European Progress in Spatial Analysis

European Progress in Spatial Analysis

Edited by R J Bennett

Pion Limited, 207 Brondesbury Park, London NW2 5JN

ISBN 0 85086 091 1

Printed in Great Britain by Page Bros (Norwich) Limited

Preface

Quantitative and theoretical geography is concerned with the application of statistical methods, mathematical models, and quantitative measurement to regional and spatial data in physical and human geography. The development of quantitative and theoretical expertise, perhaps more than any other aspect of the discipline of geography, has been dominated by Anglo-American research and scholarship. Since the late 1970s, however, increasing attention has been paid in continental European countries to the quantitative and theoretical aspects of geography. The result has been a rapid diffusion of techniques, ideas, and applications, especially in Germany, Holland, and Scandinavia. From a pattern consisting of a few isolated specialists in the early 1970s, most European countries now have a well-established base of quantitative expertise and are providing an increasing number of important publications to the body of literature in the subject. Despite this growing emergence of a communality of research approach, contact between scholars in Britain and Continental Europe has been surprisingly limited, and Anglo-American exchanges have continued to provide the major source of contact for most British workers. However, in the late 1970s this pattern has been changing rapidly and a number of international meetings, individual visits, review papers, books, and other publications have spread an increasing mutual awareness of the research developments occurring on both sides of the English Channel. This book is motivated by these developments.

The collection, it is hoped, will be found especially useful at a time of intense critical reappraisal within geography and spatial science, and it should be particularly important for geography, regional science, economics, and sociology. For the English-speaking countries it should provide a good synoptic review of the state of the art in quantitative spatial science in continental Europe and as such it should provide a valuable student text, especially in North America. For continental Europeans the book should stimulate critical debate and a greater awareness of the English-language literature. For all countries, however, it is a fervent hope that the book will stimulate more international contacts and joint research.

The aim of this volume is to assess the progress in *European* quantitative and theoretical geography. It brings together a selection of the papers presented at the *Second European Colloquium on Quantitative and Theoretical Geography* held at Trinity Hall, Cambridge between 11th and 14th September 1980. In addition, three other papers not presented at the meeting have been included to give a broader remit: chapters 1, 4, and 7, respectively, by Bennett, Marchand, and Lichtenberger. The meeting was attended by one-hundred-and-forty participants and was convened by R J Bennett on behalf of the Institute of British Geographers Study Group in Quantitative Methods. The meeting represented the second in an ongoing series which is organised every two years: the first such meeting was held in Strasbourg in September 1978, the third meeting is to be held in Munich in 1982, and future meetings will rotate among the participating countries. A summary of the proceedings of the Cambridge meeting is contained in *Area* (1981, **13** 104–108). In addition to the present volume other publications which have flowed from the meeting include a special issue of *Environment and Planning A* [1981 **13**(12)], *Munstensche Geographische Arbeiten* (1981), and a number of individual papers, further details of which are reviewed in the introductory chapter below.

The meeting in Cambridge and the publication of this volume would not have been possible without the efforts and help of a large number of people and organisations. Financial support for the meeting was provided by the Institute of British Geographers, the Social Science Research Council, and the British Council. In addition, the French, German, Dutch, and Italian participants received financial support from national and local funding agencies. The organisation was made possible only by the hard work of the network of national group organisers: G Bahrenberg for the German-speaking countries, J C Wieber for France, A Goethals for Holland, and M P Pagnini for Italy. Translation of papers was undertaken by M Blacksell, L Burmeister, and S Smith, and the participants are particularly indebted to Mark Blacksell for his help with the German translations as well as his editorial support from *Area*. Inevitably much of the burden of such meetings falls on the local department, and a special debt is also owed to the secretarial and technical staff at the Geography Department, University of Cambridge, and to D J Smith and S P Trussler who, as graduate students, helped the meeting to proceed so smoothly. Finally, the editor would like to record his special thanks to all the participants, authors, and referees whose participation, understanding, and forbearance of editorial changes has made the publication of the book possible.

R J Bennett
Secretary, IBG Quantitative Methods Study Group

Acknowledgement

The editors, *L'Espace Géographique* for permission to reproduce an English translation of an amended form of the article by J-P Marchand: "Les Constraintes physiques et la Géographie contemporaire" *L'Espace Géographique* 1980 (3), 231–240, as chapter 4.

Contributors

F Auriac	*Université de Montpellier, Place de la Voie Domitienne, BP 5043, 34032 Montpellier Cedex, France*
G Battisti	*Istituto di Geografia, piazzale Europa 1, 34100 Trieste, Italy*
R J Bennett	*University of Cambridge, Department of Geography, Downing Place, Cambridge CB2 3EN, England*
M Chesnais and the Dupont Group	*Université de Caen, Departement de Géographie, 14032 Caen Cedex, France*
F M Dieleman	*Geografisch en Planologisch Instituut, Vrije Universiteit, De Boelelaan 1105, kamer 7A4Q, Postbus 7161, 1007 MC Amsterdam, Holland*
A Douguédroit	*Université d'Aix Marseille, Institut de Géographie, 29 avenue Robert Schuman, 13621 Aix-en-Provence, France*
F Durand-Dastès	*Université de Paris VII, Section Géographie, 2 place Jussieu, Tour 34-44, 75005 Paris Cedex, France*
E Giese	*Geographisches Institut, Justus Liebig, Universität Gießen, Senckenbergstraße I, 6300 Gießen, Germany*
J P Grimmeau	*Université Libre de Bruxelles, Laboratoire de Géographie Humaine, Avenue Adolphe Buyl 87, 1050 Bruxelles, Belgium*
Y Guermond and the IMAGE Group	*Université de Haute Normandie, Institut de Géographie, Rue Lavoisier, F-76130 Mont Saint Aignan, France*
P Haggett	*Department of Geography, University of Bristol, University Road, Bristol BS8 1SS, England*
R J Johnston	*Department of Geography, University of Sheffield, Sheffield S10 2TN, England*
E Lichtenberger	*Geographisches Institut, Universität Wien, Universitatsstraße 7/V, A-1010 Wien, Austria*
G Löffler	*Universität Trier, Projekt Geschichtlicher Atlas der Rheinlande, FB III—Geschichtlicher Landeskunde, Postfach 3825, 5500 Trier, Germany*
J-P Marchand	*Université de Haute Bretagne, UER de Géographie, 6 avenue Gaston Berger, 35043 Rennes Cedex, France*

P Nijkamp	*Faculteit der Economische Wetenschappen, Vrije Universiteit, Postbus 7161, 1007 MC Amsterdam, Holland*
D Op't Veld	*Planologisch Studiecentrum TNO, Postbus 45, Delft, Holland*
M P Pagnini	*Istituto de Geografia, piazzale Europa 1, 34100 Trieste, Italy*
M L Senior	*Department of Geography, University of Salford, Salford M5 4WT, England*
A G M van der Smagt	*Geografisch en Planologisch Instituut, Katholieke Universiteit Nijmegen, Postbus 9L44, 6500 KD Nijmegen, Holland*
U Streit	*Institut für Geographie, Universität Münster, Robert-Koch-Straße 26, 44 Münster, FRG*
P J Taylor	*Department of Geography, University of Newcastle, Newcastle-upon-Tyne NE1 7RU, England*
A Turco	*Istituto di Geografia, piazzale Europa 1, 34100 Trieste, Italy*
M Vigouroux	*Université de Montpellier, Place de la Voie Domitienne, BP 5043, 34032 Montpellier Cedex, France*
A G Wilson	*School of Geography, University of Leeds, Leeds LS2 9JT, England*
G Zanetto	*Istituto di Geografia, Ca' Foscari, 30100 Venezia, Italy*

Contents

Part 3: Applications in human geography

1

Quantitative and theoretical geography in Western Europe

R J Bennett

Introduction

It would be a brave attempt indeed to summarise in the few pages available here the diversity of style of quantitative and theoretical geography in Western Europe. There are still wide gaps between the research traditions in continental Europe and those in Britain and North America, arising in part from their separate academic histories, but also related to their very different higher educational systems and their institutional, physical, and economic environments. However, considerable momentum is now emerging in European and Anglo-American interchange between quantitative and theoretical geographers, and this is likely to produce important changes in all European countries. Most important, perhaps, the deep methodological questioning of German and French geographers is increasingly likely to enrich work in the English-speaking countries, whilst the analytical technique and pragmatic approach of the English-speaking world is likely to lead to rapid development in quantitative geographical research in continental Europe.

Discussion at the *Second European Colloquium on Quantitative and Theoretical Geography* held in Cambridge in 1980, and on which this book is mostly based, showed clearly that there are a wide variety of contrasts in the outlook and position of quantitative and theoretical geography in different European countries. First there are differences in the extent to which national organisations are supportive of quantitative geography and also the extent to which quantitative workers are integrated into what can be loosely termed the geographical 'establishment' in each country. The extent of such support seems to be much less in Italy than in most other countries. A second major difference is the degree of integration between physical and human geography. For example, the French and German traditions of man–environment focus within regions, e.g. as '*Landschaft*', which has also had a major impact on British geography, has little or no significance now in Dutch geography. A third difference derives from the extent to which the spatial approach provides a unifying theme for quantitative research. Although this clearly has major importance in Britain, the French and German outlook views space as only one of a number of dimensions which must be integrated with historical, and physical, and social interrelationships. Related to this is a fourth distinction which derives from the extent to which geography dominates spatial sciences or is but one member. In Holland, in particular, geography is only one discipline which, together with economics, regional economics, and regional science, is concerned with spatial problems.

This is also linked to a fifth difference, the degree to which geography is involved with planning and other policy issues. The involvement of Dutch, British, and German geographers with such issues appears at present to be considerably greater than for the French or Italian quantitative geographers. A sixth difference relates to the degree to which quantitative and scientific approaches to geography are under challenge, question, or attack. It is evident that, whilst British, French, and German quantitative geographers have responded to the critics of positivism by incorporating new research hypotheses and purposes into their work, Dutch geography has as yet hardly tackled these issues, whilst Italian geography is still wrestling with important ideological conflicts (Bahrenberg and Streit, 1981; Bennett et al, 1981). A seventh theme of difference concerns the extent to which the different countries have strong indigenous geographical research traditions. Clearly the 'classical' French concept of *le pays*, and the 'classical' German emphasis on *Landschaft*, have both tended to inhibit the development of quantitative and theoretical geography in comparison with the Anglo-American countries. However, as will be argued below, geography will gain greatly from the closer examination of the interrelation of these indigenous Franco-German traditions with the Anglo-American tradition.

As a result of these contrasts this chapter can do no more than attempt to highlight a number of themes recurrent in the European literature, and to place the range of contributions of this volume in their wider context. However, it is attempted here also to highlight where the cross-fertilisation of national traditions is leading to a new and invigorating thrust of practical and methodological concern within quantitative and theoretical geography. As such, this book represents both a development from, and an extension to, the review of research developments in the specifically British context undertaken by the Institute of British Geographers Quantitative Methods Study Group (QMSG): *Quantitative Geography in Britain: Retrospect and Prospect*, edited by Wrigley and Bennett (1981). However, the views expressed in this chapter are those of the editor and do not necessarily reflect those of the contributors to the rest of the volume. Nevertheless, it is hoped that the contributors, and the reader, will find the attempt to look at relationships and integrative trends in European geography, evidenced in the chapters of this book and elsewhere, both fruitful and apposite.

Towards this end, the present introductory chapter is divided into four main sections. In the first of these, a brief review is given of the Anglo-American tradition of quantitative geography. This is followed by a discussion of the European developments of this tradition. This in turn leads to assessment of the extent to which a specifically European, or Anglo-European, approach is emerging, which, it is argued, derives from a series of factors: the pressure of planning needs, the reaction to methodological critique, the reinterpretation of the role of space in geography, and the reemergence of the idiographic as an item of attention.

The concluding section of the chapter then addresses the question of how the methodological developments, presaged in this and in the following chapters, will progress given the institutional constraints and other forces that will shape academic development in Europe during the 1980s.

Anglo–American quantitative and theoretical geography

Prior to the 1950s, British geography was strongly influenced by European, particularly French and German, literature. As Unwin (1979) notes in reporting the *First European Colloquium on Quantitative and Theoretical Geography*, this was evidenced by a common methodology, the 'regional method' and the study of '*Landschaft*', which gave rise to strong common grounds of approach to teaching, fieldwork, research, and publication. This was recognisable from the methodological ascendancy of the Franco–German traditions of Vidal de la Blache in France, and of Ritter, Ratzel, Maull, and others in Germany. This Franco–German tradition also had impacts on the requirement in many British universities, that geography students should have a good reading ability in German or French; it was marked by emphasis on historical and idiographic centres of concern; and it was reflected in rather specific interpretations of the physical, social, political, and economic aspects of the subject. Physical geography, social geography, political geography, and economic geography, the major 'topical subdisciplines', all had centres of concern which were different, highly specific, and rather limiting in comparison with those evident in the same subdisciplines today.

After the 1950s, however, the ascendancy of the Franco–German traditions in Britain became increasingly eclipsed by the influence of North American research. In addition Swedish research played a significant role, especially through the work of Hägerstrand (1953) (see Holt-Jensen, 1981). The development of the resulting Anglo–American research tradition has been comprehensively documented elsewhere (see especially, Burton, 1963; Johnston, 1979a; and chapter 2 in this volume). Johnston identifies the central theme of this tradition as a consensus on the role of space and on scientific method. The resulting tradition of 'spatial science' almost completely supplanted the Franco–German regional school in the Anglo–American literature.

A major source of stimulus to the development of this Anglo–American tradition, from the British point of view, was the distinctive role played by the QMSG, which was formed in January 1964, and in its early years acted as a forum for the few British quantitative geographers. In 1968 the Group became incorporated as a Study Group of the Institute. By that time quantitative methods had become an established part of geographical research and teaching. Membership of the Group increased rapidly (223 members in 1968, and 400 after 1975). The development of the group has been chronicled by Gregory (1976), Taylor (1976), and by Bennett and Wrigley (1981), and is reported annually in *Area*. What is clear from its history is that it has been a major focus for geographical research in quantitative and theoretical geography. In addition, it has also been a major instrument for developing joint research and liaison with other disciplines: itself an important means of increasing the external awareness of the work of geographers. Moreover, since 1978 the QMSG has also been a vehicle for increasing international awareness of British work in quantitative and theoretical geography. There is probably no group in any other country concerned with this area of research which has had a stronger role in encouraging research and in stimulating pedagogic development of quantitative and theoretical geography.

The Anglo-American tradition of quantitative and statistical method had increasing ascendancy in Britain until the 1970s. However, from the mid-1970s onwards a number of new 'new' geographies thrust themselves increasingly into attention. One major group of these new geographies has been concerned with what Johnston (1979a; and chapter 2 below) terms 'humanistic' approaches (see also Lee, 1974; Walmsley, 1974; Ley and Samuels, 1978). These have sought variously to redefine the regional concept (Gregory, 1978; Harris, 1978) to encompass phenomenological approaches (Tuan, 1971), idealism (Guelke, 1974), and existentialism (Buttimer, 1976; Samuels, 1978). Some of these humanistic concerns have been in part implicitly or explicitly quantitative but most have developed separate and opposed foci. Interestingly, much of the humanistic movement has been stimulated by French writers (for example, Buttimer, 1971; Brunet, 1972; Thiebault, 1972). A second major group of new geographies has been concerned with radicalism, relevance, or what Smith (1971, page 153) termed a "revolution of social responsiblity". The inception of this new approach was most clearly marked in Britain by the reaction of British participants to the Boston meeting of the Association of American Geographers in 1971 (see Prince, 1971; Smith, 1971; Berry, 1972). This proved to be a considerable stimulus to 'relevant' research and found a ready acceptance amongst many in the profession already concerned with policy-related research. The result was a stream of publications calling for socially, politically, or policy-relevant research (see for example, Hägerstrand, 1970; Chisholm, 1971; Smith, 1971; 1977; Chisholm and Manners, 1971; Eyles, 1974; Hall, 1974; Harvey, 1974a; Massam, 1974; Coppock and Sewell, 1976; Jansen, 1976). This in turn has stimulated research on the urban land market and the role of financial institutions (Harvey and Chatterjee, 1974; Boddy, 1976; Bassett and Short, 1980), on the impact of public spending and access to public goods (Smith, 1974; 1977; 1979), on the role of elections and public spending (Johnston, 1979b; 1980b; 1982), and on the interaction of public spending and taxation (Bennett, 1980; 1982), to select only a few items for attention. An emerging concern central to this work has been a geographical focus concerned with assessing 'who gets what where, and how, at what cost' by the policy actions of the State.

As a result of the joint impact of the 'humanistic' and the 'relevance' movements in geography, much attention has necessarily shifted towards social and applied research, and this has had a number of consequences. First, it has led to an increasing division between physical and human geography. In the 1950s and 1960s the common emphasis on technique and the common outlook of spatial science were strong forces binding both sides of the discipline together, and this continued their close association inherited both from *Landschaft* and from regional geography. More recently, however, human geographers have sought increasingly to justify their research by social and policy relevance, questions which are often far from central concern to physical geographers. Similarly, physical geographers have often accentuated this rift by increasingly seeking publication outlets and research liaison outside of geography: for example, in engineering, atmospheric physics, botany, ecology, and hydrology.

A further set of consequences has been recognised by Bennett and Wrigley (1981). Primary amongst these has been a shift away from 'spatial analysis' as the core subject and method of geographical enquiry. The 1950s and 1960s saw the increasing emergence of a specifically spatial quantitative focus as the core of geography. The work of Christaller, Lösch, Isard, Hägerstrand, Dacey, Berry, Curry, Olsson, Kansky, Bunge, Haggett, Chorley, and many others, seemed to point to a specifically spatial methodology. This has been reinforced by the more recent work of Cliff and Ord (1973; 1981) and Haggett et al (1977). In such studies geographical phenomena are distinguished from those in other disciplines by their spatial relations. In the 1970s, however, this view has been increasingly challenged such that many researchers would now reject that a specifically spatial view can be used as a means of defining the focus for geography as a discipline. At one extreme, for human geography, the spatial view has been characterised as a special form of fetishism which obscures more fundamental social issues. Thus, for example, Castells (1976) and Eliot Hurst (1980) state that there is no specifically spatial theory which is not part of a more general social theory; that is, that human geography is no more than a special branch of general social science. At another extreme, for physical geography, the methodology of other subjects is increasingly seen as overriding any rather weak spatial views to which geographical analysts may subscribe. With the decrease of the spatial focus, many discussants have suggested that quantitative technique and theory must also be relegated to a less significant place.

A further feature of the 1970s recognised by Bennett and Wrigley (1981) has been the reappraisal of the role of statistical inference. In the 1960s the supposed objectivity of the Neyman–Pearson theory, the seeming rigour of the methods of repeated sampling and trial, the convenience of the critical region as a criterion of choice, and the sheer intellectual beauty of the methodology itself, made the methods of inferential statistics a very attractive area for geographers. In the 1970s, however, the role of statistical inference became increasingly to be questioned in the Anglo-American literature; Gould (1970) even asked whether *Statistix inferens* was the name of a geographical 'wild goose'. In part, this resulted from a growing realisation of the difficulties involved in using classical inferential procedures, which assume independent observations, with geographical data which typically exhibit systematic ordering over space and time. In part it also reflects wider issues relating to the developing critique of positivism within geography. In addition the methods of deductive model building, phenomenological interpretation, and hermeneutics thrust themselves increasingly into attention, stimulated by the humanistic and 'relevance' movements. As a result many researchers felt the need to abandon inferential and inductive thinking, and instead to make more extensive use of prior theory, to employ techniques which were less demanding in their assumptions (for example, Bayesian statistics), and also to emphasise exploratory rather than confirmatory mode of analysis (such as using exploratory data analysis and robust estimation of the form as suggested by Tukey).

In addition to these shifts within quantitative and theoretical geography, the mode of approach based upon 'spatial science' has come under increasing attack from outside. As Johnston (1979a; and chapter 2 below) recognises, much of the Anglo-American tradition, although possessing only a weak explicit ideology, was nevertheless underlain by an implicit ideological concentration on space and science as the respective subject matter and methodology of geography. As such, Johnston and others have seen the development of the Anglo-American tradition in the 1950s and 1960s as underpinned by a *positivist* philosophy which has introduced major problems in the use and interpretation of analyses. Whilst not denying these problems and difficulties in the way of quantitative and theoretical geography, other writers have been less willing to accept the term 'positivism' without a significant number of qualifications and caveats (see especially Guelke, 1978; Hay, 1979; Bennett and Wrigley, 1981). Whatever term we accept to describe the Anglo-American tradition of this period, however, it is clear that quantitative and theoretical geography, and an emphasis on spatial analysis, was its central core.

In various ways, therefore, this Anglo-American tradition has come increasingly to be challenged. This challenge has been in part the result, and in part the cause, of the decline of the spatial focus as the integrating paradigm for the discipline. The result has been a shift in the nature of quantitative methods, their use and interpretation, but not a decline in their importance. Rather, quantitative and theoretical geography in Britain and North America has 'come of age' and is now accepted as part of the geographical orthodoxy. The challenges to it are, therefore, to be interpreted as stimuli to redirect attention, to seek new problems, and to be more 'exploratory' in modes of analysis. It is in these respects that the Anglo-American tradition has much to gain from geographical research in continental Europe, especially in the German-speaking countries; and it is towards the exploration of this set of possible international interlinkages that the rest of this chapter is directed.

Continental European developments

The 'classical' geography of France and Germany in the 1920s and 1930s became the dominant methodological influence in Britain. It also diffused to control intellectual developments in Holland, Italy, and most other European countries. Rather than providing a vigorous base of geographical interest, however, in later years these clearly-defined national schools often had the effect of stifling concern for new methodological developments. In France, in particular, the weight of the Vidalian tradition acted as a nationalistic force to exclude or at least inhibit the influences emanating from North America and elsewhere. Thompson (1975, page 349), for example, characterised French geography as reacting "with more circumspection" to these influences than did Britain, "hesitating to abandon precepts and practices which had produced a flowering in the first half of this century". Thus a break with the Vidalian tradition was seen as "a break from orthodoxy enshrined in university syllabuses, basic texts and research monographs".

In Germany the legacy of 'classical' geography was no less strong, but its influence suffered a series of major challenges to its intellectual respectability with the use of the *Landschaft* and geopolitical concepts during the Third Reich. The tradition of Ritter and Ratzel had pinned social and political geography to the understanding 'of the objects on the Earth's surface and their mutual interrelation'. When developed by Kjellan and Haushofer (see Troll, 1949), the physical features of land and deterministic ecological laws could be used to justify territorial expansion. However, the implications of Ritter's and Ratzel's work were far wider and less deterministic than allowed by their geopolitical interpretors. For example, the developments by Bobek in the 1940s and 1950s has stimulated more recent analyses which have clearly evidenced both the idealist and the phenomenal hermeneutic nature of the 'classical' geography tradition in German-speaking countries. Lichtenberger (1978c), for example, sees this evidenced particularly in the maintenance of German social geography, which remained completely distinct from the social ecology literature that dominated the Anglo-American tradition from the 1920s until the 1970s.

Up to the 1960s in both France and Germany, therefore, the tradition of 'classical' geography was maintained separately from the developments of quantitative and theoretical geography in Britain and North America. However, in the late 1960s these two national schools began to suffer challenges and reappraisal in the light of the rapid methodological and technical developments occurring outside. In France this resulted in the appearance of a number of new books and articles which placed these developments of the 'new' geography in perspective (see Beaujeau-Garnier, 1971; Clout, 1972; Isnard et al, 1976; Ciceri et al, 1977; Beguin, 1979; Durand et al, 1980; Raffestin, 1980; Groupe Dupont, 1980). In addition Haggett's 1965 edition of *Locational Analysis in Human Geography* was translated into French in 1973 (Haggett, 1973a) and has had considerable impact. Other early Anglo-American sources in translation are Cole (1969), Berry (1971), and Jones (1973). Many articles began to appear in journals: the papers by B Marchand (1972; 1974; 1978), and Racine (1974) are typical of this changing emphasis. In 1971 the prestigious journal *Annales de Géographie* published an editorial stating its intention to accord greater prominence to papers involving new methodology, and in 1972 a new journal *L'Espace Géographique* was initiated which was explicitly aimed at stimulating debate of methodological and theoretical issues. Many of the more local journals in France also stimulated innovatory initiatives, especially influential being the *Revue de Géographie de Lyon* and *Revue Géographique de l'Est* (see McDonald, 1975). In addition to these, many other groups have been influential in stimulating development of the 'new' geography: particularly important have been the 'Groupe Dupont' and 'L'Amoral Groupe', represented in subsequent chapters of this book, and the Strasbourg Laboratoire de Cartographie Thématique.

In Germany the reappraisal of 'classical' geography again occurred in the late 1960s. It was marked by a reassessment of the role of scientific theory and the development of model building and planning (see Bartels, 1968a; 1968b; Kilchenmann, 1975; 1978a; 1978b; Lichtenberger, 1978a; 1978b; 1978c). This development is summarised by Giese in chapter 6 below. Major books encouraging the development of this framework in Germany were those by Bartels (1970) and Hard (1973), and again the translation of Haggett's *Locational Analysis in Human Geography* (1973b) organised by Bartels. The influence of national and local journals was also important, especially the *Geographische Zeitschrift* nationally, and the departmental journals *Giessener Geographische Schriften, Karlsruher Manuskripte zur Mathematischen und Theoretischen Wirtschafts und Sozialgeographie*, and *Bremer Beiträge zur Geographie und Raumplanung*. The papers by Steiner (1965), Anhert (1966), Bartels (1968a; 1968b; 1979), Kilchenmann (1970; 1975), Bahrenberg (1972; 1976, 1979), Sauberer (1973), Streit (1973; 1975), Volkmann (1976), Arte and Sprengel (1977), Giese (1978), and Lichtenberger (1978a; 1978b), were some of the many influential publications, although there have also been some criticisms (such as Otremba, 1971). In Germany, to a much greater extent than in France, however, there has developed a new grouping of quantitative and theoretical geographers: the *Arbeitskreis 'Theorie und quantitative Methodik in der Geographie'* (working group on theory and quantitative methods in geography). Founded in 1974, this group, like the QMSG in Britain, is seeking to act as a means of liaison, development, and innovation in quantitative geography. Although still small and dependent on a few enthusiasts, it is also beginning to develop a pedagogic role by producing a German language equivalent of the highly successful CATMOG series published by the British QMSG (see Giese, chapter 6 below).

Outside of France and Germany the challenge to classical geography proceeded at a different pace. In some countries the classical ideas were challenged much earlier. In Sweden and Finland, in particular, the establishment of a separate set of concerns distinct from the Franco-German traditions was important before the 1950s, and in many ways the developments, particularly in Lund, enriched and stimulated the Anglo-American 'new' geography (see Granö, 1929; 1931; Ajo, 1944; Hägerstrand, 1953; 1967; Mead, 1963; 1977; Gould, 1972; Pred, 1974; Holt-Jensen, 1981). Mead, for example, argues that the work of Granö and Ajo in Finland was particularly influential in introducing mathematical ideas and quantitative techniques into Finnish historical and cultural geography. In Holland, the humanistic background and the Franco-German influence did not diminish until the early 1970s, and regional geography still remains important (see Keunig, 1969; Hauer, 1971; Dieleman and Op't Veld, chapter 10 below). However, a strong development of Dutch geographical research in the social science tradition established itself independently of 'classical geography' very early (Heinemeyer, 1977), as for example in the works of Steinmetz and von Vuuren (see Cools, 1950) and Heslinga (1974; 1980). This independent establishment occurred mainly in the *economisch-technologische* and *sociografische* institutions (see Hauer, 1971; and Dieleman and Op't Veld, chapter 10 below) and hence it was quite natural that Dutch quantitative geography should have been stimulated by, and have developed strong links with, other subjects, especially economics and regional science.

In other countries the pace of methodological development has more closely mirrored that in France. In Italy, Spain, and Portugal, for example, the methodological stimulus for change has come largely from the internal debate within French geography, rather than directly from the Anglo-American tradition (see Racine et al, 1978; Corna-Pellegrini and Brusa, 1980). As a consequence, French internal debates have mediated and modified the reactions to quantitative and theoretical concerns; and it may be argued that the specifics of the internal French methodological problems have not aided debate elsewhere. Nevertheless, in Italy there is now considerable evidence of a momentum to research in quantitative geography. In this the publications of Toschi (1941), Nice (1953), and Bonetti (1961) have been very influential. Translation of Christaller's central place study occurred in 1980, and other influential translations are Carter (1975) and Lloyd and Dicken (1979). A first Italian language handbook of quantitative geography was produced by Vlora (1979). However, in the Italian literature, as in the Anglo-American, there have been considerable conflicts both with perceptual/humanist and Marxist geographers. Moreover, the organizational fragmentation of Italian universities makes methodological innovations slow to develop; but the establishment of direct Anglo–American interactions with Italy seems to hold promise for more rapid developments in the future (see chapters 8 and 9 of this book).

This brief discussion serves to demonstrate that although the development of quantitative and theoretical geography in continental Europe began in general at a much later date than in Anglo–American geography, it is also taking a significantly different form. As Thompson (1975) has noted, perhaps the most important of these differences is that Britain and North America in the 1950s possessed no clearly-defined preexisting national school of geographical enquiry and this encouraged a greater readiness to assimilate new methodology. In contrast, the strong preexisting schools of 'regional geography' in France, and *Landschaft* in Germany, first inhibited the progress of new methodology and then led to intense and often factious debate. The consequence has been that in the Anglo–American literature 'classical geography' takes a very small place, whereas in the continental European literature quantitative geography is still largely peripheral. As a result the 'classical' and 'new' geography either exist side by side (as in the published record of *Geographische Zeitschrift* and *Annales de Géographie*, for example), or have become involved in heated debate.

This contrast between the Anglo–American and Franco–German developments is an important one, and its examination at a time of more intense methodological debate in the Anglo–American literature is now particularly apposite. Although Anglo–American geography certainly did not establish quantitative and theoretical geography without some battles (see Burton, 1963; Gregory, 1976), these were relatively short-lived and the new methodology was widely and quickly assimilated into the profession. The weakness of the French 'classical' tradition, as seen in Britain and America, was an underdeveloped framework of analysis which advanced by example rather than by explicit methodology, was idiographic, and counselled against generality (see for example Thompson, 1975) and this occasioned little defense in Britain and America. However, the battles to introduce a geography of social relevance have been far more bitter and longer lasting. The methods of quantitative and theoretical geography which are more truly indigenous to Britain and North America have been more savagely attacked and defended than was the case with the challenge to the regional method. Given this challenge, it becomes particularly apposite to look at some of the more recent European literature. In the German literature in particular we perhaps find a major stimulus to resolving the challenge of the 'relevant' to the 'quantitative'. Here, amidst the conflicting views apparent in any healthily expanding field, there is a growing integration of the socially relevant and the quantitatively proficient.

This mutual assimilation is also beginning to emerge in the British literature, and it is in this direction, it will be argued below, that the development of an Anglo-European dimension will most fruitfully lie.

The emergence of a European dimension

Stronger links between European geographers have been developing for some years. Early links were the German-French meeting of 1973 (*L'Espace Géographique*, 1974), the Anglo-French seminar in London (Claval, 1975a; 1975b; Thompson, 1975), the French-speaking meeting at Nice in 1975 (Bailly, 1976), the international meeting on spatio-temporal analysis at Montpellier in 1979 (Pumain and Saint-Julien, 1980), the NATO meeting on 'Dynamic Spatial Models' at Chateau de Bonas in 1980 (Bennett et al, 1980), and many individual contacts. The *European Colloquia on Theoretical and Quantitative Geography* (from which this book derives) initiated in Strasbourg in 1978, continued in Cambridge and Munich in 1980 and 1982, respectively (Symposium, 1979; Unwin, 1979; Bahrenberg and Streit, 1981; Bennett et al, 1981), has continued to develop these strong links. Individual and group meetings and visits between the various countries also proceed apace. From these and other meetings important lines of development for the future are emerging and are likely to continue to emerge. Moreover, it is becoming clear that these meetings are playing an increasingly important part both in the formation of research hypotheses and in the research discussions between many geographers in Britain, North America, and continental Europe (see for example Blacksell, 1981), and are particularly important for many IBG study groups (Cruickshank, 1981).

Despite these increasing international links, it is clearly impossible to seek a single emerging European view of quantitative and theoretical geography from an academic base as large and culturally diverse as Western Europe. Moreover, such an attempt may be inhibiting and hence could be undesirable. What is attempted here, therefore, is by 'exegesis' and 'osmosis' from the European literature to seek a set of latent and underlying themes which do suggest the importance of developing closer links between European geographers. It will be argued that closer linkage is important from three points of view: first, it allows a better appraisal of the strengths and weaknesses of the critique of positivism; second, it gives a wider and more social interpretation to the role of space in geography; and third, it serves to suggest the emergence of a new geographical core paradigm for the discipline as a whole. Each of these linkages is discussed in turn below.

The critique of quantitative geography as positivism

The critique of positivism in Anglo-American geography has emerged through a series of writings deriving both from Marxist/neo-Marxist and from a range of 'humanist', phenomenological, and idealist viewpoints.

Major examples are those of Harvey (1973; 1974b), Lee (1974), Walmsley (1974), Sayer (1976), Peet (1977a), and Gregory (1978; 1980) amongst a rapidly growing literature. It is the contention of these writers, in various degrees of emphasis, that quantitative geography is alienating in distracting researchers from the central question of social distribution. First, it creates a false sense of objectivity by attempting to remove the observer from the observed, which in turn lends an ability to control and manipulate society. Second, its use of computers, machines, and techniques subsumes man under a paradigm of mathematics and machinery, reifying man to an atomistic level by losing the elements of soul and humanistic concerns. Third, it is descriptive of existing behaviour and hence supports the *status quo* within society, especially of social distribution of well-being. Fourth, it allows no consideration of values and hence of the norms by which systems, and hence society, *should* be organised. Fifth, it attempts to construct models and theories of universal generality by a mode of inductive logic moving from the particular to the general.

Some Anglo-American critics have even gone beyond these criticisms of the quantitative to question the nature of the subject of geography itself. For example, Peet and Slater (1980), instead of arguing that statistical methods and empirical analysis are never appropriate, have contended that the problems of theoretical explanation are the ones to which primary attention should be directed. They then take their argument further to suggest that "Geography has no theoretical object that is specific to it, and its 'theory' is no more than a heterogenous amalgam of spatial models developed by a variety of bourgeois scholars" (Peet and Slater, 1980, pages 543-544), although it has specialised in the study of two important sets of social relations: "between social relations and their natural environments; (and) between environmental embedded social formations ... across space". Eliot Hurst (1980) even argues for a *de-definition* of geography. His case is that "geography is revealed for the most part as merely descriptive and scientifically bankrupt ... (and) is nothing more than a theoretical ideology based on technical practice alone" (op cit, page 7). Instead of geography, a truly multidisciplinary approach (dialectical and historical materialism) is required which can uncover "the true relationships between economy and society in the totality of social existence" (op cit, page 12). Geography and its concentration on space is a "fetishized domain", moreover even "the co-opted Marxism of 'radical' geography and 'Marxist Geography' ... is superficial and doctrinaire, incapable of transcending the orthodox problematic's boundaries" (op cit, page 16). Thus for some the rejection of positivism also entails the rejection of space, and hence of geography as a whole, as a valid focus of attention.

The contentions of these critics all carry some element of importance. Indeed, it cannot be disputed that much quantitative geography in the 1960s was concerned, to too great an extent, with techniques *per se*, with creating an artificial sense of objectivity, and with induction and inference rather than use of prior theory. Much of the Anglo-American tradition has even consciously sought objectivity, technical expertise, spatial description, and inductive approaches directed towards revealing universal laws which might be used to plan and control. A good example is provided by Abler et al (1971) who state that "underlying our approach ... is our belief that human geography is a social and behavioural science. We think the principles which govern human spatial behaviour are generally applicable all over the world" (op cit, page xii), and "our focus has been on the utility and goals of geographical inquiry" (op cit, page xv). This view has also affected continental European geographers; for example, Isnard (1980, page 143) states that "theory, as a provisional explanatory hypothesis, can free itself from the problems raised with the observation. Geography must not seek refuge in the abstractions of ideology".

However, it is the contention here that much of the critique of positivism in Anglo-American geography has been largely unhelpful to the development both of the discipline of geography, and to the understanding of society. It has had major importance in stimulating concern with socially relevant questions, but it has also distracted attention to unproductive methodological debate. This differs from many of the impacts of Marxist writing in continental Europe where they have often been closely linked both with spatial and with quantitative research. The Anglo-American critique of positivism is thus relatively unique in contemporary geography.

A major part of the confusion which Anglo-American geography has suffered with respect to the positivist critique seems to arise from the representation of quantitative and theoretical geography as purveyed in Harvey's (1969) *Explanation in Geography*. Although a wide-ranging and an internally contradictory book, a clear message derives from its text: namely, first, that geography *is* primarily inductive, second geography *is* searching for universal laws, and third, geography *is* an objective science. Moreover, in geography where phenomena do not seem to fit universal laws, they can be treated *as if* they do. These three aspects of Harvey's 1969 presentation all have some aspect of truth, but each is only a partial representation of the literature and ideas it seeks to describe. Despite this inadequacy, however, it is a caricature which many geographers have uncritically accepted. For example, Johnston (1979a, page 61) quotes Harvey's (1969, page 34) two routes to scientific explanation without criticism and later suggests that "the approach of positivist spatial science ... (is) codified in Harvey's *Explanation in Geography*" (Johnston, 1980a, page 404). Eliot Hurst (1980, page 3) refers to Harvey's book as a "quasi-philosophical rationalisation" of geography which "led to a distinctive practice within geography which searched for some surrogate landscape of mathematical spatial conformity". It has been accepted by Johnston (1979a; and chapter 2 below) that the Anglo-American tradition *was* positivist; Haggett et al (1977, page x) also accept this caricature and even apologise that their book is 'narrow and positivist in its philosophy'. The dangers implicit in Harvey's representation are well-evidenced by Gregory (1978). Although now modifying his views (Gregory, 1981; 1982), he makes recourse to the nineteenth-century view of positivism as defined by Comte. In particular, he seeks parallels in modern Anglo-American writing to Comte's concepts of le réel, la certitude, le précis, and l'utile. By selective quotation he is certainly able to find considerable confirmation of these views. However, the comparison is both superficial and inexact. This comparative-critical method specifies a critical entity (such as 'systems theory', 'positivism', or 'neoclassical theory'), then isolates significant characteristics of it (such as those defined by Comte). However, the representation of the source material of Anglo-American geography *solely* in terms of these characteristics is an inductive fallacy which amounts to constructing a form of 'straw man'. Rejection of the Anglo-American sources then becomes the more easily possible since these sources and 'positivism' are assumed in one-to-one correspondence. For this reason, Bartels (1974) has questioned the usefulness of this literature, and Bennett and Wrigley (1981) have termed much of the critique of positivism "at best a misrepresentative irrelevance, and at worst a fatuous distraction"; they argue that the critique has been directed at an abstract and misrepresented view of much of quantitative geography, one which attributes methods, views, and conclusions to quantitative geographers which most have never held, or if they did ever hold, have since abandoned, or if they still hold in some form, hold only in part alongside wider views. For teaching purposes and for convenience of reference it often seems useful to seek a caricature of items of academic work; for example, 'positivist', 'functionalist', or 'realist'. This is a form of shorthand. But for useful academic debate such caricatures do a disservice if they do not in *all* important respects give a fair and holistic representation of the academic work in question. Just as many writers of quantitative geography unfairly criticised regional geography as purely descriptive, idiographic, or of low academic level (see for example, Gould, 1979), so in the 1970s many 'radical' writers have invalidly criticised quantitative and theoretical geography as being purely 'positivist'.

The importance of the critique of positivism should not, however, be assessed at the level of Anglo-American debate alone. In some continental European literature, in some recent Anglo-American literature, and elsewhere, a different level of debate has been emerging. This has been concerned not with the minutiae of positivist theory, nor with the exegesis of literature assumed to be 'positivist', but instead has been concerned with the development of a critical philosophy and mode of thought aimed at better revelation of geographical phenomena, both *per se*, and as items within a wider theoretical framework. The stimulus of this literature is, then, for a new focus of geographical attention, perhaps even a new core to the discipline. One element of this, the role of *space* in geographical enquiry, is discussed below before a suggested integration of radical and quantitative notions is presented.

The role of space in geographical enquiry

It has already been stated above that space, or 'spatial science' was central to much of the quantitative and theoretical geography developed in the Anglo-American tradition up to the 1970s. So pervasive is the theme that Claval (1975a, page 345) has suggested that "British geography looks like a model for many French geographers", and Kilchenmann (1978a) has stated that German geography is "marking time" whilst it awaits the stimulus of Anglo-Swedish-American methodology. Again Isnard et al (1976) and Isnard (1980) have suggested the need for a paradigm shift towards positive science (in Popper's sense), central to which are *spatial* hypotheses. The dominance of this view of geography as spatial science has also stimulated some to axiomatise the role of space (and time) as metrics upon which phenomena are measured (see for example, Beguin, 1981). The recent NATO Advanced Studies Institute at Chateau de Bonas in France has also reemphasised the static, metric view of space (see Bennett et al, 1980). The 'spatial science' approach has therefore diffused, and is continuing to diffuse, to affect the work of many European geographers; and it is reflected in many of the following chapters (see also Bahrenberg, 1972; Cauvin et al, 1980).

Despite this widening utilisation, many important research problems in spatial science remain to be tackled. Haggett et al (1977; and in chapter 3 below) emphasise the need to clarify the special difficulties presented by spatial data in empirical analyses; for example, the effects of boundary conditions and system closure, and the difficulties of internal boundaries inducing ecological correlation (the impossibility of separating individual from group correlation) and the 'Curry effect' (the effect of map pattern on statistical estimates). Haggett sees space "as a bridge to geographical analysis" but requiring "space-proofed" analytical methods which allow the "encoding and decoding of map information". Using the more specific cases of urban planning models and decision situations, Wilson, Nijkamp, and Senior in their chapters (14, 15, and 16) also call for greater development of new forms of spatial analysis capable of encompassing differing scales of analysis, different forms of spatial representation, and successful interfacing of these with decision needs. On the one hand, then, spatial science is proceeding to generate better definitions of its problems, thus facilitating empirical research with spatially distributed data. On the other hand, spatial science is being reshaped as it evolves and comes into contact with other research traditions.

The reshaping of the spatial science view is taking many forms. Isnard et al (1976) and Isnard (1980) see spatial science as a method for reconstructing traditional regional geography: as a means for testing hypotheses concerning *le pays* and similar concepts. This methodology is strongly criticised by Racine and Bailly (1979) since it ignores the social significance of space. However, developments by Claval (1975b; 1976; 1977) suggest that the interface of society and space can be tackled by using selected elements of Marxist analysis, even though the bulk of Marxist thinking remains "ill-adapted to geographic consideration" (op cit, 1977, page 145). Others, however, have gone further and suggested that the social and spatial interrelation should be the major focus of geographical attention. Harvey (1973; 1978), for example, has developed a Marxist theory of spatial social justice. The same theme is echoed by Collectif de Chercheurs de Bordeaux (1977), Peet (1977a; 1977b), and other Marxist writers, who argue that the division of labour, the division of town and country, the role of transport in price determination, localisation rents, territorial strategies used to fight the falling rate of profit, and 'urban ground rent' all provide specifically spatial foci for social theory. Hence, they can be used to define a new, 'radical' core for geographical inquiry.

A particularly interesting view of spatial and social interactions as expressed in French geography is provided in the work of Bernard Marchand (1972; 1974; 1978; 1981) and in Auriac and Durand-Dastes (chapter 5 below). All these writers are stimulated by a linkage between the concepts of radical geography and quantitative geography and systems theory. Marchand emphasises the manner in which spatial structure expresses social structure and, since past spatial structures in part fossilise past social structures, spatial–historic structure introduces constraints and imperatives (to reform or to revolution) for current social phenomena. This gives what he terms 'a dialectic for social and spatial evolution'. Auriac and Durand-Dastes take this a stage further in their analysis of the two examples of ghetto formation and vine cultivation. They see homeostatic properties in space which serve to maintain social structures despite attempts to remove them (as in the case of small-scale vine cultivation); that "spatial inscription often serves as a 'memory' for social and economic processes". This leads them to formulate a principle of "spatialisation of systems", which may not be the principle factor in any situation, but can be decisive. Hence, space and society are seen to be irrevocably interlinked to the extent that "the spatialisation of processes will give rise to its own social and economic contradictions".

In very general terms, then, it can be stated that the tradition in spatial science is shifting its ground. On the one hand, it is emphasising the spatial properties of empirical investigation and seeking methods of overcoming the difficulties that may arise. On the other hand, it is seeking to link social and spatial concepts, using the stimulus of Marxist theory, but developing a distinct identity separate from specific Marxist dogma.

A recent development of this latter view in the American literature is that by Sack (1980). Sack argues that every society has an economic and social structure expressing underlying value positions (ideology). Part of each of these positions involves a particular conception of space: a capitalist society has one view of space, a Soviet society another, a Chinese communist society yet another, and so on. Each specific society possesses a distinct *mode of thought* and this in turn dictates a different role, meaning, value, and significance to space. In this he is close to Tuan (1971; 1977), but Sack goes further by arguing that space has two main attributes: first, the social organisations and individuals within society are in space and hence, analysis and understanding of their interrelations is a major concern; second, social organisations are territorial and hence provide a means of influence, acculturation, control, or political organisation. With space defined in this way, Sack (1980, page 197) sees it as geography's essential task "to consider the full range of meanings of space and the nature of their synthesis as part of geography's general task of understanding the earth's surface as the home of man". This task is to be tackled, Sack argues, by integrating the effect of society on modes of thought, with the conception of space deriving from any mode of thought. This involves fusing what Sack terms the subjective and objective as well as space and substance. His call is apocolyptic in scale requiring the fusing of the modes of thought characteristic of the sciences with those characteristics of the arts: the one treating symbols as objective and having no meaning attached to them, and the other treating symbols as having not only meaning but a magic and myth which are as important as the objective as foci of concern. Thus, Sack (1980, page 200) argues that the quantitative and theoretical geography of the Anglo-American tradition so characteristic of the 1960s presents enormous barriers to the introduction of a true understanding of the significance of space: "the most general structural characteristic of science itself, its discursive symbolic form, presents extraordinary problems for developing more accurate and comprehensive social scientific theories because of its inability to capture the subjective and its symbolic forms". However, he does see the need to link empirical/quantitative appraisal of space with subjective appraisal of its meaning within any mode of thought. How, then, can quantitative and theoretical geography move more closely towards capturing the subjective and symbolic of society and its spatial structure?

Is there an emerging focus?

Anuchin (1970a; 1970b) has suggested that periods of increasing generalisation within disciplines are often succeeded by periods of increasing fragmentation, which are then absorbed into a period of further generalisation, and so on in a cyclic fashion. Anglo-American geography as a whole in the 1960s underwent a period in which general concepts (derived from viewing the subject as spatial science) absorbed many of the more particular aspects of the subject. In the 1970s this generality gradually gave way to fragmentation, anarchy, and to the emergence of an increased emphasis on pluralism with consequent particularisation and specialisation within the subject. Johnston (1979a; and chapter 2 below) sees this as not only necessary, but in some ways healthy: accommodation and pluralism are preferable to conflict if consensus is impossible. Bartels (1968a; 1968b; 1978), in contrast, views pluralism and fragmentation as not only unhealthy but likely to lead to the disappearance of geography from the university curriculum by continual erosion from adjacent disciplines. This is a view echoed by many geographers from the German-speaking countries (see Lichtenberger, 1978a; and chapter 7 below), and also in Britain (Wise, 1977; Patmore, 1980).

Johnston sees increasing pluralism as largely inevitable, but certainly desirable, because of the irresolvability between what he identifies as the two central thrusts of modern geography: the 'scientific' and the 'radical'. Two aspects of these differing schools tend to be in strong oppostion; namely their differing scales and content of research, and their differing views of the interaction of the individual and society. Johnston suggests that radical theory has been concerned with a relatively small number of issues, centering attention at the level of the infrastructure of society (the mode of economic organisation) with most emphasis on long-term effects. In contrast, he argues that quantitative and theoretical geography offer the opportunity to study how particular realisations have come about *within* any *one* theory or framework of infrastructure. Johnston accepts that many of the topics so studied (voting patterns, choice of shopping centre, etc) are irrelevant to the achievement of long-term change and suggests that this points to the ideological divide between 'scientific ' and 'radical' geography. As a result, he concludes that quantitative human geography is in something of an ideological vacuum: since quantitative geography can yield solutions only to short-term problems, it may hamper change and create a false sense of legitimacy for existing societies and regimes. As a result he concludes that pluralism resulting from accommodation is preferable to conflict; consensus he views as unlikely or impossible.

In the German-speaking countries Bartels (1973; 1978) sees a parallel phase of pluralism and he identifies five major subgroups within the discipline: choreography, area studies, social geography and regional science, landscape ecology, and the ecology of man. In contrast to Johnston, however, Bartels sees a danger for the survival of the subject in acquiescing in the continuing pluralism and increasing autonomy in the separate branches of geography—a problem he sees in choreography in particular and in Anglo–American geography in general. He concludes that it is *Sozialgeographie* combined with regional (that is spatial) science which should be the theme of greatest central concern to the discipline. Most important to Bartel's conception of *Sozialgeographie* is a link between radical and scientific theories; and it is in this that the continental European literature is offering a route into the future which is distinctly in contrast to that which seems to be evolving in the Anglo–American literature.

The development of German tradition in *Sozialgeographie* in its early period of development is summarised by Hajdu (1968). Much of the foundation of this tradition derives from Bobek's criticism and extension of Ratzel: that in the study of man–environment interrelations the study of man had been largely ignored. In his subsequent developments, Bobek drew from the French concept of *géographie humaine*, developed by Vidal de la Blache, but derived a distinctive set of social rules for the way in which the social group (*Lebensformen, Kulturgemeinschaften*) affected individual decisions. Each individual was a member of numerous and overlapping social groups derived from the historical development of society, and the individual was the link between society (the groups) and the environment (Bobek, 1948; 1950). Particular social groups pursued distinctive economic activities and it was possible to define distinctive societies on this basis, for example traditional, feudal, industrial capitalist (Bobek, 1959). Regional geography then becomes the fabric of social/human groups, with regions identified by distinctive and internally similar human behaviour and interactions with the rest of the world. In this the actions of the State (national, provincial, and municipal) were important (Bobek, 1962) as was the role of international and external cultural systems such as colonialism. This conception was formalised by Hartke (1960), who concentrated attention on the discovery of social indices to allow the identification of social changes in regions, districts, or even streets of a city. In various example studies, Hartke (1952; 1963) and Lichtenberger (1977) analysed the evolution of suburbs, the gentrification process, the control of property ownership, and the role of government; and in emphasising politics, ethnic, religious, and state actions, they retained much of the wider context which was lost in most Anglo–American social area analyses. Subsequent developments of this approach by Hahn (1950), Geipel (1952), and Ruppert (1958) have demonstrated that the approach can be freed of latent emphases on social determinism (see Hajdu, 1968).

This traditional view of German *Sozialgeographie* was systematised by Ruppert and Schaffer (1969) and is 'codified' in most German textbooks (such as Maier et al, 1977; 1978). It also influenced Dutch geography (Cools, 1950). However, it is open to a number of criticisms. For example, Wirth (1977; 1978) suggests that it ignores both the scientific and the Marxist points of attention in geography. Similarly, Leng (1973) has suggested that the 'basic functions of life' of traffic, recreation, and housing analysed by *Sozialgeographie* ignore the fundamental social process of the role of territorial structure and the processes of production and reproduction within society. However, a new set of interpretations of *Sozialgeographie* have recently been given in the German literature which overcome these criticisms. These attempt to link the traditional *Sozialgeographie*, on the one hand, with the German hermeneutic philosophy of Habermas, and on the other hand, with the scientific approach of spatial science. This approach has been developed by Bartels, Lichtenberger, Hard, Wirth, and others. For example, Lichtenberger (1978a; 1978b; 1979; 1980) suggests that in German social geography there are two centres of attention: first, a deeper analysis of the subject matter of society (hermeneutics), rather than the mere ordination of phenomena so characteristic of social ecology; and second, a scrutiny of the essence of the meanings of phenomena (phenomenology), rather than mere description. Similarly, for Hard (1973; 1978) and Bartels there is the central view that analysis of 'specifically spatial behaviour' is essential to understand the functions necessary for the existence (*Daseinsgrundfunktionen*) of phenomena involving functional relations *and* values. This derives from the understanding of territory (*Lebensraum*) and spatial order (*Analyse raümlicherordnung*): this is the object of regional research (*Regionalforschung*), but is not symmetrical with 'regional science' as such, since there is a much greater concern in *Regionalforschung* with the actions of the State. A further set of issues also concern the layout of territories (*Raumordnungspolitik-Forschung*), especially the interaction between territorial space, governmental involvement, and the structure of society as expressed through its settlement systems. This draws from some of the recent Swedish time-geography research (Bartels, 1979). The two central concerns of *Raumforschung* and *Raumordnung*, as noted by Lichtenberger (1978a), have no precise equivalent in the English language and concepts; they have a focus on applied geographical research but with the central concern being settlement and territory as reflected in the actions both of society and of the State (see Bartels, 1968a; 1968b; 1970; 1973; 1978).

It seems likely that in the 1980s there must be at least a further period in which fragmentation will increase before geography is again ripe for the development of a new integrating paradigm such as that which characterised Anglo-American geography in the 1960s. However, the core of an emergency link between radical and scientific theories seems to be possible. Certainly it is suggested in the views of Bartels and in other German literature discussed above; and it is perhaps in part from this continental European literature that a new core and integrating paradigm for the discipline can develop. Admittedly the concept of *Sozialgeographie* as used by Bartels (1968a; 1968b) and Lichtenberger (1978a) is still rather ill-defined, but it does seem to offer considerable stimulus to the development of a new geographical core. It has three main strengths. First, it is integrative of the physical and human aspects of the subject. The development of quantitative technique in the 1960s was, and still is, stimulative to joint development of the two sides of the discipline (as evidenced by Streit and others in the following chapters). However, the concepts of social relevance (especially Althusserian theory) have eroded these links. In contrast, many have argued, like Marchand in chapter 4 below, that physical and human geography would benefit from closer links (see also, for example, Cooke and Robson, 1976). Marchand argues the case from the spatial point of view: the environment has a spatial role both as a constraint limiting the freedom of action (static constraints), and as part of a system of interaction between society and the environment (dynamic constraints). The view of *Sozialgeographie* like that of Sack (1980) develops these society–environment links into a set of territorial, perceptual, and cultural ties, which link individuals to their social group in a particular place. Thus Hard (1978) argues that the region is a focus of social, political, economic, *and* environmental dimensions of attention; not as *Landschaft*, but rather in terms of the social meanings deriving from social groups. Hence, the emphases in recent French and German writing on the linkage between physical and human geography offer considerable stimulus to Anglo–American geography.

A second feature of continental European tradition of *Sozialgeographie* is the emphasis on historical–spatial constraints. In the German literature this tends to be emphasised in a link between territory and society as *Lebensraum*. Similarly, the French school contains this geopolitical concept, that the ongoing system interrelations of past spatial structure and built forms constrains and modifies present society. Clearly the historical influences both of territory and of spatial constraints are important: territory because it conditions much of the meaning, nostalgia, or mystery of place; and space because it allows structural and holistic explanation of the status quo and of the mechanisms which affect future developments. These are well represented in the chapters by Marchand, Auriac and Durand-Dastes, Giese, and Lichtenberger below. These historical themes have been largely absent from the Anglo–American literature and their development offers much for the future.

A third feature of the *Sozialgeographie* tradition is its emphasis on 'applied' or 'applicable' planning needs. Many of the spatial science developments of the Anglo–American school also emphasised utility, 'instrumentalism', and the usefulness of models and theories for planning situations. However, the critics of positivism rightly castigate aspects of this literature as being ones which (implicitly) serve, by empirical analysis and description of present interrelations, to maintain a status quo, and hence in part provide the tools to facilitate the continuance of dominance and class inequities. Where the continental European literature differs in its emphasis on 'applied' research is in its emphasis on the interaction of society in place with the governmental structures which relate to and maintain that place. This is particularly well-developed in the German *Raumforschung* and *Raumordnung* mentioned above, and is briefly discussed by Lichtenberger in chapter 5 below. In this continental European literature, the interaction between economic, cultural, and social territory, the organisation of the State, the interrelations with the rest of the world, and the history of past settlement evolution all interact.

Recent British research has also focused on these aspects as part of a concern with the origins and relations of the geography of public goods (see Johnston, 1979b; 1982; Bennett, 1980; 1982). Claval (1975b; 1976; 1977) in the French literature has also emphasised the way in which governmental organisations arise as an economic, social, and territorial response to the influence and extent of individual and group externalities. However, the Anglo-American literature has not yet sufficiently grasped the historical and cultural aspects of these components which give territory and space a special and a fully social meaning.

What then are the implications of these trends for Anglo-American and continental European quantitative and theoretical geography? First, it seems that the *idiographic* must be reasserted to attention. The Anglo-American tradition has rightly rejected the highly descriptive statements of 'classical' regional geography, but in so doing has also rejected the specificity of many spatial and historical situations. The generalities of phenomena are only one aspect of analysis, and emphasis on these alone has tended to make it difficult or impossible to incorporate many social, perceptual, and political aspects. If it is a future aim of the subject to strive to include these aspects, then the idiographic (or situational) aspect, as well as the nomothetic aspect, deserves attention.

A second implication of these trends is that greater emphasis will also be required on institutional, administrative, and governmental influences. The division of powers within the State, the relation of centralist and localist models of decisionmaking, the role of institutional decisionmakers such as finance houses and local authorities, all relatively neglected topics of Anglo-American quantitative research until recently, deserve much greater attention in quantitative modes of analysis.

A third implication is that uncritical acceptance of the tenets of social theory must be rejected for a more explicit development of *geographical* social theory; although geography must use social theory, it must also contribute to it—otherwise the ability of the discipline to contribute original thought will have disappeared, and with it the raison d'être of the discipline. One example of work where this development is taking place is the use of the concept of structure (social, enviornmental, physical built form) interacting with process (desires, actions, production, consumption) to produce a set of outcomes which are spatially constrained but also space forming. They are also historically and idiographically specific, but at the same time part of more general processes. The interaction of spatial structure with spatial process has been erected by Gregory (1981; 1982) to an historical process of spatial structuration. In this he follows, in part, Giddens (1979). However, Pred (1981) takes the slightly different view that structuration is more explicitly dependent on space-time paths, dominant groups, and institutional controls, in which the role of history becomes rather more rapidly diminished by changes and transformations.

In this Pred is seeking to emphasise the historically specific, but within a more explicit generalising framework; in particular he has drawn on the stimulus of Swedish time-geography. What is important in each of these views, however, is the contention that the geographical core must itself shape social theory and not allow itself to be passively shaped by it: space itself in a major source of the situational, the idiographic, and hence of the structurational.

A final implication, therefore, of current Anglo–American and European interactions is that we must reestablish our geographical subject matter. Neither Gregory nor Pred discuss the relevance of the physical environment to social structuration, and this is a common omission in much current geographical literature. In addition, neither the focus on technique of the 1960s, nor the focus on the pure social of the 1970s, are adequate cores of subject matter alone. Moreover, space as a geometry of generalisations divorced from physical and social processes seems equally empty for providing a defensible core. What is required is a core which is clearly defined and, because it is specifically geographical, is universally defensible. This is important not only as a focus of geographical research from which scholars view and approach specific problems, but also for the very survival of the discipline in a time of expenditure cuts and administrative restructuring within the universities. This reestablishment of the subject matter of geography must have a social and humanistic core, but it must also express the specifically geographical. Whatever else that core includes, it must encompass the spatial interrelations between the situational, area differences *per se*, and the physical environment as it affects the situational—namely, the man–environment interface which gives the specific relations and constraints of place within any social or ideological framework at any particular point in time.

It is interesting that, in a recent essay, Bartels (1982) also recognises the need, in Germany, for a new geographical core which derives essentially from the linkages of researchers on man and the environment. Bartels sees geography performing an essential role in creating this linkage, first by providing a bridge between the positive physical sciences and the sociological human sciences; and second, by adding "spatial aspects of genuine importance". Hence, geography's role is both as synthesised and as operational tool to tackle spatial problems. Certainly this conclusion is not unique to German geography and all countries have much to gain by closer cooperation in establishing this important role for the spatial sciences.

Conclusion

The preceding discussion of this chapter has emphasised the emerging links and the persistent differences between Anglo–American and continental European (especially Franco–German) geography. These have important implications for quantitative and theoretical geography, which is the centre of concern here, and also for spatial science as a whole.

It has been argued that the explicit conflict between 'scientific' and 'radical'/'humanist' geography, so characteristic of the 1970s, has had beneficial effects in redirecting attention to socially significant and 'relevant' issues. However, it is also argued that much of the critique of quantitative geography as 'positivism' has been misplaced and has accepted uncritically the representation of science as positivism given by Harvey in 1969. This has had a pernicious and distractive effect on the subject in three main ways: first, it has suggested that scientific and empirical enquiry is largely socially worthless; second, it has often rejected the links between physical and human geography; and third, it has often rejected the existence of geography as a discipline at all [geography should be de-defined in Eliot Hurst's (1980) words]. In contrast, it has been argued here that the approaches of scientific and radical geography must be fused.

A possible methodological link between the fragmenting pluralism of geography, it has been speculatively conjectured, might be found in developments of the German *Sozialgeographie*. With its emphasis on *Raumforschung* and *Raumordnung* this could offer much for the future methodological development of a new geographical core in the 1980s. Such a development not only links aspects of human and physical geography, but also offers an historical-perceptual view of space and territory, and suggests the importance of integrating physical, social, spatial, and economic analysis with applied involvements with the State. It is from this area of interaction that it has been argued that not only fruitful, but central core issues will evolve. These will in significant part require quantitative and theoretical methods deriving from spatial science; although they will also require careful use and interpretation in the contexts of specific modes of thought with particular meaning and value positions. Thus, in contrast to Johnston's conclusions in chapter 2, it is argued here that pluralism, although beneficial in many respects, cannot form a methodological base for the discipline. Instead a new core must be sought, in part by empirical analyses, from the integration of environmental, social, political, and economic aspects with space, historical stimuli, and specific modes of thought and their spatial–political manifestations, and in part by the reestablishment of the geographical subject matter. Whether or not this particular conjecture proves well-founded, the subject as a whole, and quantitative and theoretical geography in particular, will benefit enormously from the renewed development of closer academic links between the Anglo–American and continental European traditions.

Acknowledgments. In writing this essay I have benefited enormously from discussions with a number of people, some of whom have also been kind enough to offer critical comments on an earlier draft. I would particularly like to acknowledge the helpful observations of G Bahrenberg, M Blacksell, R J Chorley, P Haggett, R J Johnston, W R Mead, A Pred, and O Wänerynd, all of whom, I am sure, would wish to disassociate themselves from any responsibility in the result!

References

Abler R F, Adams J S, Gould P R, 1971 *Spatial Organization: The Geographer's View of the World* (Prentice Hall, Englewood Cliffs, NJ)

Ajo R, 1944 *Tampereen Lükennealue, Publications of the Institute of the University of Helsinki 11* (mimeograph)

Anhert E, 1966 "Zur Rolle der elektronischen Rechenmaschine und des mathematischen modells in der Geomorphologie" *Geographische Zeitschrift* **54** 118-133

Annales de Géographie, 1971, Special editorial **80** 641-643

Anuchin V A, 1970a "Mathematization and the geographical method" *Soviet Geography, Review and Translation* **11** 71-81

Anuchin V A, 1970b "On the problems of geography and the task of popularizing geographical knowledge" *Soviet Geography, Review and Translation* **11** 82-112

Arte U, Sprengel U, 1977 "Zum Problem Computer und Geographie: Karlsruher Informations—und Übungskurse in Modernen Geographie—What after?" *Geographische Rundschau* **29** 307-308

Bahrenberg G, 1972 "Räumliche beitrachtungsmeise und Forschungsziele der Geographie" *Geographische Zeitschrift* **60** 8-24

Bahrenberg G, 1976 "Ein sozial gerechtes optimierungsmodell für die Standortwahl von öffentlichen Einrichtungen" *Tagungsbericht und Wissenschaftliche Abhandlungen* **40** 443-452 (Deutscher Geographentag, Innsbruck)

Bahrenberg G, 1979 "Anmerkurgen zu E. Wirth's Vergeblichen versuch einer Wissenschaftstheoretischen begrünlung der Länderkunde" *Geographische Zeitschrift* **67** 147-157

Bahrenberg G, Streit U, (Eds), 1981 *German Quantitative Geography* Papers presented at the Second European Conference on Theoretical and Quantitative Geography in Cambridge 11-14 September 1980, *Munstersche Geographische Arbeiten*, Department of Geography, University of Munster (forthcoming)

Bailly A S, 1976 "Géographie quantitative et théoretique dans les pays Francophones" *L'Espace Géographique* **5** 113-114

Bartels D, 1968a "Die Zukunft der Geographie als Problem ihrer Standortbestimmung" *Geographische Zeitschrift* **56** 124-142

Bartels D, 1968b *Erdkundliches Wissen 19: Zur Wissenschaftstheorischen Grundlegung einer Geographie des Menschen* (Steiner, Wiesbaden)

Bartels D (Ed.), 1970 *Wirtschafts- und Sozialgeographie* (Kiepenheuer und Witsch, Cologne)

Bartels D, 1973 "Between theory and metatheory" in *Directions in Geography* Ed. R J Chorley (Methuen, London) pp 23-42

Bartels D, 1974 "Sozialgeographische Grundlektüre auf Englisch?" *Geographische Zeitschrift* **62** 138-141

Bartels D, 1978 "Perspective de bas dans la Géographie Ouest-Allemande Contemporaine" *L'Espace Géographique* **3** 155-167

Bartels D, 1979 "Theorien nationaler siedlungssysteme und Raumordnungspolitik" *Geographische Zeitschrift* **67** 110-146

Bartels D, 1982 "Geography: paradigmic of functional change? A view from West Germany" in *A Search for Common Ground* Eds P Gould, G Olsson (Pion, London) forthcoming

Bassett K, Short J, 1980 *Housing and Residential Structure: Alternative Approaches* (Routledge and Kegan Paul, Henley-on-Thames, Oxon)
Beaujeau-Garnier J, 1971 *La Géographie: Methodes et Perspectives* (Masson, Paris)
Beguin H, 1979 *Methodes d'Analyse Géographique Quantitative* (Libraires Techniques, Paris)
Beguin H, 1981 "Space and time: an axiomatic approach" in *Dynamic Spatial Models* Eds D Griffith, R D MacKinnon, Proceedings of NATO Advanced Studies Institute (Sijthoff and Noordhoff, Alphen aan den Rijn)
Bennett R J, 1980 *The Geography of Public Finance: Welfare under Fiscal Federalism and Local Government Finance* (Methuen, London)
Bennett R J, 1982 *Central Grants to Local Governments: The Political and Economic Impacts of the Rate Support Grant* (Methuen, London)
Bennett R J, Haining R P, Thornes J, 1980 "Dynamic spatial models" *Area* **12** 284-286
Bennett R J, Wrigley N, 1981 "Introduction, quantitative and theoretical geography: retrospect and prospect" in *Quantitative Geography in Britain: Retrospect and Prospect* Eds N Wrigley, R J Bennett (Routledge and Kegan Paul, Henley-on-Thames, Oxon) chapter 1
Bennett R J, Blacksell M, Cliff A D, Cox N, Gatrell A C, Harris R, Senior D, Wrigley N, 1981 "Second European Colloquium on quantitative and theoretical geography" *Area* **13** 104-108
Berry B J L, 1971 *Géographie des Marches et du Commerce de Detail* (Armand Collin, Paris)
Berry B J L, 1972 "More of relevance and policy analysis" *Area* **4** 77-80
Blacksell M, 1981 "A future in Britain—or Europe?" *Area* **13** 1
Bobek H, 1948 "Stellung und Bedeuntung der Sozialgeographie" *Erdkunde* **2** 118-125
Bobek H, 1950 "Aufriss einer vergleichenden Sozialgeographie" *Mitteilungen der geographischen Gesellschaft Wien* **92** 34-45
Bobek H, 1959 "Die Hauptstufen der Gesellschaft- und Wirtschaftsentfaltung" *Die Erde* **90** 259-298
Bobek H, 1962 "Über den Einbau der Sozialgeographischen Betrachtungsweise in die Kulturgeographie" *Verhandlungen des deutschen Geographentages* **33** 148-166
Boddy M J, 1976 "The structure of mortgage finance: building societies and the British social formation" *Transactions, Institute of British Geographers* New Series **1** 58-71
Bonetti E, 1961 *La Teoria della Localizzazione* Publication 5, Università di Studi-Facultà de Economia e Commercio, Instituto di Geografia, Trieste
Brunet R, 1972 "Les nouveaux aspects de la recherche géographique: rupture ou raffinement de la tradition?" *L'Espace Géographique* **1** 73-77
Burton I, 1963 "The quantitative revolution and theoretical geography" *The Canadian Geographer* **7** 151-162
Buttimer A, 1971 *Society and Mileau in the French Geographic Tradition* (Association of American Geographers, Washington, DC)
Buttimer A, 1976 "Grasping the dynamicism of lifeworld" *Annals, Association of American Geographers* **66** 277-292
Carter H, 1975 *La Geografia Urbana. Teoria e Methodi* (Zanichelli, Bologna)
Castells M, 1976 *The Urban Question* (Edward Arnold, London)
Cauvin C, Reymond H, Hirsch J, 1980 "Valuation de l'information autocorrélation et analyse spectrale: examples de la theorie des places centrales" Rapport final, Tome 2C, Annexe E9 *Cartographie Informatisée et Géographie Humaine* (CNRS, Laboratoire de Cartographie Thématique, Strasbourg)
Chisholm M D I, 1971 "Geography and the question of relevance" *Area* **3** 65-68

Chisholm M D I, Manners G, 1971 *Spatial Policy Problems of the British Economy* (Cambridge University Press, Cambridge)
Ciceri M, Marchand B, Rimbert S, 1977 *Introduction à l'Analyse de l'Espace* (Masson, Paris)
Claval P, 1975a "Geography in Britain: A French view" *Geographical Journal* **141** 345-349
Claval P, 1975b "Contemporary human geography in France" *Progress in Geography* volume 7, Eds C Board and others (Edward Arnold, London)
Claval P, 1976 "La Géographie et les phénomènes de domination" *L'Espace Géographique* **5** 145-154
Claval P, 1977 "Le Marxisme et l'Espace" *L'Espace Géographique* **6** 145-164
Cliff A D, Ord J K, 1973 *Spatial Autocorrelation* (Pion, London)
Cliff A D, Ord J K, 1981 *Spatial Processes: Models and Applications* (Pion, London)
Clout H D, 1972 "Considérations sur les tendencies de la recherche en géographique humaine en Grande-Bretagne dans les années 1960" *L'Espace Géographique* **1** 49-53
Cole J P, 1969 *L'URSS* (Armand Collin, Paris)
Collectif de Chercheurs de Bordeaux, 1977 "A propos de l'article de P. Claval 'Le Marxisme et l'Espace" *L'Espace Géographique* **6** 165-177
Cooke R U, Robson B T, 1976 "Geography in the United Kingdom 1972-1976" *Geographical Journal* **142** 3-22
Cools R H, 1950 "Die Entwicklung und der heutige Stand der Sozialgeographie in den Niederlanden" *Erdkunde* **4** 1-5
Coppock J T, Sewell W R D, 1976 *Spatial Dimensions of Public Policy* (Pergamon Press, Oxford)
Corna-Pellegrini G, Brusa C (Eds), 1980 *La Ricerca Geografica in Italia 1960-1980: Convegno Svottosi sotto gli auspici del Consiglio Nationale della Ricerche* (Associazione dei Geografi Italiani and Ask Edizioni, Milan)
Cruickshank J G, 1981 "Study groups rule, O.K.?" *Area* **13** 18-21
Durand M-G, Le Berre M, Chamussy M, 1980 *Modèle Régional des Alpes du Sud* (Institut de Géographie Alpine, Grenoble)
Eliot Hurst M E, 1980 "Geography, social science and society: towards a de-definition" *Australian Geographical Studies* **18** 3-21
L'Espace Géographique, 1974 special issue "La Géographie Allemande Contemporaine" 7(3) 153-231
Eyles J, 1974 "Social theory and social geography" in *Progress in Geography, volume 6* Eds C Board and others (Edward Arnold, London) pp 27-88
Geipel R, 1952 *Rhein-Mainische Forschungen, 38, Soziale Struktur und Einheitsbewusstsein als Grundlagen geographischer Gliederung* (W Kramer, Frankfurt)
Giddens A, 1979 *Central Problems in Social Theory: Actions, Structure and Contradictions in Social Analysis* (University of California Press, Berkeley)
Giese E, 1978 "Kritische Anmetkungen zur Anwendung faktorenanalytischer Verfahren in der Geographie" *Geographische Zeitschrift* **66** 161-182
Gould P R, 1970 "Is *Statistics inferens* the geographical name for a wild goose?" *Economic Geography* (Supplement) **46** 439-448
Gould P R, 1972 "Reino Ajo and contemporary geography" *Terra* **84** 64-66
Gould P R, 1979 "Geography 1957-1977: the augean period" *Annals, Association of American Geographers* **69** 139-151
Granö J G, 1929 "Reine Geographie" *Acta Geographica* 2(2)
Granö J G, 1931 "Die Geographische Gebiete Finnlands" *Fennia* **52**(3)
Gregory D E, 1978 *Ideology, Science and Human Geography* (Hutchinson, London)
Gregory D E, 1980 "The ideology of control: systems theory and geography" *Tijdschrift voor Economische en sociale geografie* **71** 327-342

Gregory D E, 1981 "Human agency and human geography" *Transactions, Institute of British Geographers* New Series **6** 1-18
Gregory D E, 1982 *Social Theory and Spatial Structure* (Hutchinson, London) forthcoming
Gregory S, 1976 "On geographical myths and statistical fables" *Transactions, Institute of British Geographers* New Series **1** 385-400
Groupe Dupont, 1980 *Montpellier 1954-1978: Processes de transformation de l'Espace et developpment du bati urbain* 2 volumes, Groupe Dupont, Faculté des Lettres et Science Humaines, Avignon
Guelke L, 1974 "An idealist alternative in human geography" *Annals, Association of American Geographers* **64** 193-202
Guelke L, 1978 "Geography and logical positivism" in *Geography and the Urban Environment, Volume 1* Eds D T Herbert, R J Johnston (John Wiley, Chichester, Sussex) pp 35-61
Hägerstrand T, 1953 *Innovations for loppet ur Korologisk Synpunkt* (Gleerup, Lund)
Hägerstrand T, 1967 "On Monte Carlo simulation of diffusion" *Northwestern University, Studies in Geography 13* pp 1-32
Hägerstrand T, 1970 "What about people in regional science?" *Papers, Regional Science Association* **24** 7-21
Haggett P, 1965 *Locational Analysis in Human Geography* First edition (Edward Arnold, London)
Haggett P, 1973a *L'analyse spatiale en géographie humaine* (Armand Collin, Paris)
Haggett P, 1973b *Einführung in die Kultur- und sozialgeographische Regionalanalyse* (De Gruyter, Sammlung Göschen, Berlin)
Haggett P, Cliff A D, Frey A, 1977 *Locational Analysis in Human Geography* Second edition (Edward Arnold, London)
Hahn H, 1950 *Bonner Geographische Abhlandlungen, 4. Der einfluss der Konfessionen auf die Bevölkerung und Sozialgeographie des Hunsrücks* Bonn
Hajdu J, 1968 "Toward a definition of post-war German social geography" *Annals, Association American Geographers* **58** 397-410
Hall P, 1974 "The new political geography" *Transactions, Institute of British Geographers* **63** 48-52
Hard G, 1973 *Die Geographie: Eine Wissenschaftstheoretische Einführing* (De Gruyter, Sammlung Göschen, Berlin)
Hard G, 1978 "Noch einmal: Die Zukunft der physischen Geographien. Zu Ulrich Eisels Demontage lines Vorshlags" *Geographische Zeitschrift* **66** 1-23
Harris R C, 1978 "The historical mind and the practice of geography" in *Humanistic Geography* Eds D Ley, M S Samuels (Maaroufa Press, Chicago) pp 123-137
Hartke W, 1952 *Rhein-Mainische Forschungen, 32, Die Zeitung als Funktion Sozialgeographische Verhältnisse im Rhein-Main Gebiet* (W Kramer, Frankfurt)
Hartke W, 1960 *Denkschrift zur Lage der Geografie* (Steiner, Wiesbaden)
Hartke W, 1963 *Frankreich* (Diesterweg, Bonn)
Harvey D, 1969 *Explanation in Geography* (Edward Arnold, London)
Harvey D, 1973 *Social Justice in the City* (Edward Arnold, London)
Harvey D, 1974a "What kind of geography for what kind of public policy" *Transactions, Institute of British Geographers* **63** 18-24
Harvey D, 1974b "Population, resources, and the ideology of science" *Economic Geography* **8** 239-255
Harvey D, 1978 "Espace et Justice Sociale" *L'Espace Géographique* **7** 300-310
Harvey D, Chatterjee L, 1974 "Absolute rent and the structuring of space by governmental and financial institutions" *Antipode* **6** 22-36

Hauer J, 1971 "Een poging tot plaatsbepaling von de Nieuwe geografie" in *Mathematische methoden in de sociale geografie* (Stichting Interuniversitair Institut voor Social-welenschuppelijk Onderzoek, Amsterdam)
Hay A, 1979 "Positivism in human geography: response to critics" in *Geography and the Urban Environment: Progress in Research and Applications, Volume 2* Eds D T Herbert, R J Johnston (John Wiley, Chichester, Sussex) pp 1-26
Heinemeyer W F, 1977 "De sociaal-geografische bemoeienis met stad en stedelijk systeem. Aperçu van der Nederlandse stadsgeografie sinds 1950" *Geografische Tijdschrift* **11** 259-272
Heslinga M W, 1974 "Over de Vooys en de geografie" in *Een Sociaal-Geografisch Spectrum* Eds J Hinderink, M de Smidt (Geografisch Instituut, Rijksuniversiteit, Utrecht) pp XI-XLIII
Heslinga M W, 1980 "Sociale geografie in meervoud" *Geografische Tijdschrift* **14** 177-181
Holt-Jensen A, 1981 *Geography: Its History and Concepts* (Harper and Row, London)
Isnard H, 1980 "Methodologie et Géographie" *Annales de Géographie* **89** 129-143
Isnard H, Reymond H, Racine J B, 1976 *Une Problematique de la Géographie* (Presses Universitaires de France, Paris)
Jansen A-C M, 1976 "On the theoretical foundation of policy-oriented geography" *Tijdschrift voor Economische and Social geografie* **67** 342-351
Johnston R J, 1979a *Geography and Geographers: Anglo-American Geography since 1945* (Edward Arnold, London)
Johnston R J, 1979b *Political, Electoral and Spatial Systems* (Oxford University Press, Oxford)
Johnston R J, 1980a "On the nature of explanation in human geography" *Transactions, Institute of British Geographers* New Series **5** 402-412
Johnston R J, 1980b *The Geography of Federal Spending in the United States* (John Wiley, Chichester, Sussex)
Johnston R J, 1982 *The Geography of the State* (Macmillan, London) forthcoming
Jones E, 1973 *Villes et cités* (Mercure de France, Paris)
Keunig J H, 1969 *De Denkwijze van de Sociaal-geograaf* (Het Spectrum, Utrecht)
Kilchenmann A, 1970 *Statistischen-analytische Arbeitsmethoden in der regional-geographischen Forschung. Untersuchungen zur Wirtschaftsentwicklung von Kenya und Versuch einer Regionalisierung des Landes auf Grund von thematischen Karten* (University of Michigan Press, Ann Arbor, Michigan)
Kilchenmann A, 1975 "Zum gegenwäntigen Stand der 'Quantitativen und Theoretischen Geographie'" *Giessener Geographisches Schriften* **32** 194-208
Kilchenmann A, 1978a "Dokumentation über Forschungsprojekte aus dem Bereich Theorie und quantitative Methodik in der Geographie" *Karlsruher Manuskripte zur Mathematischen und theoretischen Wirtschafts und Sozialgeographie 24* (mimeograph)
Kilchenmann A, 1978b "Une analyse subjective de la Géographie humaine, en Allemagne Fédérale: avec quelques perspectives pour l'avenir" *L'Espace Géographique* **3** 169-177
Lee D R, 1974 "Existentialism in geographic education" *Journal of Geography* **73** 13-19
Leng G, 1973 "Zur München Konzeption der sozialgeographie" *Geographische Zeitschrift* **61** 121-134
Ley D, Samuels M S (Eds), 1978 *Humanistic Geography* (Maaroufa Press, Chicago)
Lichtenberger E, 1977 *Die Wiener Alstadt* (Deuticke, Vienna)
Lichtenberger E, 1978a "Quantitative geography in the German-speaking countries" *Tijdschrift voor Economische en Social Geografie* **69** 362-373
Lichtenberger E, 1978b "Klassische und theoretisch-quantitative Geographie im deutschen Sprachraum" *Berichte zur Raumforschung und Raumplanung* **22** 9-21

Lichtenberger E, 1978c "Regional science-social systems. A paradigmatic approach" in *Festschrift Karl A. Sinnhuber zum 60 Geburstag* Wirtschafts-geographische studien 4 (Herausgegeben von der Österreichischen Gesellschaft für Wirtschafts-raumforschung, Hirt, Wien)

Lichtenberger E, 1979 "The impact of political systems upon geography: the case of the FRG and the GDR" *Professional Geographer* **31** 201-211

Lichtenberger E, 1980 "Zur Standortbestimmung der Universitätgeographie Reflexion über die institutionelle Situation in der BRD und Grossbritannien" *Mitteilungen der Österreichischen Geographischen Gesellschaft* **122** 3-48

Lloyd P E, Dicken P, 1979 *Spazio e Localizzazione* (Angeli, Milan)

Maier J, Poesler R, Ruppert K, Schaffer F, 1977 *Sozialgeographie* (Bramschweg)

Maier J, Poesler R, Ruppert K, Schaffer F, 1978 "Sozialgeographie—zum diskussions Beitrag von E. Wirth in den Geographischen Zeitschrift 1977" *Geographische Zeitschrift* **66** 262-275

Marchand B, 1972 "L'usage des statistiques en géographie" *L'Espace Géographique* **1** 79-100

Marchand B, 1974 "Quantitative geography: revolution or counter-revolution" *Geoforum* **17** 15-23

Marchand B, 1978 "A dialectical approach in geography" *Geographical Analysis* **10** 105-119

Marchand B, 1981 "Methods for studying change in an urban spatial structure" in *Dynamic Spatial Models* Eds D Griffith, R D MacKinnon, Proceedings of NATO Advanced Study Institute (Sijthoff and Noordhoff, Alphen aan den Rijn, Holland)

Massam B, 1974 "Political geography and the provision of public services" *Progress in Geography* **6** 179-210

McDonald J, 1975 "Current trends in French geography" *Professional Geographer* **27** 15-18

Mead W R, 1963 *The Geographical Tradition in Finland* (H K Lewis, London)

Mead W R, 1977 "Research developments in human geography in Finland" *Progress in Human Geography* **1** 361-375

Nice B, 1953 *Geografia e Pianificazione* Memoria di Geografia Economica, IX, Instituto di Geografia di Università, Naples

Otremba G, 1971 "Zur Anwendung quantitativer Methoden und mathematischer Modelle in der Geographie" *Geographische Zeitschrift* **59** 1-22

Patmore J A, 1980 "Geography and relevance" *Geography* **65** 265-283

Peet R, 1977a "The development of radical geography in the United States" *Progress in Human Geography* **1** 240-263

Peet R, 1977b *Radical Geography: Alternative Viewpoints on Contemporary Social Issues* (Maaroufa Press, Chicago)

Peet R, Slater D, 1980 "Reply to the Soviet review of 'Radical Geography'" *Soviet Geography* **21** 541-545

Pred A R, 1974 *An Evaluation and Summary of Human Geography Research Projects* (Statens Råd för Samhällsforsking, Stockholm)

Pred A R, 1981 "Production, family, and free-time projects: a time-geographical perspective on the individual and societal change in nineteenth-century U.S. cities" *Journal of Historical Geography* **7** 3-36

Prince H C, 1971 "Questions of social relevance" *Area* **3** 150-153

Pumain D, Saint-Julien T, 1980 "L'analyse spatio-temporelle en géographie (Montpellier 1979)" *L'Espace Géographique* **9** 55-56

Racine J B, 1974 "Modèles de recherches et modèles théoriques en géographie" *Bulletin, Association de Géographie Français* **51** 51-56

Racine J B, Raffestin C, Ruffy V, 1978 *Territorialita e Paradigma centro-periferia: la Svizzera de la Padana* (Edizioni Unicopli, Milan)

Racine J B, Bailly A S, 1979 "La géographie et l'espace géographique: à la recherche d'une épistémologie de la géographie" *L'Espace Géographique* 8 283-291
Raffestin C, 1980 *Pour une Géographie du Pouvoir* [Libraires Technique (LITEC), Paris]
Ruppert K, 1958 *Spalt: Ein methodischer Beitrag zum Studium der Agrarlandschaft mit Hilfe der Kleinräumlichen Nutzflachen- und Sozialkartierung und zur Geographie des Hopfenbaus* (M L Kailmünz, Munchener Geographische Heft 14, Regensburg)
Ruppert K, Schaffer F, 1969 "Zur Konzeption der Sozialgeographie" *Geographische Rundschau* **6** 205-214
Sack R D, 1980 *Conceptions of Space in Social Thought: A Geographic Perspective* (Macmillan, London)
Samuels M S, 1978 "Existentialism and human geography" in *Humanistic Geography* Eds D Ley, M S Samuels (Maaroufa Press, Chicago)
Sauberer M, 1973 "Anwendungsversuche der Faktorenanalyse in der Stadtforschung-Sozialräumliche Gliederung Wien" *Seminarberichte der Gesellschaft für Regionalforschung 7* Heidelberg
Sayer R A F, 1976 "A critique of urban modelling: from regional science to urban and regional political economy" *Progress in Planning* **6** 189-254
Smith D M, 1971 "Radical geography—the next revolution" *Area* **3** 153-157
Smith D M, 1974 "Who gets what where and how: a welfare focus for human geography" *Geography* **59** 289-297
Smith D M, 1977 *Human Geography: A Welfare Approach* (Edward Arnold, London)
Smith D M, 1979 *Where the Grass is Greener* (Croom Helm, London)
Steiner D, 1965 "Die Faktorenanalyse: ein modernes statistisches Hilfsmittel des Geographen für die objecktive Räumgliederung und Typenbildung" *Geographica Helvetica* **20** 20-34
Streit U, 1973 "Ein mathematisches Modell zur Simulation von Abflussganglinien (Am Beispiel von Flüssen des Rechtsrheinischen Schiefergebirges)" *Giessener Geographische Schriften* **27**
Streit U, 1975 "Zeitreihensimulation mit Markov-Modellen, dargestellt an Beispielen aus der Hydrologie" *Giessener Geographische Schriften* **32** 165-180
Symposium, 1979 *Symposium on Theoretical and Quantitative Geography, September 28-30, 1978* Répertoire des particiants (Laboratoire de Cartographie Thématique, Strasbourg)
Taylor P J, 1976 "An interpretation of the quantification debate in British geography" *Transactions, Institute of British Geographers* New Series **1** 129-142
Thiebault A, 1972 "L'analyse des espaces régionaux en France depuis début du siècle" *Annales de Géographie* **81** 129-170
Thompson I B, 1975 "Geography in France: a British view" *Geographical Journal* **141** 349-354
Toschi U, 1941 *La Teoria Economica della Localizzazione delle Industrie Secondo Alfredo Weber* (Macri, Bari)
Troll C, 1949 "Geographic science in Germany during the period 1933 to 1945" *Annals, Association of American Geographers* **39** 128-135
Tuan Y-F, 1971 "Geography, phenomenology and the study of human nature" *Canadian Geographer* **15** 181-192
Tuan Y-F, 1977 *Space and Place* (Edward Arnold, London)
Unwin D J, 1979 "Theoretical and quantitative geography in North-West Europe" *Area* **11** 164-166
Vlora N R, 1979 *Città e Territorio* (Mursia, Milan)
Volkmann H, 1976 "Quantitative, empirische Methoden in der Didaktik der Geographie" *Geographische Rundschau* **28** 295-296

Walmsley D J, 1974 "Positivism and phenomenology in geography" *The Canadian Geographer* **18** 95-107
Wirth E, 1977 "Die deutsche Sozialgeographie in ihrer theoretischen Konzeption und in ihrem Verhältnis zu Soziologie und Geographie des Menschen" *Geographische Zeitschrift* **65** 161-187
Wirth E, 1978 "Zur wissenschaftstheoretischen Problematik der Länderkunde" *Geographische Zeitschrift* **66** 241-261
Wise M J, 1977 "Geography in universities and schools" *Geography* **62** 249-258
Wrigley N, Bennett R J (Eds), 1981 *Quantitative Geography in Britain: Retrospect and Prospect* (Routledge and Kegan Paul, Henley-on-Thames, Oxon)

Part 1

Developments in methodology and theory

Ideology and quantitative human geography in the English-speaking world

R J Johnston

The growth of quantitative geographical work in the English-speaking world during the last three decades has been substantial. As early as the late 1960s, La Valle et al (1967) were able to report on widespread numeracy requirements among North American graduate schools, and similar requirements at the undergraduate level have since been reported for the United Kingdom (Unwin, 1978). The plethora of research studies using quantitative methods is widely appreciated, and the many advances and applications by British geographers are chronicled in a recent impressive volume (Wrigley and Bennett, 1981). Numerous textbooks have been produced, both to prepare the would-be researcher and to induct the tentative student. Nearly every British and North American geography undergraduate since the mid-1960s will have been required to take at least one quantitative course, and the majority of substantive research papers report some usage of quantitative procedures.

This apparent dominance of Anglo-American geography (in particular of human geography, which is the topic of concern here) has not come about without a series of academic struggles. Its introduction was strongly contested in some quarters (Burton, 1963; Taylor, 1976), and although grudging admission to course structures was readily achieved in most institutions—to avoid open conflict at least—there were (and still are) many who saw no lasting value to the subdiscipline [see, for example, Stamp (1966)]. And in the last decade or so, antiquantitative movements have been launched, generating substantial (sometimes bitter) debate over the value of what in the 1960s was hailed as the 'new geography'.

Any approach to an academic discipline—and perhaps especially to a social science discipline—is ideologically influenced. Thus much of the debate over the value of the quantitative approach has been ideologically phrased. In reviewing the debates, and in evaluating the present and future contribution of quantitative work in human geography, this ideological base must be examined. This is the purpose of the present essay. First, however, a definition of the term ideology as it is to be used in this paper is needed, if for no other reason than that "Ideology is perhaps one of the most equivocal and elusive concepts one can find in the social sciences" (Larrain, 1979, page 13). In general, the concept is used to denote a set of belief systems, a world view which governs the beliefs, attitudes, and actions of an individual, group, or class (Bell, 1977). As Larrain (especially page 172ff) points out, however, ideology can have either a positive or a negative meaning. The former is the "system of opinions, values and knowledge which are connected with certain class interests" (page 172), whereas the latter is "distorted knowledge" set up in opposition to "true knowledge", which is science (page 173). To many—especially critics of certain radical approaches to social science (see below)—the antithesis of science and ideology is stressed. Here however, the positive meaning of ideology is applied: the approach to science is developed within certain class interests—although those interests are generally implicit and frequently subconscious. In this way, attitudes to science, and to human geography in particular, are part of attitudes to society as a whole. Thus debate between social scientific approaches becomes a debate between 'world views'.

In presenting this review of ideological debates over quantification in human geography, all work in the discipline is classified into one of four paradigms. A paradigm—as the first user of the term in this context indicated (Kuhn, 1977)—can itself be a world-view or an ideology. It is used here as a shorthand description of the four main types of work practised in Anglo-American human geography since c. 1945. (Clearly, the classification of some pieces is problematic.) No argument is presented here in favour of Kuhn's (1962) paradigm-succession model, in which each period of single paradigm-dominance (a period of 'normal science') is ended by a scientific revolution and conversion of the academic community to a new paradigm. Rather, a period of increasing pluralism is recognised, with the relative strengths of the competing paradigms being accounted for by factors which are of no importance to the present argument (Johnston, 1978; 1979).

Quantification, spatial science, and regional geography

The roots of quantitative human geography reach back at least into the 1930s. (Here it should be stressed that quantification refers to the use of formal mathematical modelling and statistical analyses and not simply to the arithmetic manipulation of data.) But it was only in the mid and late 1950s that the major developments that were to launch the 'quantitative revolution' took place in the United States. Before that, human geography was dominated by the regional approach, which stressed the uniqueness of places.

The original 'quantifiers' made two major criticisms of the work of their peers. The first was expressed most forcibly by Schaefer (1953) in a posthumous paper attacking what he identified as the exceptionalist view of geography codified by Hartshorne (1939) in his classic, *The Nature of Geography*. According to Schaefer, any science should seek to explain that which it studies; and such explanations require laws. Geographers investigate spatial patterns, and so

> "... geography has to be conceived as the science concerned with the formulation of the laws governing the spatial distribution of certain features on the surface of the earth" (page 227).

The impact of Schaefer's paper is unclear, though it has been widely praised by Bunge (1968; 1979). Nevertheless, other workers were formulating similar opinions, and were seeking to develop human geography as a member of the sciences rather than of the humanities.

The second criticism of regional geography referred to its methods, which were generally considered sloppy. Although undoubtedly expressed verbally at the time, such beliefs were not widely published (perhaps because they were unacceptable to conservative editors). They have recently been stated forcibly by Gould (1979), one of

> "... a new generation, one that was both sick and ashamed of the bumbling amateurism and antiquitarianism that had spent nearly half a century of opportunity in the university piling up a tip-heap of unstructured factual accounts ... it was practically impossible to find a book in the field that one could put in the hands of a scholar in another discipline without feeling ashamed" (pages 140–141).

As with most critiques, the attack on the regional school contained some excesses. Later work has suggested that Hartshorne and Schaefer did not differ substantially on many fundamental points (Guelke, 1977), and the Berkeley school, led by Carl Sauer, had a clear belief in the primacy of one causal factor—culture (Duncan, 1980). Nevertheless, quantitative geography was a novel approach to human geography, which was launched in opposition to the current orthodoxy and it soon attracted many adherents.

As this new approach grew, so it clearly (at least in hindsight) was fumbling for a focus and rationale; its ideology was implicit and only partially understood, in large degree because many (perhaps most) of the early converts had no strong basis in the philosophy of science which their ideology demanded. (That philosophy is often argued to be logical positivism.) Much stress was placed on the precision brought by quantification (see Cole, 1969), and although the need to develop generalisations and laws was seen as paramount, the means to achieve that end were poorly understood. Indeed, the first detailed (and far from complete) statement of the philosophy was only provided more than a decade after the initiation of quantitative work in human geography [Harvey (1969): that those working at Iowa were aware of the philosophical base is indicated by King (1979a)]. At the same time, the focus of a quantitative human geography was being sought. Eventually, the consensus view was that, within the corpus of the social sciences, human geographers should focus on space (or distance) as a variable influencing human behaviour (Cox, 1976). This view was codified in a number of texts, notably the influential volumes by Haggett (1965), Morrill (1970a), and Abler et al (1971): interestingly, much of the stimulus for the spatial approach came from the other social sciences (Pooler, 1977).

Much early quantitative human geography was naive, both technically and substantively. Its contribution, even to the 'new' literature, was minimal, though much of it remains in the published record. Relatively few axioms were used to develop theory (undoubtedly the most popular of which was Christaller's central place theory) and the deductive models proved poor approximations of the 'real world'. In the mid-1960s, however, a stronger inductive element was introduced, based on investigations of behaviour in space. By the late 1970s, an impressive volume of research had been completed (Haggett et al, 1977), and the technical sophistication of much of this was indicative of extremely rapid advances.

As already indicated, the implicit ideology of this initial development of quantitative work was weakly understood and expressed. Basically, the goal was to develop human geography as a science, equivalent to the 'harder' social sciences (notably economics) if not to the natural sciences. The belief underpinning this goal [as best expressed in Stewart's social physics—Stewart (1956)—later converted into macrogeography—Stewart and Warntz (1958)] was that the spatial organisation of human society is highly ordered and the function of human geography is to identify that order. [This did not imply that man, as an individual, lacked free will but that, in the aggregate, he appeared to obey certain spatial laws (Jones, 1956).]

The ideology of the natural sciences—and in particular of logical positivism—is expressed in a large number of ways, and in almost as many variations. As far as can be identified from the published literature and references, human geographers adopted no one ideological statement. They apparently accepted, and felt no need to defend, what Mulkay (1979) has called 'the standard view of scientific knowledge':

> "The natural world is to be regarded as real and objective. Its characteristics cannot be determined by the preferences or the intentions of its observers. These characteristics can, however, be more or less faithfully represented. Science is that intellectual enterprise concerned with providing an accurate account of the objects, processes and relationships occurring in the world of natural phenomena. To the extent that scientific knowledge is valid, it reveals and encapsulates in its systematic statements the true character of this world ... accepted scientific knowledge, because it has satisfied ... impersonal, technical criteria of adequacy, is independent of those subjective factors, such as personal prejudice, emotional involvement and self-interest, which might otherwise distort scientists' perception of the external world" (pages 19–20).

Human geographers believed that these criteria could be applied to the world of human behaviour as well as to that of 'the natural world'. Thus Harvey (1969, page 29) sought to examine the relevance of the '*standard model* of scientific explanation', associated in particular with physics (italics in original), to geography: the whole tenor of his book is that the standard model is applicable, and that the main route to explanation in geography is by developing "general theory ... examining the interactions between temporal process and spatial form" (page 483).

This orientation is reflected in several texts of the 1960s/1970s. Amedeo and Golledge (1975), for example, state that their purpose is "to present an introduction to scientific reasoning in geography ... [which is] the reasoning of logical positivism" (page iii), and their description (cf page 36) closely follows the standard model. Similarly, Abler et al (1971) assert that "The scientific method is singular" (page 54) and it is applied by them to "The distinctively geographical question ... '*why are spatial distributions structured the way they are?*' " (page 56; their italics): "our view ... is that the explanation of classes of events by demonstrating that they are instances of widely applicable laws and theories is the function of geography" (page 87). Finally, the first section of the first chapter of Haggett's classic *Locational Analysis in Human Geography* (1965) was entitled "The search for order" and the organisation of the entire book stressed the logical positivist methodology. Twelve years later, the revised edition maintained that structure; it assumed mathematical sophistication among undergraduates but was slightly apologetic for being possibly 'narrow and positivist in its philosophy' (Haggett et al, 1977, page x).

During the 1960s and 1970s the applications of the standard model of the natural sciences to human geography became technically more sophisticated and substantively less naive. These developments are not reviewed here. One extra feature of them is, however. One of the attractions of the natural science model to many workers is that explanation brings with it the ability to control; thus advances in scientific understanding generate social and economic progress, as the natural world becomes more amenable to man's manipulations. A logical consequence of this is that, if the natural science model is applicable to human geography, then so is the correlate: an advance in social science application leads to an advance in man's ability to control society. Thus a scientific human geography is valuable not only as a provider of information but as an input to planning, particularly spatial planning.

This was not the first utilitarian view of human geography, but it was the first presentation of the discipline as a predictive/forecasting activity rather than as a data-compiling and presenting agency. Thus the links between geography and spatial planning become closer (as exemplified, for example, in the works of Wilson, 1974, and of Batty, 1976; 1978). One partial consequence of the incorporation of control (via planning) into the ideology was a more pragmatic selection of research topics, especially if the research needed funding and thus had to be justified to sponsors. (The ideology did not, of course, present the role of human geography in planning as one of social control. The standard model sees science as objective and neutral. Thus geographers-as-planners were disinterested professionals, providing expert advice on technical issues with no reference to 'personal prejudice, emotional involvement and self-interest': geographers were merely joining the band of scientists who were helping to create a better world for everybody.) Wilson, for example, presented his book as "a body of theory on cities and regions. Important, but subsidiary, aims are to show how the resulting mathematical models can be built empirically, and how they can be used in planning studies to help solve problems" (Wilson, 1974, page vii). Batty, too, focused on the link between social science theory and planning practice. He recognised that no social science theory can ever be fully validated, but argued (Batty, 1978, page 77) that

> "There seems no way out for the use of the scientific method in social science, but if science is not to be used, this is a prescription for doing nothing, for there is nothing to replace it."

Finally, Bennett and Chorley (1978) gave their book *Environmental Systems* the subtitle *Philosophy, Analysis and Control*, and stated as one of their aims

> "It is hoped to show that the systems approach provides a powerful vehicle for the statement of environmental situations of ever-growing temporal and spatial magnitude, and for reducing the areas of uncertainty in our increasingly complex decision-making arenas" (page 21).

To these authors, as to many others, analysis of how the world works leads naturally to applications of research results to improve the world. (But see Gregory, 1980.)

The challenges to quantitative geography

The ideology of natural science has never reigned supreme in human geography, although in the mid-1960s it was undoubtedly subscribed to by a majority of members in the discipline, and certainly by a very large majority of its youngest generation. The older generation remained sceptical, however (Taylor, 1976), especially those who were cast as the 'field workers' who would collect the information needed to evaluate the models and test the hypotheses of the theoreticians (Johnston, 1979, pages 70–72). By the late 1960s they had converted others to their scepticism, and increasingly, in the decade to follow, the challenge to quantitative human geography became louder and stronger. This challenge came from two major ideological sources.

The humanistic challenge

This first challenge attempted to rebuild the human geography that had preceded the quantitative and theoretical revolution, but on a firmer philosophical foundation. Thus the critique included not only an evaluation of the validity of the standard model of natural science (Guelke, 1971; 1978) and a defence of regional geography as synthesis (Harris, 1971; 1978) but also an exploration of other epistemologies, notably those of phenomenology (Tuan, 1971; Relph, 1970), idealism (Guelke, 1974), and existentialism (Buttimer, 1976; Samuels, 1978).

The main focus of the humanistic challenge to the natural science approach was the model of man that it embodied. Ley (1977), for example, argued that in social geography

> "... a materialist treatment drawing upon a physical science tradition [encouraged] deterministic thinking. Social relations were suppressed in the interest of spatial facts, and social geography became preoccupied with man's material works and the irresistable objectivity of the map" (page 500).

To him, a positivist approach to locational studies carried with it an "implicit economic determinism, with a pale spectre of man" (page 501); it should be replaced by an attitude that

> "To understand social process one must encounter the situation of the decision-maker, which includes incomplete and inconsistent information, values and partisan attitudes, short-term motives and long-term beliefs" (page 502).

Such a man-centred geography requires a phenomenological, not a positivist, approach. In similar vein, Buttimer (1979) asked whether positivism (concerned, she claimed, with 'know how' rather than 'know what') tends

> "... to bury rather than promote individual creativity in scholarship, and open, caring engagement in discussion with people actually involved in environmental issues" (page 24).

Thus, she concludes

> "A great deal of the direction and volume of applied research today is shaped by the practical and technical interests of sponsors, which in turn tend to favor the promotion of existing (ideological and economic) conditions ... the actual intellectual exploration of [problems] does not promote dialogue among different stances or feedback between planners and those for whom they plan" (page 26).

As Ley and Samuels (1978) express it, positivism reduced man to an "image of nature ... a mere product of an environment" (page 7): quantitative human geography is dehumanised.

The radical challenge

This second challenge developed into a much wider-ranging one than the first. Initially, it was a relatively mild critique, launched at the end of the 1960s, not of the scientific approach per se but of the topics which human geographers sought to investigate with it. At a period of considerable social and economic unrest, much apparently 'applicable' human geography was deemed irrelevant to the real issues of the day, and thus was considered trivial, it not naive. More 'relevant' work was called for, with reference to the problems of, for example, poverty and inequality (Smith, 1973; Coppock, 1974; Steel, 1974; White, 1972); 'geography and public policy' became an important phrase. Increasingly, however, some realised that this relatively minor change of direction was insufficient (Peet, 1977): to some, the conclusion was one of despair (for example, Zelinsky, 1975); to others, it was that the reigning ideology must be overthrown.

A considerable number of ideological onslaughts was published. Harvey, for example, pointed out the irrelevance of quantitative human geography to the solution of major world problems, many of which have a spatial component (Harvey, 1973); he also illustrated how the proposed solution to a particular perceived problem—overpopulation—was ideological (Harvey, 1974a), and questioned the impacts, if not the motives, of the activities of geographers in the 'corridors of power' (Harvey, 1974b). Others have claimed, for example, that: the content of geographical inquiry is determined by data availability rather than by theory; that the available theories are mechanistic abstractions from socioeconomic reality; that description triumphs over explanation of process; that the relationships between spatial patterns and political economy, especially the class relations of capitalist society, are ignored; that consensus rather than conflict is advanced as the cement of society; and that available theories cannot incorporate qualitative change. (For these criticisms see, for example: Eyles, 1974; Slater, 1975; Gregory, 1978a.)

For the radical critics, therefore, quantitative human geography is, at best, sophisticated description of patterns over space and time, which have no explanatory power. As Sayer (1979a, page 1059) puts it

> "... it is often quite unnecessary to look for empirical regularities amongst large data sets in the hope of finding basic relationships ... . Such regularities sometimes highlight what has to be explained, but in themselves they explain nothing."

Further, it is claimed that in adopting the standard model of normal science, human geographers assume that the generalisations they have derived are universals:

> "... positivist science ... expects to find invariant, universal empirical regularities of human behaviour. Any regularities discovered at a particular point in time are hopefully expected to hold true for the future" (Sayer, 1979b, page 30).

Such an assumption is contrary to the dialectical view of social change presented by Marxists and other radicals (Marchand, 1978): each observation is peculiar to the specific place and time (a conjuncture) and cannot be generalised from (Massey and Meegan, 1979).

Many of the critics see the dialectical process operating between two levels (Cosgrove, 1979). These carry various names. One pair is the infrastructure and the superstructure: the former represents the operating process, and the latter the particular realisation (or outcome) of the process (see Gregory, 1978a). Many of the radicals have an ideology which is oriented to fundamental change, which sees no separation betwen theory and action, and rejects the objective neutrality model as relevant to social science: to them facts about human society are not neutral but are value-laden, according to the ideology within which they are both collected and interpreted. Thus their focus is not on the particular realisation of a process—on the detail of the spatial pattern—but on the process itself (Gregory, 1978b; 1978c; Johnston, 1980a). In addition, some claim that a detailed explanation of a particular realisation is not possible, because a large number of contingent (space-time specific) relationships are acting on the general process, and the interactions are not readily unravelled. Thus, to them, most—if not all—quantitative human geography is irrelevant. Furthermore, it is counter-productive; its existence buttresses the status quo within society, supporting planning activity which will maintain the current unequal distribution of power and life-chances.

Whereas the ideology of the radical challenge is clear in terms of its ends, the means are poorly specified. Revolution—in terms of an overthrow of the existing power structure—is argued for by only a few advocates, and there have been only a small number of arguments for achieving fundamental change via social democracy and participation (Blowers, 1974). Harvey's (1973) position suggests an 'ivory-tower approach': the proper role of the scholar is to think and to develop theory, which presumably will then permeate the rest of the population via the education system. Gregory (1978a), like many of the humanistic geographers, wants a critical science which will emancipate people, releasing them from their current alienated state and giving them control over their existence; again, it must be presumed that this ideological end is to be achieved via the education process.

The quantifiers' response

The response to these two major challenges has been a varied one. The humanistic critique has received least attention from the quantifiers, in part because it presents a complete antithesis to the natural science model and in part because it is both relatively muted and prepared for coexistence. Thus Tuan (1977) states that

> "... we measure and map space and place, and acquire spatial laws and resource inventories for our efforts. These are important approaches, but they need to be complemented by experiential data that we can collect and interpret in measured confidence because we are human ourselves" (page 5).

The radical challenge has been considered much more important and the debate has often been strident (see Johnston, 1979, chapter 6). Many of the most basic criticisms have been recognised and accepted. Description for description's sake via multivariate techniques—notably factorial ecology —is much less popular now than it was a decade ago, and the 'fetishism of space' has been realised: relatively few claims are made now that geography is a 'discipline in distance'. (The major philosophical criticisms of this fetishism were made by May, 1970, and Sack, 1972; 1974, neither of whom is allied with the radical group.) Similarly, much of the naive belief in the efficacy of geographical techniques for the solution of social problems is gone, as exemplified by Taylor and Gudgin's (1976) explosion of the myth of nonpartisan cartography, as claimed by Haggett and Chorley (1969) in their treatment of the redrawing of boundaries to electoral districts.

Some of the more fundamental criticisms have been accepted, too, as some members of the vanguard of the quantitative movement reflect on its achievements (King, 1976; 1979b). The role of the infrastructure is realised, and the need to study the major processes of political economy accepted. But the full radical ideology cannot be accepted (Morrill, 1970b; 1974): social change is required, but not fundamental social change. Nor can a focus solely on the infrastructure be supported: analysis of the superstructure—the particular realisations on spatial patterns—is considered both desirable and necessary (as well as satisfying). What, then, is the future for quantitative human geography?

Quantitative human geography in a pluralist discipline

Neither of the challenges to quantitative human geography outlined here has succeeded: spatial or locational science is perhaps healthier now than it has ever been; it is certainly as lively and as stimulating, even though the research frontier is leaving many struggling at the customs post (Johnston, 1980b). But the challenges will not disappear either. Human geography in the next decade is likely to be more pluralistic in its approaches than heretofore. This pluralism may be characterised by bitter confrontations; it may be characterised by accommodations (Hay, 1979).

Assuming that accommodation—or at least mutual respect—is feasible, what will be the role for quantitative human geography? As already indicated, the complementary situation relative to humanistic geography has been recognised. The major issue concerns the interaction between the natural scientific and the radical ideologies.

Two major points seem of importance here. The first concerns the scale of study and the substantive content. Much radical theory is concerned with two topics only: spatial inequalities at an international and interregional level; and the manipulation of space for capitalist gain within urban areas. The theories are written at the level of the infrastructure. But many geographers are interested in the superstructure, in accounting for the particular realisations (in part because they appreciate that such accounts may stimulate emancipation more rapidly than will general theories). Few theories written at the infrastructural level are deterministic, presenting a mechanistic view of man and society comparable to some of the early, naive spatial science (though see Ley, 1978). They allow for autonomy. Thus within a general theory, whether it be of the state or of the world economy (Johnston, 1981a; Taylor, 1981), it is possible to study, using quantitative procedures (regression—Johnston, 1981b; factor analysis—Taylor, this volume), how particular realisations have come about. No claim is made for the universality of the findings over time and space; they are simply descriptions of a particular situation set in an infrastructural theoretical matrix.

In addition, many of the topics studied by quantitative human geographers are concerned with the exercise of individual autonomy within the constraints imposed by the infrastructure. Generalisations are sought about these choices, and of the characteristics differentiating people electing for particular selections: the choices may be of places to visit, of shops to purchase from, of political parties to vote for, of homes to purchase, of social services to spend tax income on, of crops to grow, and so on. All of these take place within a particular conjuncture (the length of a conjuncture has not been defined in the geographical literature), and quantitative generalisations about them are valid.

The response from the radical school to the argument in the preceding paragraph may well be that the topics listed there (and the many others studied by quantitatively-inclined human geographers) are trivial: choice of shopping centre is irrelevant to the achievement of long-term social change. Indeed, but this highlights a major ideological divide between the two groups. The radicals' goal is a long-term one. That of the quantifiers is more short-term; when and where their expertise is applied it is to achieve immediate goals, involving the solution of problems (some serious, some relatively trivial—except to the sufferer, perhaps) for the current world population, rather than the future one.

This dualism immediately raises other issues. The solution of the short-term problems may, because it creates a false consciousness among the proletariat, hamper, delay, even prevent, achievement of the long-term goals (to which many of the quantifiers may be committed). This poses a moral problem, clearly identified in an analogy by Wallerstein (1979) of an individual faced with the following problem. A speedboat race is taking place near the coast; 1000 people are swimming in the water and liable to drown, because several of the boats come too close inshore. He has several courses of action available: (1) swimming out and rescuing 2 or 3, but leaving the other 997 to drown; (2) locating a lifeline, with which 50 are saved but 950 drown in the interim; (3) preventing the boats coming near to shore, thus saving the other 950 but letting the 50 saved under plan (2) drown; and (4) getting the law changed, so that no swimmers will be put in danger in the future. Wallerstein's choice is plan (3) in the short-term and plan (4) in the long-term. But he realises the moral dilemma:

> "I can understand how others might choose differently. I can respect these other choices, if made with clarity of vision and understanding of the consequences and of who it is that will benefit from each of the various solutions to the dilemma. The basic problem is one concerning

the structure of inequality of the present world-system. As long as it persists, swimmers will drown. But it is not inevitable that swimmers drown" (page 131).

This is not to claim that all quantitative human geographers have developed the clarity of vision and understanding that will influence their choice of what to study, how, and why: indeed, some accept that it is inevitable that some swimmers drown and others prefer to tackle only the small problem. But the ideological challenge of the last decade has made quantifiers aware of the implicit choices that they are making and of the consequences of their actions. Their reactions may have been, like David Harvey's, to renounce the natural science model and to decide that where evidence is required, sophisticated technology is irrelevant (Harvey, 1974c). Or they may have reacted by sharpening their research, so as to improve understanding of the present conjuncture and to enhance emancipation. Or, they may have decided to 'do their own thing' arguing (falsely) that ideology is not for them.

Conclusions

Quantitative work in human geography in the English-speaking world began in something of an ideological vacuum. As a result, much of the initial research was naive, both in its conception and implementation and in its application. During the last decade, however, the challenges to the implicit ideology have been more forceful and fruitful than those raised in the 1950s and 1960s. As a result, the ideological status of quantitative human geography is much more widely realised, and current work is better informed.

The thrust of some of the humanistic challenge and almost all of the radical challenge has been that quantitative human geography is ill-conceived, irrelevant, and counter-productive to the achievement of both a better and fairer society and an improved quality of life for the individual. Ideologies have been placed in opposition: consensus and accommodation seem unlikely. The possibility of accommodation has been proposed here, however. A role for quantitative work has been outlined which is not antithetical to the major theoretical tenets of the radical challenge, though it may inhibit achievement of its practical goals. The accommodation, if achieved, will be uneasy, but quantitative work will continue. If there is no accommodation, dissensus may lead to entrenched positions, to the detriment both of the discipline and of society.

Acknowledgements. Continued discussions on the general issues raised here with Derek Gregory, Alan Hay, and Peter Taylor are acknowledged; they are, of course, not to be associated with these views, which they have not commented upon. The stimulating contributions of Gerhard Bahrenberg and Bob Bennett at the Colloquium are also gratefully acknowledged.

References

Abler R F, Adams J S, Gould P R, 1971 *Spatial Organization: The Geographer's View of the World* (Prentice-Hall, Englewood Cliffs, NJ)
Amedeo D, Golledge R G, 1975 *An Introduction to Scientific Reasoning in Geography* (John Wiley, New York)
Batty M, 1976 *Urban Modelling* (Cambridge University Press, Cambridge)
Batty M, 1978 "Urban models in the planning process" in *Geography and the Urban Environment*, volume 1, Eds D T Herbert, R J Johnston (John Wiley, Chichester, Sussex) pp 63-134
Bell D, 1977 "Ideology" in *The Fontana Dictionary of Modern Thought* Eds A Bullock, O Stallybrass (Fontana Books, London) pp 298-299
Bennett R J, Chorley R J, 1978 *Environmental Systems* (Methuen, London)
Blowers A T, 1974 "Relevance, research and the political process" *Area* **6** 32-36
Bunge W, 1968 "Fred K Schaefer and the science of geography" *Harvard Papers in Theoretical Geography*, Laboratory for Computer Graphics and Spatial Analysis, Cambridge, Mass
Bunge W, 1979 "Perspectives on *Theoretical Geography*" *Annals, Association of American Geographers* **69** 169-174
Burton I, 1963 "The quantitative revolution and theoretical geography" *The Canadian Geographer* **7** 151-162
Buttimer A, 1976 "Grasping the dynamism of lifeworld" *Annals, Association of American Geographers* **66** 277-292
Buttimer A, 1979 "Erewhon or nowhere land" in *Philosophy in Geography* Eds S Gale, G Olsson (D Reidel, Amsterdam) pp 9-38
Cole J P, 1969 "Mathematics and geography" *Geography* **54** 152-163
Coppock J T, 1974 "Geography and public policy: challenges, opportunities and implications" *Transactions, Institute of British Geographers* **63** 1-16
Cosgrove D, 1979 "Ron Johnston and structuralism" *Journal of Geography in Higher Education* **3** 106-112
Cox K R, 1976 "American geography: social science emergent" *Social Science Quarterly* **57** 182-207
Duncan J S, 1980 "The superorganic in American cultural geography" *Annals, Association of American Geographers* **70** 181-198
Eyles J, 1974 "Social theory and social geography" in *Progress in Geography*, volume 6, Eds C Board et al (Edward Arnold, London) pp 27-88
Gould P R, 1979 "Geography 1957-1977: the augean period" *Annals, Association of American Geographers* **69** 139-151
Gregory D, 1978a *Ideology, Science and Human Geography* (Hutchinson, London)
Gregory D, 1978b "The process of industrial change 1730-1900" in *An Historical Geography of England and Wales* Eds R A Dodgshon, R A Butlin (Academic Press, London) pp 291-312
Gregory D, 1978c "Social change and spatial structures" in *Making Sense of Time* Eds T Carlstein, D N Parkes, N J Thrift (Edward Arnold, London) pp 38-46
Gregory D, 1980 "The ideology of control: systems theory and geography" *Tijdschrift voor Economische en Sociale Geografie* **71** 327-342
Guelke L, 1971 "Problems of scientific explanation in geography" *The Canadian Geographer* **15** 38-53
Guelke L, 1974 "An idealist alternative in human geography" *Annals, Association of American Geographers* **64** 193-202
Guelke L, 1977 "Regional geography" *The Professional Geographer* **29** 1-7
Guelke L, 1978 "Geography and logical positivism" in *Geography and the Urban Environment*, volume 1, Eds D T Herbert, R J Johnston (John Wiley, Chichester, Sussex) pp 35-61

Haggett P, 1965 *Locational Analysis in Human Geography* (Edward Arnold, London)
Haggett P, Chorley R J, 1969 *Network Models in Geography* (Edward Arnold, London)
Haggett P, Cliff A D, Frey A E, 1977 *Locational Analysis in Human Geography* (Edward Arnold, London)
Harris R C, 1971 "Theory and synthesis in historical geography" *The Canadian Geographer* **15** 157-172
Harris R C, 1978 "The historical mind and the practice of geography" in *Humanistic Geography* Eds D Ley, M S Samuels (Maaroufa Press, Chicago) pp 123-137
Hartshorne R, 1939 *The Nature of Geography* (Association of American Geographers, Lancaster, Pa)
Harvey D, 1969 *Explanation in Geography* (Edward Arnold, London)
Harvey D, 1973 *Social Justice and the City* (Edward Arnold, London)
Harvey D, 1974a "Population, resources, and the ideology of science" *Economic Geography* **50** 256-277
Harvey D, 1974b "What kind of geography for what kind of public policy" *Transactions, Institute of British Geographers* **63** 18-24
Harvey D, 1974c "Class-monopoly rent, finance capital and the urban revolution" *Regional Studies* **8** 239-255
Hay A M, 1979 "Positivism in human geography: response to critics" in *Geography and the Urban Environment*, volume 2, Eds D T Herbert, R J Johnston (John Wiley, Chichester, Sussex) pp 1-26
Johnston R J, 1978 "Paradigms and revolutions or evolution: observations on human geography since the Second World War" *Progress in Human Geography* **2** 189-206
Johnston R J, 1979 *Geography and Geographers* (Edward Arnold, London)
Johnston R J, 1980a "On the nature of explanation in human geography" *Transactions, Institute of British Geographers* New Series **5** 402-412
Johnston R J, 1980b "After 150 years ..." *The Geographical Magazine* **52** 601-603
Johnston R J, 1981a "Political geography in Britain since Mackinder" in *Political Studies from Spatial Perspectives* Eds A D Burnett, P J Taylor (John Wiley, Chichester, Sussex)
Johnston R J, 1981b *The Geography of Federal Spending in the United States* (John Wiley, Chichester, Sussex)
Jones E, 1956 "Cause and effect in human geography" *Annals, Association of American Geographers* **46** 369-377
King L J, 1976 "Alternatives to a positive economic geography" *Annals, Association of American Geographers* **66** 293-308
King L J, 1979a "Areal associations and regressions" *Annals, Association of American Geographers* **69** 124-128
King L J, 1979b "The seventies: disillusionment and consolidation" *Annals, Association of American Geographers* **69** 155-157
Kuhn T S, 1962 *The Structure of Scientific Revolutions* (University of Chicago Press, Chicago, Ill.)
Kuhn T S, 1977 "Second thoughts on paradigms" in *The Structure of Scientific Theories* Ed. F Suppe (University of Illinois Press, Urbana) pp 459-482
Larrain J, 1979 *The Concept of Ideology* (Hutchinson, London)
La Valle P, McConnell H, Brown R G, 1967 "Certain aspects of the expansion of quantitative methodology in American geography" *Annals, Association of American Geographers* **57** 423-436
Ley D, 1977 "Social geography and the taken-for-granted world" *Transactions, Institute of British Geographers New Series* **2** 498-512
Ley D, 1978 "Social geography and social action" in *Humanistic Geography* Eds D Ley, M S Samuels (Maaroufa Press, Chicago) pp 41-57

Ley D, Samuels M S (Eds), 1978 "Contexts of modern humanism in geography" in *Humanistic Geography* (Maaroufa Press, Chicago) pp 1-18
Marchand B, 1978 "A dialectical approach in geography" *Geographical Analysis* **10** 105-119
Massey D, Meegan R A, 1979 "The geography of industrial reorganisation" *Progress in Planning* **10** 155-238
May J A, 1970 *Kant's Concept of Geography* Research Publication 4, University of Toronto Department of Geography, Toronto
Morrill R L, 1970a *The Spatial Organization of Society* (Wadsworth, Belmont, Calif.)
Morrill R L, 1970b "Geography and the transformation of society" *Antipode* 2(1) 4-10
Morrill R L, 1974 "Review of D. Harvey *Social Justice and the City*" *Annals, Association of American Geographers* **64** 475-477
Mulkay M J, 1979 *Science and the Sociology of Knowledge* (Allen and Unwin, London)
Peet J R, 1977 "The development of radical geography in the United States" *Progress in Human Geography* **1** 240-263
Pooler J A, 1977 "The origins of the spatial tradition in geography: an interpretation" *Ontario Geography* **11** 56-83
Relph E, 1970 "An inquiry into the relations between phenomenology and geography" *The Canadian Geographer* **14** 193-201
Sack R D, 1972 "Geography, geometry and explanation" *Annals, Association of American Geographers* **62** 61-78
Sack R D, 1974 "The spatial separatist theme in geography" *Economic Geography* **50** 1-19
Samuels M S, 1978 "Existentialism and human geography" in *Humanistic Geography* Eds D Ley, M S Samuels (Maaroufa Press, Chicago) pp 22-40
Sayer R A, 1979a "Philosophical bases on the critique of urban modelling: a reply" *Environment and Planning A* **11** 1055-1067
Sayer R A, 1979b "Epistemology and conceptions of people and nature in geography" *Geoforum* **10** 19-43
Schaefer F K, 1953 "Exceptionalism in geography: a methodological examination" *Annals, Association of American Geographers* **43** 226-249
Slater D, 1975 "The poverty of modern geographical enquiry" *Pacific Viewpoint* **16** 159-176
Smith D M, 1973 *The Geography of Social Well-being in the United States* (McGraw-Hill, New York)
Stamp L D, 1966 "Ten years on" *Transactions, Institute of British Geographers* **40** 11-20
Steel R W, 1974 "The Third World: geography in practice" *Geography* **59** 189-207
Stewart J Q, 1956 "The development of social physics" *American Journal of Physics* **18** 239-253
Stewart J Q, Warntz W, 1958 "Macrogeography and social science" *Geographical Review* **48** 167-184
Taylor P J, 1976 "An interpretation of the quantification debate in British geography" *Transactions, Institute of British Geographers New Series* **1** 129-142
Taylor P J, 1981 "Political geography and the world-economy" in *Political Studies from Spatial Perspectives* Eds A D Burnett, P J Taylor (John Wiley, Chichester, Sussex)
Taylor P J, Gudgin G, 1976 "The myth of non-partisan cartography" *Urban Studies* **13** 13-25
Tuan Yi-fu, 1971 "Geography, phenomenology and the study of human nature" *The Canadian Geographer* **15** 181-192
Tuan Yi-fu, 1977 *Space and Place* (Edward Arnold, London)

Unwin D J, 1978 "Quantitative and theoretical geography in the United Kingdom" *Area* **10** 337-344
Wallerstein I, 1979 *The Capitalist World-Economy* (Cambridge University Press, Cambridge)
White G F, 1972 "Geography and public policy" *The Professional Geographer* **24** 101-104
Wilson A G, 1974 *Urban and Regional Models in Geography and Planning* (John Wiley, Chichester, Sussex)
Wrigley N, Bennett R J (Eds), 1981 *Quantitative Geography in Britain: Retrospect and Prospect* (Routledge and Kegan Paul, Henley-on-Thames, Oxon)
Zelinsky W, 1975 "The demi-god's dilemma" *Annals, Association of American Geographers* **65** 123-143

3

The edges of space

P Haggett

Introduction

In this essay, I wish to review the continuing impact on contemporary work in quantitative and theoretical geography of the fact that geography continues to be, in some significant part, a spatial science. By contemporary is included work of the last two decades but with increasing emphasis on the last five years.

Two points should be noted at the outset. First, many of the advances in the most recent history of quantitative geography have stemmed from the importation of concepts or techniques originally developed in other sciences. Thus we can point to the introduction of the many standard statistical techniques, now common in geographical research, which were first developed from the probability calculus used in the study of biological populations; to factor analysis, brought from psychology; or to maximum entropy models, first used in physics. Second, increasing experience of the use of such techniques by geographers has shown 'spatial side effects' to be present. Such side effects may be both positive and negative. The welcome, positive aspects are the potential for extending the original nonspatial models, enriching them to exploit a fuller range of geographical situations. We can illustrate this from the trip-generation models developed by A G Wilson. Here the use of a statistical mechanics language brought major empirical gains as well as conceptual insights when compared to the gravity models that preceded them. The whole family of models constructed on the Wilson foundations—production constrained, attraction constrained, and doubly constrained—have permitted new levels of accuracy in estimating and forecasting flows within cities and regions.

Against such gains must be set the unwelcome, negative side effects. These stem in the main from the problems of applying models in which space is either ignored or assumed to be very simple (for example, continuous, isotropic, and unbounded), to the case of a geographical space which can be immensely complex. Again, we can illustrate the effects by reference to trip-generation models and the problems in interpreting distance terms in such models (see the later discussion in this essay).

The existence of spatial side effects is one thing, their significance is another. Experience to date suggests that gains and losses cannot be easily categorized. In some cases the spatial complexity is of a low order, and where problems arise they are trivial in nature, easily recognized, and as readily solved; in other cases the effects are more subtle and pervasive, and practical answers have yet to be worked out. In this essay we shall look first at the negative side effects and then at the positive advantages.

Geographical space as a barrier to quantitative analysis

Three characteristics of geographical space (by which we mean an area of interest on the earth's surface) cause difficulties for quantitative analysis: first, the fact that it has a variable dimension; second, that it has an external boundary in all but the limiting global case; and third, that it frequently has internal boundaries or partitions. Before illustrating each problem in order, we must define more closely the measurement characteristics of geographic space itself.

Types of geographical space

Until now we have left the nature of geographical space itself unexplored. Insofar as the space within which we work (the earth's surface and its subdivisions) affects the significance of our side effects, we need to give a more precise definition of the type of space being studied.

Work in quantitative geography usually demands at some stage that we assign symbols or numbers to our observations of geographical space. In accordance with conventional measurement theory it is possible to recognize four levels of geographical space—two at the categorical level and two at the continuous level. The details with examples are set out in table 3.1. In studying the table it is important to recall that it refers to single categories; in a given problem, space may be a hybrid of these categories.

The same four-level typology that we have applied to space, *S*, also applies to the measurement of time, *T*, and to the phenomena being studied (namely the data, *D*). We may combine these into a three-dimensional *STD* cubic matrix with 64 cells (figure 3.1) ranging from

Table 3.1. Types of geographical space.

Type	Level	Defining geographical relations	Examples
Categorical space	Nominal (S_1)	Equivalence (1)	Name identification of locations (for example, city names)
	Ordinal (S_2)	(1) plus relative position (nearer than, further than)	Locations (for example, nodes) identified on a topological graph indicating rank position in space
Continuous space	Interval (S_3)	(1), (2), plus absolute position in terms of graticule with an arbitrary origin point in space (3)	Locations identified by spatial coordinates (for example, the National Grid System)
	Ratio (S_4)	(1), (2), (3) plus absolute position in terms of a graticule with a natural origin point in space (4)	Locations identified by spatial coordinates (for example, distance and direction from CBD of identified census tracts)

$S_1T_1D_1$, in which space, time, and data are all measured in categorical terms at the nominal level, to $S_4T_4D_4$, in which all are measured in continuous terms at the ratio level. The different levels have implications for the analytical methods which may appropriately be used, with 'higher' levels having a wider range than 'lower'. It does not necessarily follow that the most primitive cell, $S_1T_1D_1$, is lacking in interest; indeed, some geographers would argue that our most basic understanding must stem from careful set-analysis at the nominal level (Chapman, 1980).

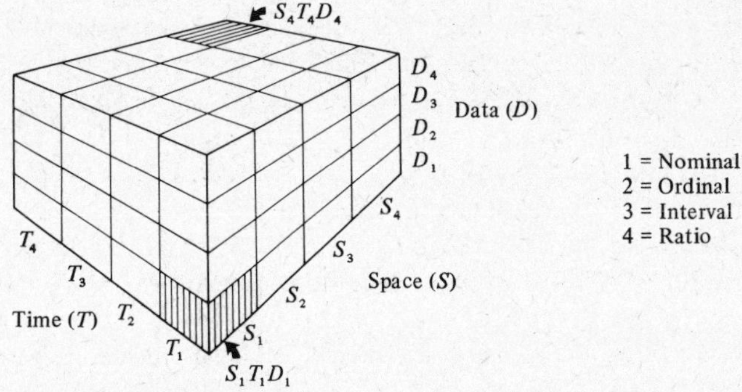

Figure 3.1. *STD* matrix showing the interrelationships of space, time, and data at each of four measurement levels.

The dimensional barrier

Geographical analysis may take place at many different scales, from that of the earth itself down through a staircase of intermediate magnitudes to the level of a single individual. Linking quantitative analysis at the different scales poses a problem, conventionally labelled the 'ecological correlation problem', which has been extensively analyzed in the geographical and related literature [see the original statement by Robinson (1950) and the recent review by Clark and Avery (1976)] and will not be treated further here. It is worth noting, however, that the problem of integrating information at different scales is not one unique to geography. Economists encounter acute problems in integrating macro and microeconomic models, while the biologist has to range over a very wide set of spatial scales in linking findings at the level of the organism, the cell, the biochemical helix. For each discipline it remains essential to recognize the dimensions at which the analysis is being conducted and to make inferences from one scale to another only with great care. It remains an open question whether in geography, as in certain other fields, there are 'windows' within the scale spectrum where research is likely to be particularly fruitful, and dark areas where structures are confused and probably unresearchable.

External edges and boundary effects

Few of the numerical models introduced into the geographical literature are global in extent and only those in numerical climatology concern themselves with the continuous surface of the sphere itself (see Holloway, 1958). For the most part, geographers have concentrated their research on spatial models of small and confined segments of the earth's surface. Such segments of interest may conventionally be thought of as two-dimensional and bounded by sharp or fuzzy limits. We look here at the effects of such limits both on one-dimensional and two-dimensional models.

Boundary effects and one-dimensional models

The fact that geographical space is bounded means that the frequency distribution of distances between any pair of points within the space must also be bounded. For simple shapes the frequency distribution rises to a single peak from its zero-distance minimum back to zero again at a maximum distance equivalent to the long axis of the space. Where shapes are punctured or fragmented, there may be more than one peak on the distribution curve.

Such finite frequency distributions which describe distance relations within an area are important in understanding distance-related behaviour. Migration differences between two geographical areas may result from (a) real differences in migration behaviour, and (b) differences in the boundaries of the two areas. People changing homes within a long, thin area are *ceteris paribus* more likely to cross its boundaries (and be recorded as external movers) than those within a compact round area. Since the effects of (a) and (b) may be confounded, it is important to know the expected pattern of frequencies so that the second effect can be separated out. That this effect is nontrivial is shown by Taylor (1971), who estimated that 60% of the differences of migration patterns from small areas within nineteenth-century Liverpool could be attributed to differences in their shape.

Boundary conditions may also influence the distance–decay functions commonly used to compare population-density patterns within large urban areas. Edmonston and Davies (1978) illustrate this by comparing median distances, M, for unbounded and bounded models. The median distance from the city centre is that value which divides the number of people in the city into equal numbers of residents. At the median value, one-half of the population of the total urban area is located at distances beyond the value, thus assuming an infinite boundary

$$-(\alpha M+1)\exp(-\alpha M)+1 = 0{\cdot}5\,, \tag{3.1}$$

where α is the density gradient in the negative exponential density gradient.

But actual metropolitan areas do not extend without limit over space. If we assume a finite boundary c from the city centre, where $c > 0$, the median value M for a population distribution with a finite boundary will be

$$\frac{-(\alpha M+1)\exp(-\alpha M)+1}{-(\alpha c+1)\exp(-\alpha c)+1} = 0{\cdot}5\,. \tag{3.2}$$

Equation (3.2) can be solved by the methods appropriate for calculating the roots of a nonlinear function.

Figure 3.2 presents values of the median distance for twelve US urbanized areas in 1970 studied by Edmonston and Davies (1978, page 240). Comparison of the finite boundary model, equation (3.2), with the infinite boundary model, equation (3.1), shows the latter consistently overestimates the distances. Thus in the case of Washington, DC, the average distance to its urbanized area boundary from the city centre was 12·48 miles. Under the finite boundary model, the median lay at 6·5 miles; with the infinite boundary model, at 9·3 miles. In the case of three large low-density cities (Denver, Fort Worth, and Jacksonville), the median with the infinite boundary model lies—illogically—beyond the average distance of the boundary itself.

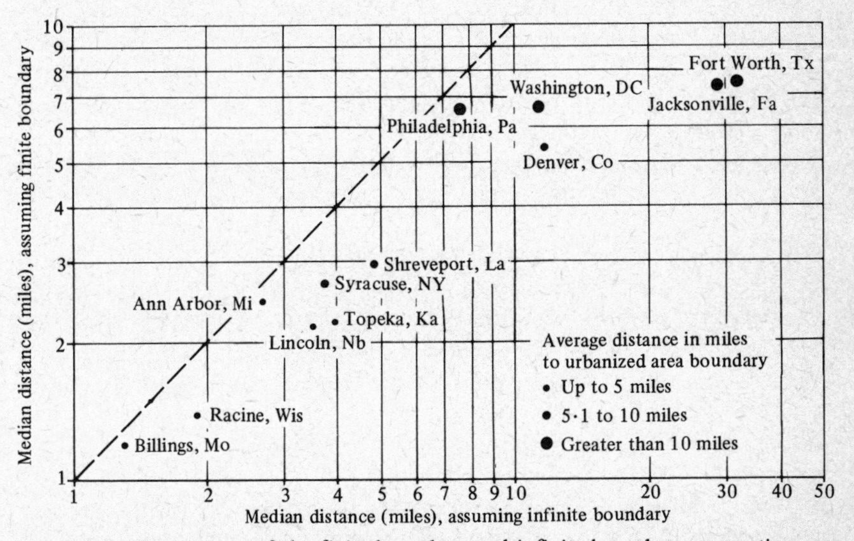

Figure 3.2. Comparison of the finite-boundary and infinite-boundary assumptions on the median distance of population distribution in twelve United States urbanized areas in 1970, using the negative exponential density model. Source: Edmonston and Davies (1978, table 2, page 240).

Boundary effects and two-dimensional models: 1. Surface models

A powerful battery of quantitative techniques has been developed to estimate 'best fit' surfaces for describing spatially-distributed observations as contour maps. As with the negative exponential models, the existence of external boundaries influences the form of the surface.

In the earlier models, surfaces were generated by moving-average processes applied to grid cells of different sizes. The size of the filter determined the width of the 'dead' zone around the perimeter of the map and for this zone no generalized contours could be drawn—the larger the filter size, the wider the dead boundary zone. Even the most recent of local mapping processes for contour smoothing, Tobler's (1979) pycnophylactic interpolation, is not independent of the boundary condition, and Tobler illustrates how two different assumptions about values in the boundary zones (the so-called Dirichlet and Neumann assumptions) affect the shape of the contour map.

In later models, surfaces were generated by taking all data points within an area and estimating a map-wide equation. Krumbein (1966) has examined the problems to be met in extrapolating a trend-surface map beyond the control limits (where data are known) into the peripheral boundary zone (where data are unknown). To do this, a rectangular section from the McClure, Pennsylvania, topographic map was sampled to give 90 data points recording elevation and location. Although not shown here, this control area is traversed diagonally from SW to NE by a long ridge which extends laterally beyond the mapped area. Two types of trend-surface map were fitted to the elevation data. First, the commonly-used polynomial model which structures the data in such a way that a succession of fitted surfaces, linear, quadratic, cubic, etc, can be constructed. Second, the less-common Fourier model which structures the data to produce a series of harmonic surfaces, which have wave forms of diminishing length as the order of the surface rises.

Figure 3.3 shows the application of the two models to the test area. In each case the equation is estimated for the data points lying within the rectangular control area but extended into the surrounding boundary zone. The degree of fit achieved within the control area is given for each model in terms of sums of squares as percentages and is in all cases more than one half. For both models, the fit increases as the order of the surface (and the complexity of the generating equation) increases. Thus in the polynomial model, the quartic surface with 15 coefficients in the equation gives better fits than the cubic surface with 10, or the quadratic surface with 6. In the Fourier model the equations consist of blocks of cosine and sine terms; the higher degree of fit (compared to the polynomial) partly reflects the larger number of terms, with 9, 25, and 49 coefficients, respectively, for the three surfaces shown.

Most interest focusses on the performance of the models in the boundary zone. From the inspection of figure 3.3, two generalizations seem apparent. First, the low-order quadratic map gives a more satisfactory representation of the continuing diagonal ridge than its higher order equivalents where the extrapolated values rapidly increase or decrease beyond reasonable numerical values (shaded areas). Paradoxically, the better the fit of a polynomial surface within the control area the poorer it is as an extrapolatory model outside the map boundary. As Krumbein (1966, page 39) suggests the distance over which extrapolation is reasonable contracts as the order, and internal fit, of the model increases. Second, the Fourier models differ markedly from the polynomial in that they simply display periodic repetition of the pattern within the control area when extrapolated.

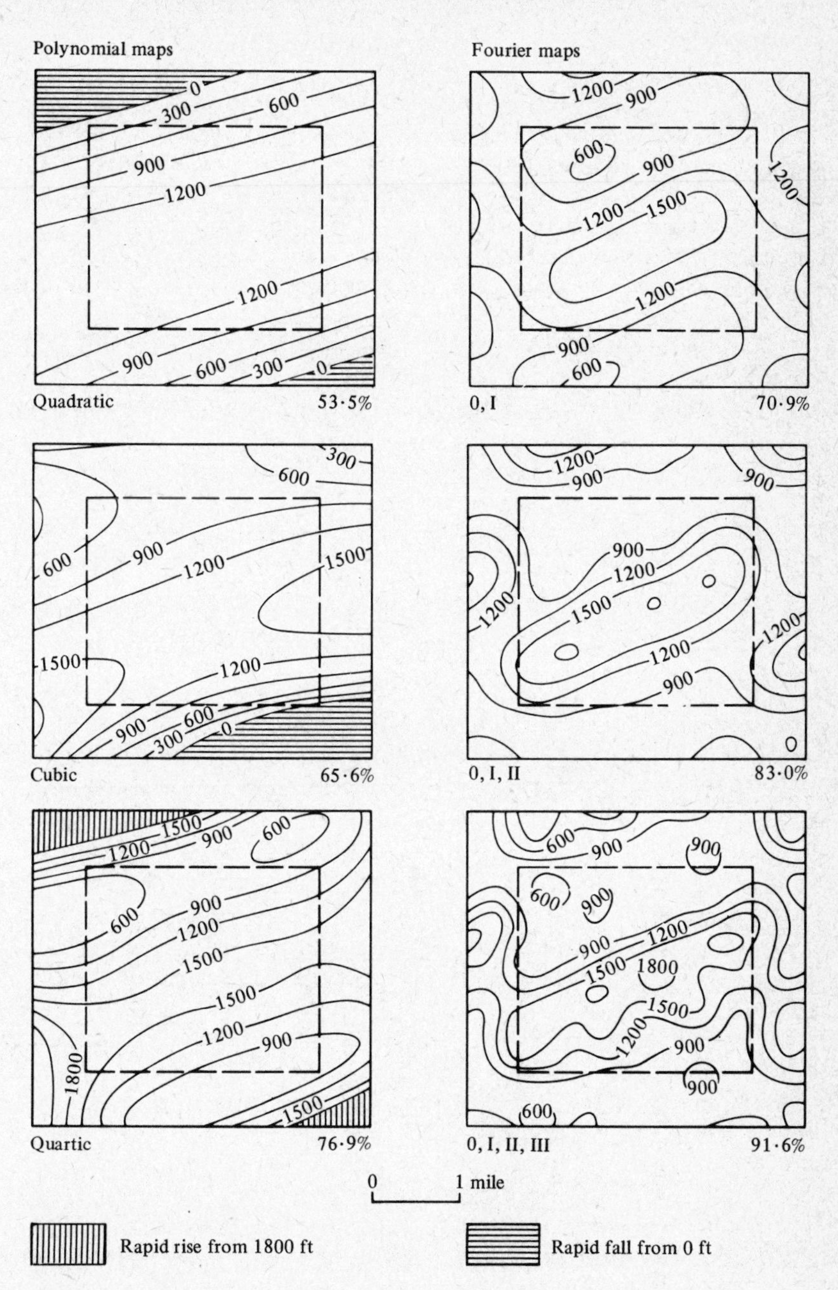

Figure 3.3. Performance of trend-surface models in boundary areas. Contrasts between the estimates of polynomial and Fourier models in extrapolating contours beyond the edges of the control area. Source: Krumbein (1966, figures 8 and 9, pages 35, 37).

This generalization applied to Fourier maps of all orders so that extrapolation of a Fourier surface, even one of low order, seems ruled out beyond the edge of the control map.

Boundary effects and two-dimensional models: 2. Nearest-neighbour models
A large number of distance-based measures of spatial point patterns have been proposed in the literature (see the review by Haggett et al, 1977, pages 439–446). One of the earliest and still one of the more widely used is the first nearest-neighbour statistic, given by the ratio of observed and expected mean distances between first nearest neighbours.

Such an index is sensitive to boundary conditions in two senses. First, the index value is directly dependent on the dimensions of the reference area under study. For a given pattern of points, the value of the nearest-neighbour index will fall *ceteris paribus* as the reference area increases.

A second sensitivity to boundary conditions is shown in figure 3.4. Here each point i is the centre of a circle with a radius $d(\min)_i$. Where the perimeter of each circle lies wholly within the area of interest (bounded

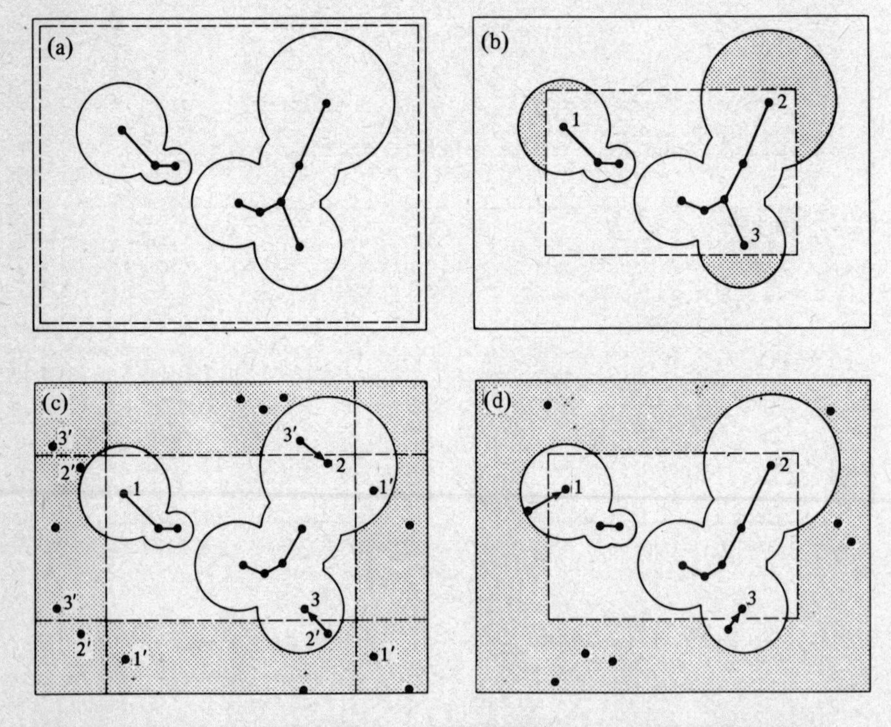

Figure 3.4. The impact of boundary assumptions on the estimated value of the first nearest-neighbour statistic. The values of the statistic for the four maps are (a) 0·560, (b) 0·915, (c) 0·748, and (d) 0·831. The four assumptions are discussed in the text. Source: Haggett (1980, figure 4).

by the dashed line), as in map (a), the value of the first nearest-neighbour statistic, R, is independent of any boundary assumptions. However, where the perimeter intersects the boundary a working rule has to be made about information in the space which lies outside it. Three common rules encountered are: (1) to confirm from empirical evidence or to assume that the area outside the boundary is empty, as in map (b); (2) to map the area of interest onto a torus so that the point distribution is replicated in the adjacent areas as in map (c); and (3) to take advantage of the known distribution of points outside the boundary of interest. Clearly the third option implies more information being available than do the first two. The distances between first nearest neighbours are shown by links in figure 3.4. Under assumption (2), point 2 finds a new nearest neighbour in point 3 and vice versa, so that both new links are of the same length. Under assumption (3), the new links formed by points 1 and 3 have lengths independent of each other. It will be evident from inspection of the maps that assumptions (2) and (3) yield R values below that of assumption (1), in this case $R = 0{\cdot}748$ and $R = 0{\cdot}831$, respectively, as opposed to $R = 0{\cdot}915$. How significant such boundary assumptions will be in varying the value of R will depend on the proportion of the n points whose $d(\min)$ circles intersect the edge of the area of interest. Theoretical aspects in relation to planar sampling are considered by Miles (1974).

Internal edges and the impact of map infrastructure

The internal edges which split up an area of interest into subareas may be thought of as the map infrastructure. Commonly, different systems of partition are used for different categories of data—electoral districts for voting data, census districts for demographic data, and so on. Attempts to standardize the infrastructure by adopting grid squares or standard regions have met with limited success. For example, in the United Kingdom central government has set up eleven standard regions for which many official statistical series are published. Unfortunately, since World War II, while the number has generally remained constant the exact boundaries have changed. Thus Rees (1979, pages 2–7) in a study of population migration is forced to use an ingenious series of conversion matrices. These convert the proportional distribution of the population of 'before' regions into 'after' regions.

Even when shifts over time are not an issue, the impact of internal boundaries on quantitative measures may be severe. Figure 3.5 illustrates some recent findings by Openshaw and Taylor (1979). In a study of the association between the Republican vote and older populations for the state of Iowa, USA, they obtained correlation coefficients ranging from a high of $+0{\cdot}862$ to a low of $+0{\cdot}265$ when the ninety-nine counties that make up the state were grouped, using different criteria, into six conventional regions. When many other six-region partitions were tested, the range of possible values for the correlation coefficient widened still further to include high negative as well as high positive associations! The implications of spatially-grouped data for quantitative analysis is considered further by Williams (1978).

Variations in map infrastructure also affect quantitative measures of spatial interaction, such as journey-to-work flows and migration (see Smith, 1977). Curry (1972) has argued that the attenuating effect of distance, as measured by spatial interaction models, confounds two distinct components: (a) a *behavioural* component describing how trips may be modified by the time or cost of increased separation, and (b) a *map pattern* component describing the particular bounded configuration of locations between which trips are occurring. In practice the distance attenuation effect is measured by a single term, β, in the standard, doubly-constrained entropy model,

$$T_{ij} = A_i B_j O_i D_j \exp(-\beta c_{ij}) \,, \tag{3.3}$$

where

T_{ij} are the trips between locations i and j,
A_i and B_j are balancing factors for origin i and destination j,
O_i is the total number of trips generated by origin i,
D_j is the total number of trips terminating at destination j,

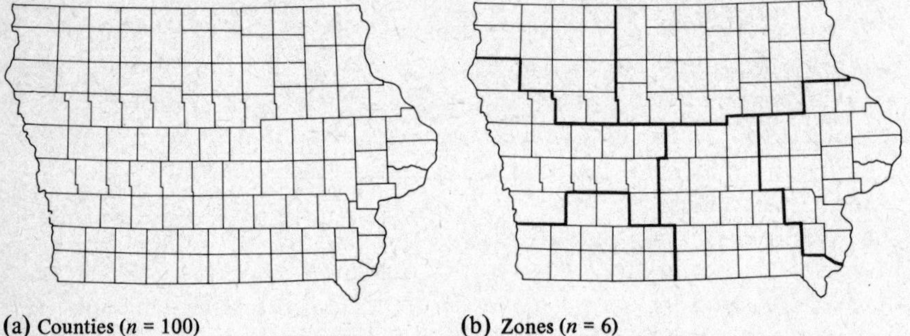

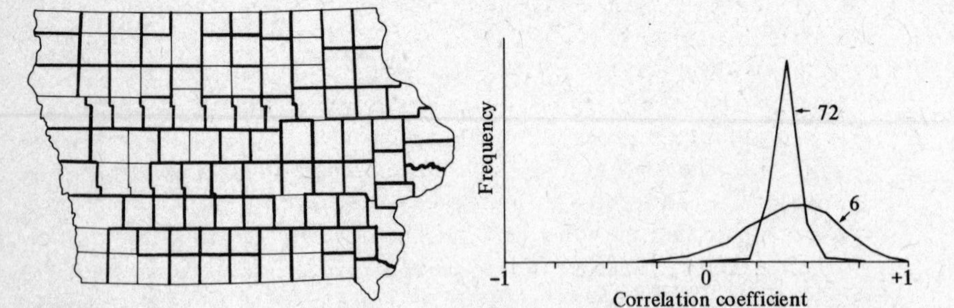

Figure 3.5. Frequencies of correlation coefficients at different scales for two variables within Iowa. The maps show a single example of the very large number of boundary configurations at each spatial scale. Source: Openshaw and Taylor (1979, figure 5.1, page 131).

β is the distance–decay parameter, and
c_{ij} is a measure of the distance or cost of travel between origin i and destination j (Wilson, 1970).

Since Curry's original paper, a lively debate has occurred in the geographical literature on the exact interpretation of the term (summarized in Haggett et al, 1977, pages 35–36). A recent paper by Griffith and Jones (1980) confirms, from a study of journeys-to-work in twenty-four Canadian cities, that the shape of the bounded urban area and the location of origins and destinations within it did indeed affect the β term.

Geographical space as a resource in quantitative analysis

If the problems of estimating and correcting for spatial effects were the only concern of quantitative geography, then it would rank as a useful but somewhat sombre intellectual activity. Fortunately there is a reverse side to the coin. For although the awkward, multilevel, anisotropic world we study may pose problems for our sister disciplines in making space-free generalizations, those same problems provide a resource for geography itself. A world without boundary or autocorrelation problems would be one with little geographic interest (Gould, 1970).

To illustrate the role of space as a bridge to geographical analysis, I select examples from one aspect of a conventional but central part of the field of map analysis. In this we look, first, at the encoding process by which real-world information is converted into map form; this is then complemented by a second process of decoding, by which inferences may be drawn from maps about the real world. Both form the two halves of a mapping cycle. The second process of map interpretation leans heavily on spatial autocorrelation and is most fully explored by Cliff and Ord (1973; 1981); it is not pursued further here.

Surface estimation

Establishing the degree of confidence one is able to place in a map pattern is a crucial stage in geographical analysis. Consider figure 3.6, which shows the incidence of a blood disease (toxoplasmosis) in thirty-six cities in the Central American country of El Salvador. Measured rates are expressed as deviations from the national incidence (the average of the rates for all cities), so that a city with $-0{\cdot}04$ has a rate 4% lower than the country as a whole [figure 3.6(a)].

One obvious map to draw from these data is shown in figure 3.6(b). The incidence value for each city is taken as the best estimator and used to generate an isarithmic map for the country. The choice of an isarithmic map, giving a continuous surface from discontinuous data, is itself a matter of debate but, leaving this cartographic question to one side, we may legitimately inquire whether this is the best isarithmic map we can draw? If we go back to the original data we find that tests were carried out for toxoplasmosis on roughly five thousand people in El Salvador, but that

(1) these were not drawn equally from the thirty-six cities, (2) the cities showed considerable contrasts in the variability of their results (as measured by the standard deviation), and (3) the extreme-value cities, both very high and very low, were also those with large standard deviations and were based on small populations. Given these facts, the map in figure 3.6(b) takes on a less confident shape and we can no longer be sure that the highs and lows are a result of real variations in the disease rather than an artifact of the sampling process.

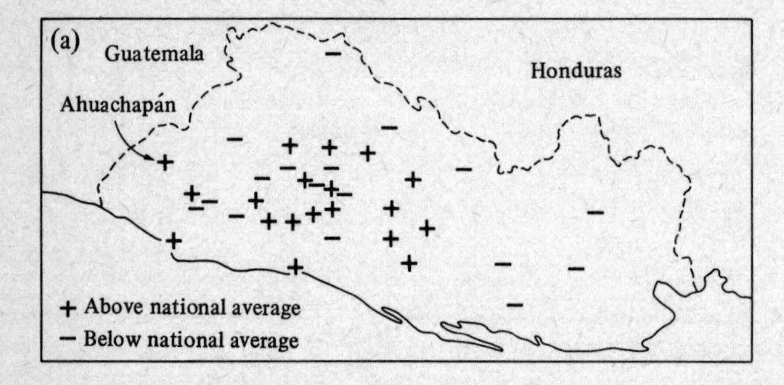

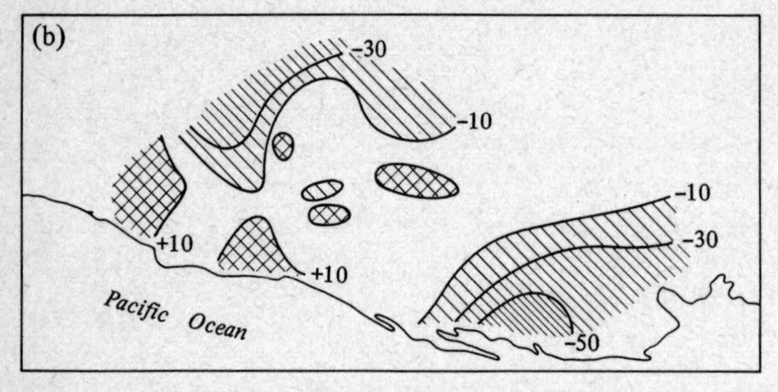

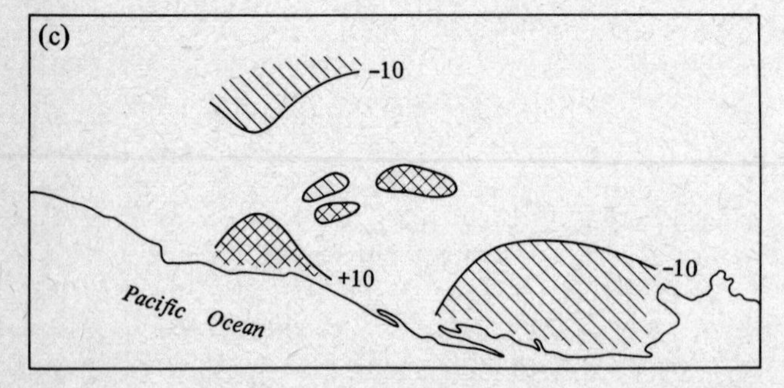

Figure 3.6. Use of Bayesian estimators in mapping. (a) Distribution of toxoplasmosis in thirty-six locations in El Salvador. (b) Contour maps of incidence based on local mean values. (c) Modified contour maps based on James-Stein estimators in which the variability at each location has also been taken into account. Source: (a) based on Efron and Morris (1977, page 125).

Efron and Morris (1977) have suggested that a Bayesian approach to the mapping problem may allow the mean values and the variability to be taken into account. They propose a James–Stein estimator which 'shrinks' the mean value for each city towards the overall mean for the country as a whole. However, the shrinking factor is not the same for all cities, the shrinkage being greatest for cities with the largest standard deviations. Figure 3.6(c) shows the revised map based on the James–Stein estimators. In some cases the change in values is considerable; thus the western city of Ahuachapán, ranked highest with a value 29% above the national value, is demoted by the James–Stein estimator to twelfth-ranking position only 5% above the national value.

The establishment of more accurate map patterns from spatially-distributed data has a long history in quantitative geography going back to Choynowski (see the review by Cliff and Haggett, 1980). Recent work on statistical estimation theory (Efron, 1975) promises to provide a new family of Bayesian estimators (of which the James–Stein statistic is an example) on which more accurate maps can be drawn.

Space estimation

In map construction, the accurate estimation of the vertical (z) values represents the first of a two-stage process. We need further to consider whether the horizontal dimensions (xy) correctly describe the locations of data points. I shall illustrate the approach to the problem at two contrasting magnitudes: the global and the local.

A global example

The attempt to portray the curved surface of the earth on flat paper is one of the classic problems of map-making. Although complete solutions are clearly impossible, it is feasible to retain certain desired characteristics (such as equal area), and research has proceeded by the invention of a consistent mathematical transformation which maps one surface onto another. One of the significant developments in this field in the last two decades is the use of high-speed computers to develop special-purpose or empirical map transformations.

One simple example at the global level is provided by the nine cities shown in figure 3.7. Each city is shown in its 'correct' position with respect to the relevant map transformation though clearly not in its correct position on the globe. Let us assume that we wish to map as accurately as possible the airline distances between the nine cities (values above the diagonal in table 3.2). Then we can rank the projections in figure 3.7 by comparing the scaled-up map distances with the real distances and computing an error or stress statistic. The error values, in table 3.3, show clearly that the polar case of the Azimuthal Equidistant projection centred on 90°N is the most accurate, with an error of 15·46%.

The question asked by modern cartographers is whether this is the 'best' map that can be drawn to represent the data in question. One approach, initiated by Tobler (1966), is to devise an empirical map projection in which the data structure determines the spatial structure of the map. Although various algorithms for such nonlinear mapping exist (Golledge, 1972), all consist of a convergence routine in which locations are iteratively

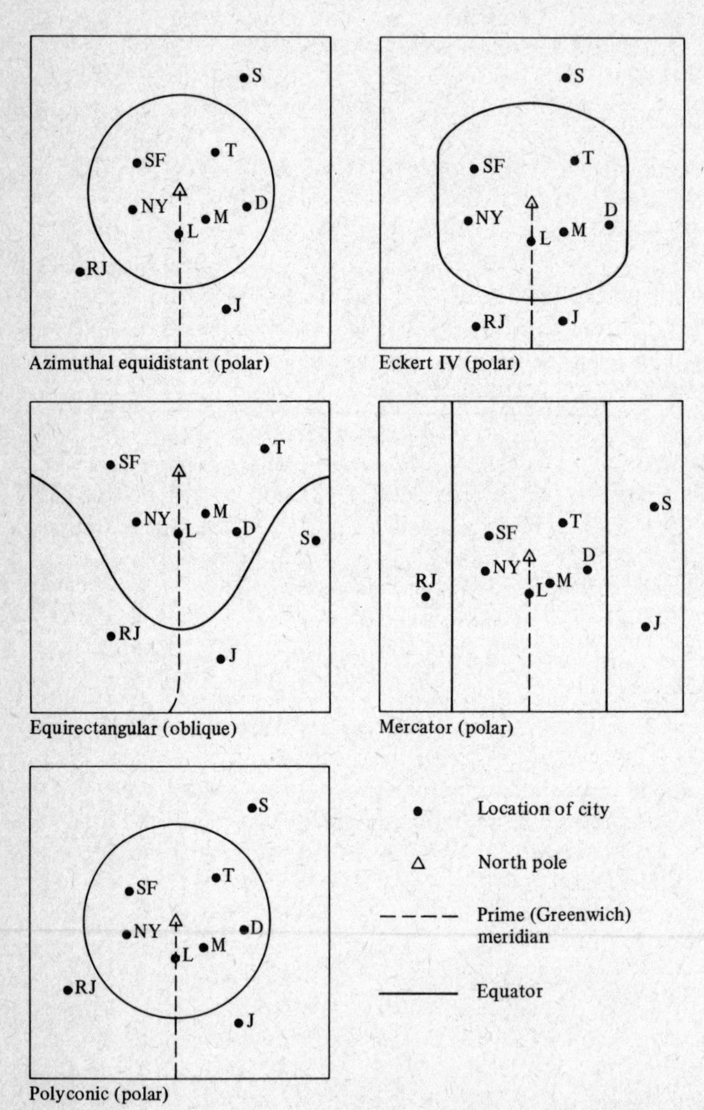

Figure 3.7. The location of nine world cities as plotted on six conventional map transformations. The prime Greenwich meridian and the equator are plotted for reference but continental outlines are omitted. The lines connecting the cities represent the minimum spanning time as based on the great-circle distances in table 3.2: these are arbitrarily drawn as straight lines rather than the different curved paths that the lines would actually follow on the six graticules.

moved so as to increase the goodness-of-fit between real distances and the equivalent map distances. If we apply a nonlinear mapping to the distances between the nine cities given in table 3.2, then the resulting map (figure 3.8) shows this information in a slightly more accurate manner than the best of the conventional map transformations.

The relatively modest gain that empirical map projections give to distances results largely from the high consistency between them in the sense that each can be related to the other through some elementary spherical trigonometry. But what happens when the separation between the cities is measured by something much less consistent, namely the prevailing lowest cost flight in either direction in the Summer 1980? The cost values, shown below the diagonal in table 3.2, reflect not only consistent distance effects but also inconsistencies related to competition, fare policy, route loading, etc. The irregular cost relations between the nine cities are much more difficult to portray accurately on a map: as the

Table 3.2. Great-circle distances and airline fare costs between nine world cities.

	Del	Joh	Lon	Mos	N-Y	R-J	S-F	Syd	Tok
Delhi, India	0	84	67	44	117	151	124	104	59
Johannesburg, SA	72	0	91	92	128	71	170	110	136
London, UK	80	94	0	25	55	92	86	170	196
Moscow, USSR	62	108	48	0	75	111	92	163	75
New York, USA	53	77	13	46	0	77	41	160	180
Rio de Janeiro, Brazil	129	70	105	140	58	0	106	135	186
San Francisco, USA	106	115	22	77	24	69	0	119	83
Sydney, Australia	101	96	67	146	154	158	72	0	78
Tokyo, Japan	76	145	139	111	75	113	49	147	0

Great-circle distances are shown *above* the diagonal in hundreds of kilometers. (Source: *IATA Air Distance Manual*, Cointrin-Geneva, 1977.) Lowest cost single fares on scheduled airline in either direction are shown *below* the diagonal in tens of US dollars. (Source: *ABC World Airways Guide*, June 1980.)

Table 3.3. Accuracy of six conventional map transformations in showing the inter-city data in table 3.2.

Map transformation	Case	Map centre	Map accuracy (% error)	
			Great-circle distance	Cost (7/80)
Azimuthal equidistant	Polar	90°N	15·46	33·59
Azimuthal equidistant	Equatorial	90°W, 0°N	27·25	28·21
Eckert IV	Polar	90°N	23·35	34·73
Equirectangular	Oblique	0°E, 45°N	24·31	35·16
Mercator	Polar	90°N	21·14	33·32
Polyconic	Polar	90°N	16·54	34·52

last column of table 3.3 shows, the error term for the six projections ranges from 28% to 35%.

Cost relations between locations can again be used to generate an empirical map projection by using a nonlinear mapping algorithm. In figure 3.8, the set of best-fit locations for the distance matrix is used as a starting point, and the locations are progressively adjusted to bring them into line with the cost-separation data. The trajectories followed by the city locations and the positions after 50, 100, and 150 adjustment cycles are shown. Convergence was achieved after 180 iterations, when the error level had fallen to 14·93%. It is important to note that the mapping algorithm used aims to reduce the *overall* error in the map, that is to make costs and map distances as highly correlated as possible for all 36 dyads. Clearly this is not the same as optimizing the map for a single city. Figure 3.9 shows how the error term for four of the cities changes during the iteration process. In the case of Johannesburg, the error for the individual city changes monotonically in line with general improvements for the city system as a whole; the best map for this city is after 180 iterations. This pattern stands in contrast to San Francisco, where the lowest error is found after 90 iterations. Sydney, Australia, shows a still more complex pattern.

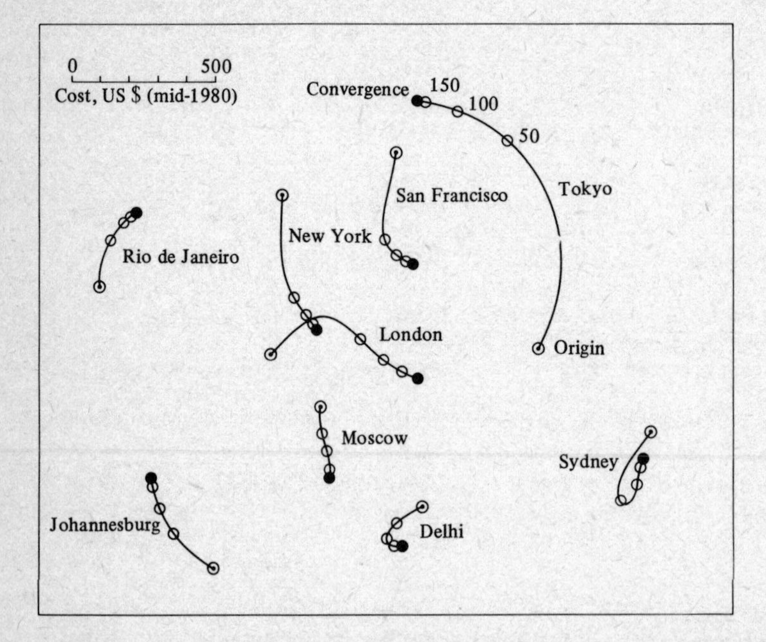

Figure 3.8. Global nonlinear mapping. Convergence trajectories of nine world cities in cost space. The origin points (⊙) indicate the best-fit location in terms of the matrix of Great-circle distances (table 3.2). The position after 50, 100, and 150 cycles is shown (○), together with the final convergence position after 180 cycles (●) in cost space.

Even after convergence has been obtained, it is clear that the final map leaves some cities more accurately placed than others. As in the case of the regular map transformations shown in figure 3.7, there is a characteristic error pattern for each map. For nonlinear maps it tends to be irregular rather than regular.

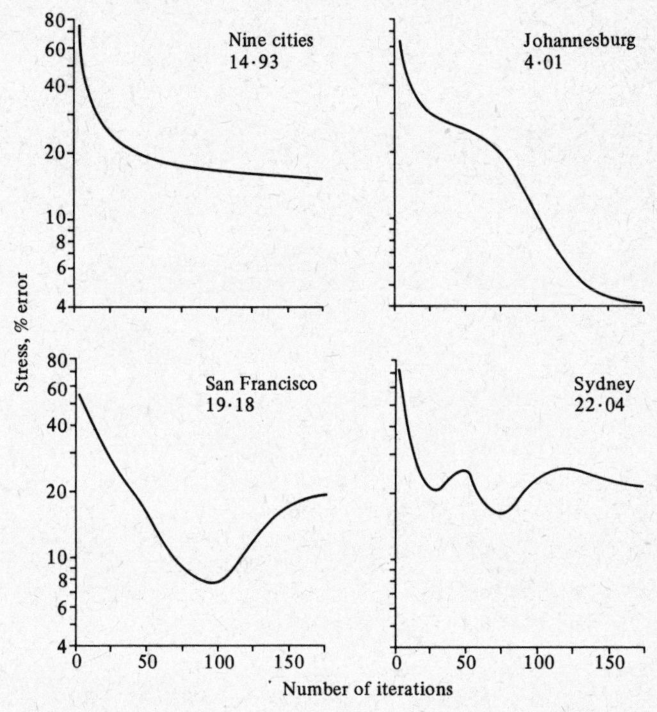

Figure 3.9. Accuracy curves for the set of cities and for three individual cities plotted in figure 3.8.

A local example

A contrasting local example of nonlinear mapping is shown in figure 3.10. The four maps show the locations of nine halts around a small Austrian lake as connected by a state-owned motor-vessel service in the summer of 1980. In addition to the conventional map based on geographic distance [figure 3.10(a)], we can draw a series of other maps based on forecast, time of journey, or frequency of service. In each case the mapping algorithm is the same; only the input data of connections between places are changed.

To sum up. Nonlinear mapping provides one example of the ways in which maps can be encoded to exploit the essential spatial nature of a data matrix. They allow a breakaway from the conventional geographical map based on a physical distance separation and allow it to be replaced with any other relevant appropriate metric. I have given very simple examples

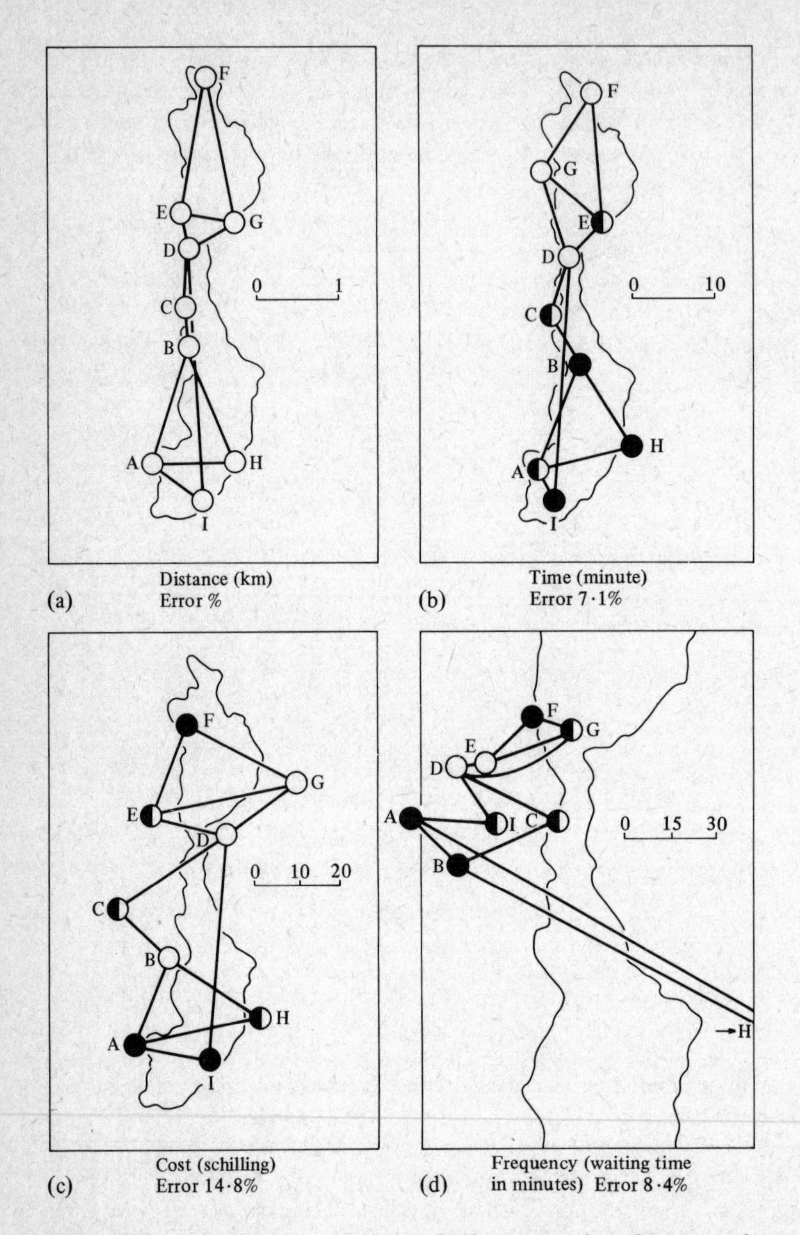

Figure 3.10. Local nonlinear mapping. Different spatial configurations for a transport network connecting nine locations around an Austrian lake (Wolfgangsee).
(a) Distance. (b) Time of journey. (c) Cost of journey. (d) Frequency of service. Whereas (a) shows a conventional linear mapping, maps (b), (c), and (d) are produced by a nonlinear algorithm. The lake outline is included for comparative purposes in all four maps.

based on cost, time, and service. Such maps are likely to be particularly valuable where they show changes over time. Self-evidently the method can incorporate behavioural or perceptual metrics based on scaled values of knowledge, attitude, information, etc. The literature already includes important work in these areas (Golledge, 1972).

Conclusions

To write about "the edges of space" nearly five hundred years after the Columbus voyage might appear quixotic. Nonetheless it can be argued that most geographic research centres on smaller, nonglobal worlds where space is bounded and where edges—sometimes sharply cut, sometimes fuzzy—enter into the reckoning.

It has been argued in this presentation that the existence of such discontinuities, both around and within our study regions, leads to barriers to quantitative analysis. Where space acts as a liability there appear to be three continuing tasks for geographers:

First The precise identification of spatial effects separating the significant from the trivial.

Second Development of 'space-proofed' analytical methods wherever such corrections are needed and feasible.

Third Flagging of those areas which (a) indicate errors may be large and (b) indicate solutions may not be feasible.

The beginning of the 1980s has shown encouraging signs of renewed interest in these problems both from inside geography (for example, Griffith, 1980; Haining, 1980; Smith, 1980; Cliff and Ord, 1981), and from outside (for example, Draper and Guttman, 1980).

Balancing the space-correcting tasks within quantitative geography are those which exploit spatial effects as one of the central planks of geographical research. These call for the identification and exploitation of those areas where the spatial aspect contributes insights both to the encoding and to the decoding of map information.

References

Chapman G P, 1980 "*Q* analysis" in *Quantitative Geography in Britain: Retrospect and Prospect* Eds N Wrigley, R J Bennett (Routledge and Kegan Paul, Henley-on-Thames, Oxon) chapter 23

Clark W A V, Avery K L, 1976 "The effects of aggregation in statistical analysis" *Geographical Analysis* 8 428-438

Cliff A D, Haggett P, 1980 "Mapping respiratory diseases" in *Scientific Foundation of Respiratory Medicine* Ed. J G Scadding and others (Heinemann Medical, London) chapter 4

Cliff A D, Ord J K, 1973 *Spatial Autocorrelation* (Pion, London)

Cliff A D, Ord J K, 1981 *Spatial Processes: Models and Applications* (Pion, London)

Curry L, 1972 "A spatial analysis of gravity flows" *Regional Studies* **6** 131-147

Draper N R, Guttman I, 1980 "Incorporating overlap effects from neighbouring units into response surface models" *Applied Statistics* **29** 128-134

Edmonston B, Davies O, 1978 "Interpreting the negative exponential density gradient" *Journal of the Royal Statistical Society, Series A* **141** 235-241

Efron B, 1975 "Biased versus unbiased estimation" *Advances in Mathematics* **16** 259–277
Efron B, Morris C, 1977 "Stein's paradox in statistics" *Scientific American* **236**(5) 119–127
Golledge R G, 1972 *Multidimensional Scaling: Review and Geographical Applications* TP-10, Commission on College Geography, Association of American Geographers, Washington
Gould P R, 1970 "Is *Statistix Inferens* the geographical name for a wild goose?" *Economic Geography* **46** 439–448
Griffith D A, 1980 "Towards a theory of spatial statistics" *Geographical Analysis* **12** 325–339
Griffith D A, Jones K G, 1980 "Explorations into the relationship between spatial structure and spatial interaction" *Environment and Planning A* **12** 187–201
Haggett P, 1980 "Boundary problems in statistical geography" in *Die Bedeutung von Grenzen in der Geographie* Ed. H Kishimoto (Kummerley and Frey, Zurich)
Haggett P, Cliff A D, Frey A E, 1977 *Locational Analysis in Human Geography* second edition (Edward Arnold, London)
Haining R P, 1980 "Spatial autocorrelation problems" in *Geography and the Urban Environment* Eds D T Herbert, R J Johnston (John Wiley, Chichester, Sussex) pp 1–44
Holloway J L, 1958 "Smoothing and filtering of time series and space fields" *Advances in Geophysics* **4** 351–389
Krumbein W C, 1966 "A comparison of polynomial and Fourier models in map analysis" ONR Task No. 388-078, Contract 1228 (36), TR-2, Office of Naval Research, Geography Branch, Northwestern University, Evanston, Ill.
Miles R E, 1974 "On the elimination of edge effects in planar sampling" in *Stochastic Geometry: A Tribute to the Memory of Rollo Davidson* Eds E F Harding, D G Kendall (John Wiley, Chichester, Sussex)
Openshaw S, Taylor P J, 1979 "A million or so correlation coefficients: three experiments on the modifiable areal unit problem" in *Statistical Applications in the Spatial Sciences* Ed. N Wrigley (Pion, London) pp 127–144
Rees P H, 1979 *Migration and Settlement: the United Kingdom* International Institute for Applied Systems Analysis, Laxenburg, Austria
Robinson W S, 1950 "Ecological correlations and the behavior of individuals" *American Sociological Review* **15** 351–357
Smith M J de, 1977 "Distance distributions and trip behaviour in defined regions" *Geographical Analysis* **9** 332–345
Smith T E, 1980 "Additive measures of perceived accessibility" *Environment and Planning A* **12** 829–841
Taylor P J, 1971 "Distances within shapes. An introduction to a family of finite frequency distributions" *Geografiska Annaler* **B53** 40–54
Tobler W R, 1966 "Medieval distortions: the projections of ancient maps" *Annals of the Association of American Geographers* **56**(2) 351–360
Tobler W R, 1979 "Smooth pycnophylactic interpolation for geographical regions (with Comment and Rejoinder)" *Journal of the American Statistical Association* **74** 519–536
Williams I N, 1978 "Some implications of the use of spatially grouped data" in *Towards the Dynamic Analysis of Spatial Systems* Eds R L Martin, N J Thrift, R J Bennett (Pion, London) pp 53–64
Wilson A G, 1970 *Entropy in Urban and Regional Modelling* (Pion, London)

4

Physical constraints and contemporary geography

J-P Marchand

Introduction

The recent evolution of geographical studies has been characterised by a questioning, not only of the methodology, but also of the problematic of the discipline itself. The arrival of new techniques, taken from the new as well as the exact sciences, or stemming from familiar technology, has been the starting point for an epistemological mutation tending to overturn the postulates of the French school of geography. In these upheavals 'physical' geography has tended to become more and more distinct from 'human' geography.

Physical geography has rarely been defined, or at least its definition has not led to analyses such as that of human geography by Demangeon (1940). Just as a rural area is too often defined as 'nonurban', physical geography tends to be determined by contrast. "Physical geography is the study of the epidermis of a unique entity: the Earth. It deals with natural landscapes such as would have been seen by an observer travelling all over the globe before the intervention of man", wrote Pierre Birot (1968) in the introduction to his précis on physical geography.

Nevertheless, physical geography has always had a principle role. Before 1922, the French history–geography degree consisted of four parts: a written commentary, a latin translation, a historical dissertation, and a study on physical geography, and up until 1939, according to Meynier (private communication), "many morphologists believed themselves to be the only geographers". For decades past and even today, geographical unity has been the subject of discussion which speaks of a science studying man living in a natural environment, but with a conceptual equality with regard to the values of the two component parts of geography. In theory, no one part is subordinate to the other, even if, according to certain authors, the one or the other of these aspects has long been privileged: the French approach to regional geography was the ultimate translation of this university tradition.

Ten years ago, the unity myth was shattered (see, for example, Meynier, 1969). This split has often been justified by the necessary specialisation of the researcher. In addition, however, the increasingly important role of the economy, developments in epistemology and knowledge of Anglo-Saxon and Soviet geography have also influenced the split in geographical thought. After the split, which already existed in morphology, physical geography has become the object of an initiative trying to establish it as an autonomous discipline.

At a time when opponents face each other over this dichotomy, this chapter seeks to analyse a debate that is by no means over yet. It also aims to propose a concept of physical constraints, which would facilitate a clarification of the nebulous notion of 'environment and natural landscape'.

The natural environment in human geography

From the formal equality between man and his environment, there follows a positive natural order as stated by Brunet (1972, page 650), for whom regional studies "aligned economy or demography in an established order after the physical environment, thus presupposing that the latter is the true source of change. Yet how could this be any different, as geography is, above all, based on the study of the interrelationships between man and his environment." This tradition has evolved into an approach that I would term neoclassical in the sense that it presents a different view of these interrelationships, but without changing the problematic. Although the functional equality between man and his environment is always present, this interdependence may be translated into economic terms where it may be heightened or overcome by technological or financial intervention. Beaujeu-Garnier (1971, page 13), for example, would like to see studies where "man is the agent of transformation of the physical environment", and, in her opinion, "no constraints are insurmountable if one puts a price on them". According to Béthemont (1977, page 31), the translation of water into economic terms "can come about through irrigation or drainage set against stable constraints". Hence, the relationships between man and his environment are viewed, not as interdependent or as complementary, but rather as economic confrontations.

Alongside this concept a 'new geography' is developing in which the socioeconomic environment and society are paramount within the framework of the discipline. By 'new geography' I mean investigations, often deriving from the stimulus of Anglo-Saxon geography, but also influenced by economic theories as diverse as those of Marx and Keynes. One, and perhaps it is the only one, of the aspects common to these new approaches is the rejection of the discipline's classical concepts. Geography is becoming "the social science of spatial phenomena" (Auriac, 1978, page 2), "a differentiation of social space" (Pailhe, 1977). "As space is organised by society, it is a social science" according to Chamussy et al (1977, page 20), who add "the study of natural phenomena only belongs to geography if it is subordinate to this proposition". According to these authors, the degree of radicalisation varies but three concepts seem to emerge; two of these will be analysed below and the third in the third part of this chapter.

Concept 1. The natural environment is unimportant or irrelevant

It may appear paradoxical to find a convergence of views between followers of Marx and those of Keynes; certainly their approaches are different, favouring history and economics, yet in both cases the physical environment is rejected. Thus, Chaline (1966) deals with natural moderations and limits within the United Kingdom in six pages and according to Reynaud (1974, page 25), "what need is there to consider the difference in climate between Lorraine and Alsace when 90% of the population do not work in agriculture". A speaker at the 1978 'Colloque Geopoint' declared "the landscape is not a geographical concept, as it is monopolised by biogeography".

The influence of Althusser

The works of this philosopher have influenced many French geographers, for example the editors of the journal *Espace-Temps*. Thus, according to Grattaloup (1978), when referring to the Sahel region it is unnecessary for the geographer to study recent climatic variations; although these could be useful in understanding phenomena that have a dramatic social impact, this does not justify "the existence of a discipline that is both natural and social" and "the least rational attitude is to combine the study of natural phenomena and social analysis under the pretext that they are interrelated". Hence, the physical framework and its variability are not the concern of geography, and "natural resources and physical constraints primarily concern the economy".

The theoretical justification for this lies in the study of Althusser and highlights another myth of the link of geography with history. This myth is obviously still with us today. Pailhe (1977) explains: according to Althusser, the Greeks discovered the 'continent' of mathematics, Freud that of the subconscious, and Marx that of history (itself divided into different 'regions: 'traditional' history, economics, and social sciences—of which geography is a subdivision). Hence the place of geography is not in an 'eclectic multidiscipline' since it is a "system for the differentiation of social space".

The influence of the neoliberal economy

Through its weight and sociocultural framework, Keynesian theory has influenced the resurgence of geographical thought, firstly across the Atlantic, and subsequently on the continent. Over the past ten years, Anglo-Saxon geography has been a fundamental element in the resurgence of geographical thought in France, but this has served to highlight its ideological foundation, the neoliberal economy.

This economic school considers the physical environment to be an exogenous variable, that is to say at best as a constant and at worst as a hazard, a parameter to be rejected. Samuelson (1965), winner of the Nobel Prize for economics, admitted, as a heuristic principle, that "nature is neutral or benevolent" (page 1172). Morgenstern (1967), another Nobel prizewinner, adds that if we accept this postulate, "we should continue with the same methodology" but, if not, "our studies would become particularly difficult", this evidently not being favourable towards the position of physical constraints within econometric models. Hence, according to Morgenstern, man is confined within pressure and temperature barriers, in a chemical and physiological equilibrium. Here we may recall the ideas dealt with by Max Sorre in his 'Biological Foundations of Human Geography (*Fondements biologiques de la géographie humaine*), where he was one of the first to develop an ecology of man, stressing, in particular, medical and physiological constraints. Claval (1972) took the same view. For him, the "new geography has developed from economic theory" (page 4); but man must adapt to climatic conditions which in turn affect the spatial distribution of diseases, hydrological and thermal conditions, and metabolic differences (Claval, 1974).

The paradoxical convergence of liberal and Marxist views extends beyond theoretical justification. Their approaches no longer favour 'socionatural' equilibrium, but are founded upon the concepts of growth, of progress, and on a model fashioned by history. Are natural disasters to be seen as 'acts of God' in American Law? For the most part, the physical environment appears to be defeated from the outset. We find numerous examples of this, whether it be Roosevelt's New Deal, influenced by Keynes and symbolised by the Tennessee Valley Authority, or oecumenic progress such as the pioneering frontiers and the development of Soviet Asia.

Concept 2. The natural environment as a 'frame'

The word 'frame' is used here as defined by the Petit Robert dictionary, namely as a "border enclosing a picture". According to traditional human geography, the relationship of relief to historical time, and of climate to seasonal rhythm is regarded as a constant. Viewed within this frame there would be a difference in the nature of the socioeconomic complex, whose existence is now acknowledged but which is more the object of description than of serious analysis: fear of determinism preventing any synthetic initiative. Reynaud (1974) proposed a sentimental but far from unrealistic explanation. "Geography has provided us with food for our dreams and imaginations. How many geographers themselves are not basically drawn to physical geography to satisfy their love of the landscape" (page 34). Yet another geographical myth exploded by the always iconoclastic Reynaud, that of the 18th century when exploration was the very reason for the existence of geography. Geographers with a love of different landscapes, having discovered the world in the footsteps of Jules Vernes or the Fenouillard family, had only to turn to the books of their childhood! This can be taken further, and for Brunet a region is a structure where the forces at work are those of labour, capital investment, communications, and local resources; these "are inert, and remain so unless (other energies) mobilise them. These are what are also called natural potentialities" (Brunet, 1972, page 658).

Theoretical explanations of the geographical framework are less common, but they are based on a difference in nature between the physical and the socioeconomic environments, and the influence of neoliberal thought may equally be felt at this level. For Morgenstern (1967), 'nature' cannot be considered as a stochastic (aleatory) phenomenon, when human behaviour can display the same characteristics. "We are unable to offer statistical predictions on the behaviour of an enemy; but with regard to nature, we believe that current statistical hypotheses are applicable". Against the statistical uncertainties of 'nature', he sets those of social communities, which result from socioeconomic behaviour patterns.

Following from the work of Prieto, Raffestin (1978) put forward an explanation that echoes the conclusions of liberal economics with regard to essential differences. For him, human geography is not on the same level as physical geography. The latter is only "knowledge" which results from the study of the Earth from various points of view, whereas the human approach is the "understanding of that knowledge", describing the way in which the subject deals both with the socioeconomic and with the physical environments.

In addition to theoretical justifications, the study of geography in France stems naturally from the framework concept. When at the outset it claimed to isolate interactions, cartographical analysis, for example, was merely the French university interpretation of the classical concept. Today we have a physical environment directly fixed to an historical scale and undergoing variations from Precambrian to Villafranchian and Würm, from whence is derived an impression of stability linked to structural factors and reinforced by an evergreen forest.

In fact, this framework is a mere deviation from the classical concept which is based on the regional geography of a stable rural world, founded on the agrarian situation current in the years 1870–1940 (Le Berre, 1980), and influenced by the historical contribution of the *Annals* school (economy, society, civilisation), that provided the essential works of agrarian history until the 1960s (Goubert, Le Roy, Ladurie). Since 1950, the rural world has undergone radical change, and into geography have been incorporated industrial development, the concept of the town as a spatial centre of activity, and a reversal of the 'town–country' relationship in the mind of the geographer. However, since we seek to retain what we believe to be the essence of geography, we continue to write chapters on physical geography; chapters which are often merely a descriptive framework for human activity. Furthermore, regional geography is still based on fundamental ideas of the physical environment (seasonal rhythm, relief forms, etc). Whatever our concept of geographical study, for the rural environment, for example, it comprises climatic variation, surface geology, and soil erosion. Here the geographer is tempted to seek refuge in the 'framework' concept, such action being justified by the increasing complexity of techniques to be mastered by the researcher.

Today, if we read the introduction to *Tableau de la géographie de la France* (1903), Vidal de la Blache appears to have been a precursor of the framework concept. "A geographical unity does not simply result from geological and climatic influences. It is not something predetermined by nature. We must rid ourselves of such ideas, as against these there exist a reservoir of dormant energies laid down by nature but whose use depends upon man" (page 8). This is Brunet's (1968; 1972) definition of the region and its natural potentialities, but seventy years earlier. It is only now becoming apparent that the interrelationship between man and his environment is the fundamental basis for geography, with Demangeon (1940) defining human geography as "the study of human settlements and their relationships with the geographical environment" (page 27) that combines the effects both of nature and of man.

Classical geography, with its emphasis on the formal equality between man and his environment, maintained against wind and tide, carries its own internal contradiction and can relate only to a physical environment which inhibits socioeconomic activity. This almost automatically results in a split into two antagonistic branches, the physical and the human geographer each studying their particular problems and, ill-equipped to bring about a synthesis, they too often forget human–physical interactions altogether.

Physical geography as an autonomous discipline?

It is difficult to ascertain in which direction physical geography will develop, with geomorphologists appearing to favour a regrouping of allied disciplines, and climatologists feeling closer to the 'new geography'. Whatever the outcome, man would not appear to be of particular interest to the physical geographer, who seems unwilling to consider within his discipline the role of the utilisation of space by society.

During the 1970s, geomorphology has tended to establish itself as an autonomous discipline. Indeed the choice of a number of geomorphologists for the Earth Sciences section of the National Centre for Scientific Research has been seen as official recognition of the autonomy of this discipline. However, other geographers would appear to favour a regrouping around the concept of the geosystem so that, in the words of the French Association for Physical Geography, the discipline "is able to demonstrate its unity". It is, however, strange to see a discipline founded upon groups of scientists having neither the same basic interests nor the same problems. Geography has been viewed as "a mediator between certain natural disciplines and a certain type of social analysis" (Bertrand, 1978), yet physical geography, which is after all a subdivision of the discipline, has never chosen to adhere to a social methodology (see, for example, *Bulletin de l'Association Française de Géographie Physique*, 1977; 1979).

From the geographical study of the Third World, a 'naturalist' school with a strongly ecological bias has evolved for which physical geography represents a synthesis of ecology and geography. However, if the geosystem is to serve as the basis for a new-found unity in physical geography, its relations with the socioeconomic environment remain ambiguous.

According to Bertrand and Beroutchachvili (1978) and Bertrand and Dollfus (1973), "ecological facts are directly integrated with economic and social analysis". Yet the laws which govern the geosystem are of a natural order whereas the rest of geography is subject to socioeconomic laws, thus highlighting the difference in nature between the two, and the apparent autonomy of physical geography.

Together with geomorphology and biogeography, climatology represents the third traditional aspect of 'physical' geography and here, even more so than in geomorphology, we can envisage an abstract, autonomous study area distinct from mainstream geography. However, phenomena such as agricultural disaster and droughts reinforce the fact that climatology is an integral part of the socioeconomic environment. In effect, climatology, as studied by the meteorologist, was conceived to analyse climate and not the socioeconomic environment, but the droughts in the Sahel region would appear to have resulted in a rethink amongst climatologists, with the relationship between economy and climate developing into a new area of study.

The rejection of the dominant role of physical factors within geography by those scientists primarily concerned with the socioeconomic environment (see, for example, George, 1970) has led to the increasing autonomy of physical geography as a discipline. The classical concepts of geography still have considerable influence on physical geography but, far from analysing the relationship between man and his environment (even when the latter is viewed as a framework), physical geography continues to explain the form, nature, origin, and metamorphosis of the environment in isolation; which is akin to looking at a great painting yet appreciating only the craftsmanship of the frame.

Physical constraints as agents of spatial organisation

"Time is continuously modifying the environment",
Elisee Reclus (*L'homme et la terre*, 1905, page 114).

The preceding discussion has highlighted the feature that the formal equality between man and his environment contains opposing deviations. For some people, physical environments are the domain of natural science and for others they are primarily the concern of the economy. The new geographies have attempted to avoid the two-fold natural and social heritage.

In 1905, the work "Man and the Earth" (*L'homme et la terre*) by Reclus was published. In this extensive survey of historical geography, the author used history to defend the primary role of man in the organisation of his environment. This view of the physical environment is certainly close to that which we have today. He distinguishes between static, essentially physical, environments and dynamic environments resulting from historical evolution and human action (being very close here to the "framework" concept). However, unlike Vidal de la Blache, he ventured towards the idea of interactions: "human history can only be explained by the addition over the centuries of environments with 'composite interests', but we must also appreciate to what extent these environments have themselves developed through general evolution and have modified their action as a consequence." (page 113). At that time, reciprocal relationships were difficult to picture owing to the lack of suitable means, particularly statistical ones, to analyse the complexity of the physical environment.

Today, apart from developments in quantitative geography, systems analysis should enable us to analyse the interactions between society and the natural environment. Brunet (1972) and Bertrand (1972; 1978) refer to this in particular; in relation to the region on the one hand, and to the geosystem on the other. However, Bertrand holds that the difference in nature between physical and socioeconomic laws tends to prevent a liaison between two systems that are evolving separately, although they are able to utilise the same analysis methodology. If systems analysis proves to be a useful means of studying these interactions, it will also be necessary to define a relationship that would allow not a partial, but a complete integration of the two systems into a single, more comprehensive, geographical system.

To achieve this, the classical definition of geography must be abandoned as it leads us nowhere; if geography is the discipline that studies spatial organisation by society, then we must consider physical environments to intervene as organisational factors that are nonprivileged, nondominant, but which are not without significance. Hence a study area appears that is totally concerned with neither the reciprocal interaction of physical forces (the geosystem), nor uniquely with those of society, but rather with a combination of constraints interacting between the geosystem and the socioeconomic environment, governed by physical, social, and economic laws. Therefore there is no longer confusion between the existence of the relationship and the law that explains it.

Systems analysis allows the formulation of an overall spatial perception, in particular with regard to the role of physical phenomena within this perception. However, physical phenomena may only be studied in this way subject to the condition that we always remember that we are dealing with a socioeconomic organisation that utilises space and that is, therefore, subject to and protects itself from the actions of the natural environment.

This viewpoint is close to that formulated by Chamussy et al (1977, page 20): "natural phenomena are only of geographical importance in the sense that they are taken into account by society in its organisation of space". Charre (1977, page 218) has added, "in our opinion, natural elements are supplementary and invariable, and they are the concern of other sciences; they are observed, modified and subdued by society, yet they also belong to geographical research." However, in my opinion, an invariant natural element may be a geographical object, even if, in effect, the study of it has nothing to do with geography in the sense it is used here. Society moves and evolves within a spatial framework that is differentiated and that, even outside the limits of the oecumene, possesses invariable elements that are geographical objects, particularly when these elements represent obstacles in the relationships between societies in different parts of the framework. They are not only reducible to elements in a perception of planetary space, but they can be a legitimate part of geographical research.

Hence, through the nature of its manifestations, whether constant or not, the physical environment appears to be a spatial component in the same sense as political organisation or the economic system. For example, Charre (1977) described the concept of drought on the basis of the situation in Sahel from 1969 to 1973: "the limits of the drought are both temporal and spatial. They are situated within the pluviometric variable, but it is economic, social and political characteristics that determine their position with respect to this variable." (page 218).

Reclus (1905) had already distinguished between 'environment-space' and 'environment-time'. In addition, on the basis of the notion of temporal variability, the concept of the 'physical constraint' was defined. In the same way that Bertrand (1978) faced with the "abundant and multiple reality" of nature, defined the concept of the geosystem, so we must define a concept enabling us to identify the influences and actions of a natural environment in a space organised by society.

Concept 3. Interaction between the physical environment and a socioeconomic space

By a constraint, we generally mean a 'limit to freedom of action'. In mechanics the word has a more precise meaning being "a magnitude characterising the intensity of superficial contact forces", and it is in this sense that I use the word here. A physical constraint is defined as the interaction between an element of the natural world (climatic, geological, geomorphological, or botanical) and a space utilised by man. The physical constraint which measures the intensity of forces between the two is situated on a scale of values in the same way as social, mental, political, and economic constraints, and, like these, is a component of spatial organisation. Whatever its internal mechanism, it becomes an element of the regional system.

Reclus (1905) differentiated between static and dynamic environments. Brunet (1968) did the same for geographical discontinuities. In their footsteps, I aim to distinguish between static and dynamic constraints, the difference stemming from their temporal evolution.

Static constraints

These interpret the interactions between the socioeconomic environment and the physical environment as a framework (the static environment of Reclus). The latter offers only a simple resistance—a threshold that cannot be crossed, such as a mountain for example. However, static does not mean fixed; thus in climatology, seasonal rhythm is one of these constraints that cannot be shaken off.

Moreover, the critical threshold of a constraint may well not be reached within the existing socioeconomic context. There is therefore a margin, and developments may occur which extend utilisation capacity within the existing framework (such as technical progress, agrarian reform, demographic evolution, or the improved resistance to frost of seed varieties). Even if, for some reason, the threshold of a constraint is extended (polders, hydrological schemes, tunnels, etc), it may still be considered to be static. In effect, if in the previous case the constraint was constant over a period of time, here it remains constant until the moment of economic mutation. As soon as this takes place, even if its threshold has been lowered, the constraint remains constant from that moment on.

Hence, a static constraint would essentially appear to be characterised by temporal stability. It may appear either as a constant function or as a function increasing with time, and society adapts to it *ex post*.

Dynamic constraints

These constraints are not subject to time in the same fashion; they present neither stability nor temporal regularity. They may result from anthropological (podzolization or soil erosion) or nonanthropological (climatic variation) action, and they may or may not be irreversible (landslide or early frost) in their environmental consequences. Whatever their form, however, they cannot be regarded as constant over periods of time. Thus, they are prone to variation that we may express, if not in terms of probability, then at least in terms of frequency.

Through having too often confused seasonal and interannual variation, classical and modern geography have both classified a number of dynamic constraints as static, and have thus reinforced the framework effect. This is particularly clear in climatology. For example, Béthemont (1977, page 18) wrote: "Interannual climatic variations are without importance, as they end up as averages or more or less constant frequency values. Only seasonal rhythms or secular rhythms able to bring about a reversal in tendency are of real significance."

We must now consider the relative evolutionary rates of these two components. Generally we tend to believe that the socioeconomic complex evolves the fastest and that it is susceptible to more profound mutations than the physical environment. This is certainly the case: political reform and technological development act on a geosystem which is itself considered as an 'average' and, in general, they do not exceed their potential utilisation threshold. By their very structure, static constraints are illustrations of this opinion.

Dynamic constraints offer more complex interactions. For example:

(a) They result from human action and, as the process is not irreversible physically speaking, a mutation through human action is always possible.

(b) We may consider a political or socioeconomic system to be stable in the short term. Within this framework (which, if not completely stable, is at least stabilised and whose evolutionary possibilities we know) the variability of certain physical constraints presents a greater or lesser probability characteristic. The consequences of the constraints upon a relatively homogeneous socioeconomic environment are subject to spatial variation: the drought in North-West Europe in the summer of 1976 was felt to be a disaster in Britanny and in South-West England, but in Ireland the harvest was exceptional.

If probability is high, that is to say if the physical constraint acts with a certain regularity, this results in a quasi-static constraint. In this case, society either has, or has not, the technical means or the desire to fight against the adversity (the collapse of embankments on the Loire, or the need for snow clearance in the mountains).

If probability is low, two facts should be noted. We may find society totally unable to control the environment, as is the case with snow and ice in Britanny. In addition, the resultant economic consequences may be overcome with adequate equipment, but this may be considered to be financially unprofitable.

We are able to find more complex economic adaptations with regard to phenomena whose dates of occurrence are themselves variable. For example, the time of the first snowfalls, a late spring, or the date of the end of the dry season have profound effects on tourism and food supplies. In turn these effects lead to higher prices and stockpiling, and this is in a stable socioeconomic system that utilises climatic variations and that adapts in an *ex ante* manner to the economic market, that is to say, it makes forecasts on the basis of current market conditions.

The solution to such situations is political. But although this is undoubtedly correct, if we do not improve or even transform the socio-economic system, interactions with the natural environment may be different but they will not be automatically overcome. Today, the geographer studies space as organised by society, but as it is and not as it should be.

References

Auriac F, 1978 *Table Ronde Européenne de Géographie Quantitative; Rapport scientifique de l'auteur* 2 page typescript, Centre National de la Recherche Scientifique, Strasbourg

Beaujeu-Garnier J, 1971 *La géographie, methodes et perspectives* (Masson, Paris)

Bertrand G, 1972 "La science du paysage" *Revue de Géographie des Pyrennées et du Sud-Ouest* avril 127-292

Bertrand G, 1978 "La géographie physique contre nature?" *Hérodote* numéro 12 (Oct-Dec) 77-96

Bertrand G, Dollfus O, 1973 "Le paysage et son concept" *L'Espace géographique* 2(3) 161-163

Bertrand G, Beroutchachvili N, 1978 "Le 'géosystème' ou 'système territorial naturel'" *Revue de Géographie des Pyrennées et du Sud-Ouest* avril 167-180

Béthemont J, 1977 *De l'eau et des hommes* (Bordas, Paris)
Birot P, 1968 *Précis de géographie physique générale* second edition (Colin, Paris)
Brunet R, 1968 *Les phénomènes de discontinuité en géographie* Editions du Centre National de la Recherche Scientifique, Paris
Brunet R, 1972 "Pour une théorie de la géographie régionale" in *La Pensée géographique française contemporaine.* Mélanges offerts à A Meynier (Presse Universitaire de Bretagne, St-Brieuc) pp 649-662
Bulletin de l'Association Française de Géographie Physique, 1977 and 1979 Paris, ronéotypés
Chaline C, 1966 *Le Royaume-Uni et la République d'Irlande* (Presse Universitaire de France, Paris)
Chamussy H, Charre J, Durand M-G, Le Berre M, 1977 "Espace que de brouillons commet-on en ton nom!" *Brouillon Dupont* numéro 1 15-30
Charre J, 1977 "A propose de sécheresse" *Revue géographique de Lyon* 2 215-226
Claval P, 1972 "La rèflexion théorique en géographie et les méthodes d'analyse" *L'Espace géographique* **1** 7-22
Claval P, 1974 *Eléments de géographie humaine* (Génin, Paris)
Demangeon A, 1940 "Une définition de la géographie humaine" in *Problèmes de géographie humaine* (Colin, Paris) pp 25-34
Dorize L, 1977 "Economie et climat" *Revue d'économie politique* **6** 867-891
George P, 1970 *Les méthodes de la géographie* (Presse Universitaire de France, Paris)
Grattaloup C, 1978 "Géographie physique, écologie, espace social. Les enfants du divorce" *Espace-temps* **9** 113-123
Le Berre M, 1980 "Heur et malheur de la géographie régionale" *Travaux de l'Institut de Géographie de Reims* numéros 41-42 3-20
Meynier A, 1969 *Histoire de la pensée géographique en France* (Presses Universitaire de France, Paris)
Morgenstern O, 1967 "L'attitude de la nature et la comportment rationnel" in *Les fondements philosophiques des systèmes économiques. Textes en l'honneur de Jacques Rueff* (Payot, Paris) pp 131-141
Pailhe J, 1977 "La géographie, procès sans sujet" *Espace-temps* **7** 33-35
Phlipponneau M, 1976 *Géographie et action* (Colin, Paris)
Raffestin C, 1978 "Les construits en géographie humaine: notions et concepts" *Colloque Géopoint 78* (Lyon), Groupe Dupont, Avignon, pp 55-74
Reclus E, 1905 *L'homme et la terre* Volume 1 of 6 (Edition universelle, Paris)
Reynaud A, 1971 *Epistémologie de la géomorphologie* (Masson, Paris)
Reynaud A, 1974 "La géographie entre le mythe et la science" *Travaux de l'institut de géographie de Reims* numéros 18-19
Samuelson P, 1965 "Professeur Samuelson on theory and realism: a reply" *The American Economic Review* numbers 4-5 1164-1172
Vidal de la Blache P, 1903 *Tableau de la géographie de la France* (Hachette, Paris)

5

Reflections on some recent developments in systems analysis in French geography

F Auriac, F Durand-Dastès

The concept of the system is not unknown in French geography. The term was used many years ago in geographical texts, but only in relation to 'culture systems' and 'erosion systems'. In addition, the term was also widely used with regard to a series of rather undistinguished notions, and the correct logic of system theory has only been sporadically utilised.

It is only fairly recently that efforts have been made to utilise systems analysis in a more coherent manner, mainly in physical and regional geography[(1)]. It is our intention here to pose a certain number of questions on the subject of systems analysis, from two viewpoints: first, on the use of the systems model in geographical analysis, and then on its ideological and theoretical status. We intend to pose these questions on the basis of two examples of localisation problems formulated in terms of systems, and to follow with general observations.

Two examples of the utilisation of systems analysis

Both examples concern 'paradoxical localisations' of groups of geographical data, one could even say of geographical 'individuals'.

The black ghettoes situated in the centre of American cities can in effect be considered as 'paradoxical' in the sense that there we find a poor population inhabiting often dilapidated buildings in areas close to the city centre, yet these buildings have extremely high real-estate value [a paradox underlined particularly by Rose (1972) and Ward (1968)]. In the same or in a similar way, the large-scale cultivation of vines on the Languedoc plain in the South of France, which has undergone decades of crisis (Auriac, 1979) and which is continuously condemned by distinguished economic analysts, is still able to resist constant initiatives on the part of local government to reduce its size. It is no exaggeration to say that since 1907, a year which saw serious troubles in the region following a severe crisis, vine cultivation here has been considered to be condemned by economic evolution. Despite this, it is still with us.

We feel that it is possible to show that these two geographical examples correspond to systems built up over many years, during which time their component elements were first established. This was followed by a fairly rapid phase of systemogenesis, of system formation, which subsequently acquired a certain stability owing to its homoeostatic properties. (On the

(1) Barel (1973); Brunet (1973); Bertrand (1978); Dauphiné (1979); Dollfus and Durand-Dastès (1975).

formation and maintenance of systems, see Durand-Dastès, 1978.) We shall present these two systems in the form of diagrams (see figures 5.1 and 5.2).

On the top of the two diagrams, we have shown the appearance of the elements of the future system; these often become established long before the appearance of the system itself. Hence, in the case of Languedoc, the tight urban framework may be considered to be Roman, whereas the accumulation of business capital in the towns began in the Middle Ages. In contrast, the rapid development of vine cultivation occurred much later, during the nineteenth century. However, even after this rapid development, we could not have spoken of a viticultural 'system' owing to one main contradictory factor, of social order, which extended over the entire sphere of vine cultivation in Languedoc; after the 1907 phylloxera crisis, we might have supposed that this contradiction was being aggravated, with the phylloxera crisis increasing capitalist intervention. However, 1907 marked a profound change, a phase of rapid 'systemogenesis' during which a series of interrelationships were formed which were to assure the functioning of a true system. These interrelationships are illustrated on the lower section of the diagram. We can see that there exist three elements: first, three types of holdings (medium-sized holdings, large capitalist holdings, and smaller, part-time holdings); second, the region's towns, which are sources of capital and employment; and third, the cooperatives. Between these elements we are able to discern the movement

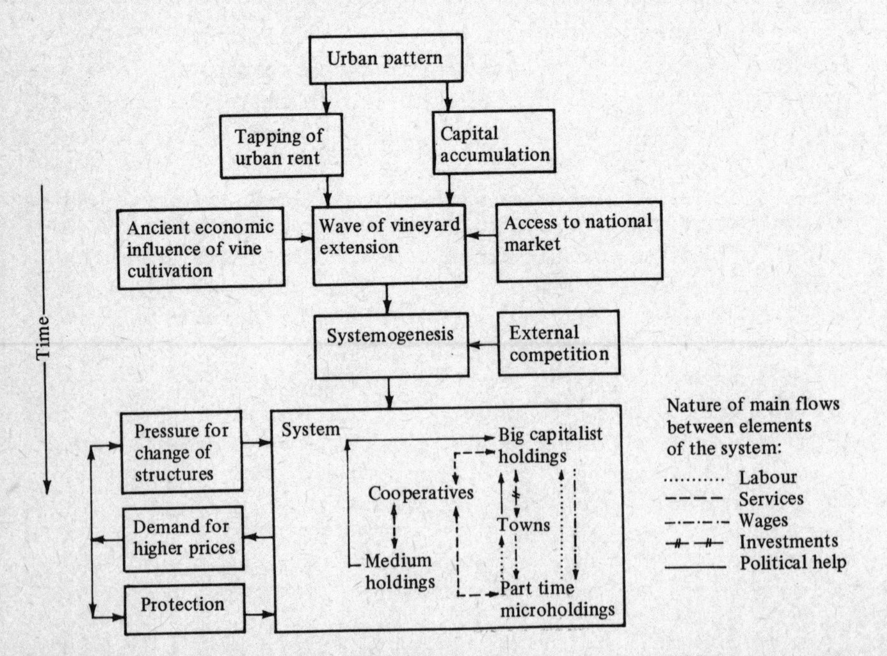

Figure 5.1. The Languedoc vineyard system.

of manpower, employment, and services, in addition to the interrelationships between the forces acting upon them. The combination of movement towards and away from an element ensures its survival in the face of adverse conditions. Thus, the very smallest holdings survive in part owing to the employment offered by the large holdings and by the urban markets, and in part owing to the services provided by the cooperatives. Conversely, excess labour from the smaller holdings guarantees a sufficient workforce for the large holdings, which in turn enables them to realise a sufficient level of activity. We are able to proceed in the same way for each element of the system, which is in fact normal in the logic of a system.

Over the years, these elements have undergone certain changes with regard to their importance, their function, and their 'place' in the overall picture. Thus the cooperatives, born from an authentic original socialist ideal, are perfecting their technical vine-growing function and their spatial organisation; it was in the 1930s that this organisation was achieved through a network covering *all* productive land. At the same time, the

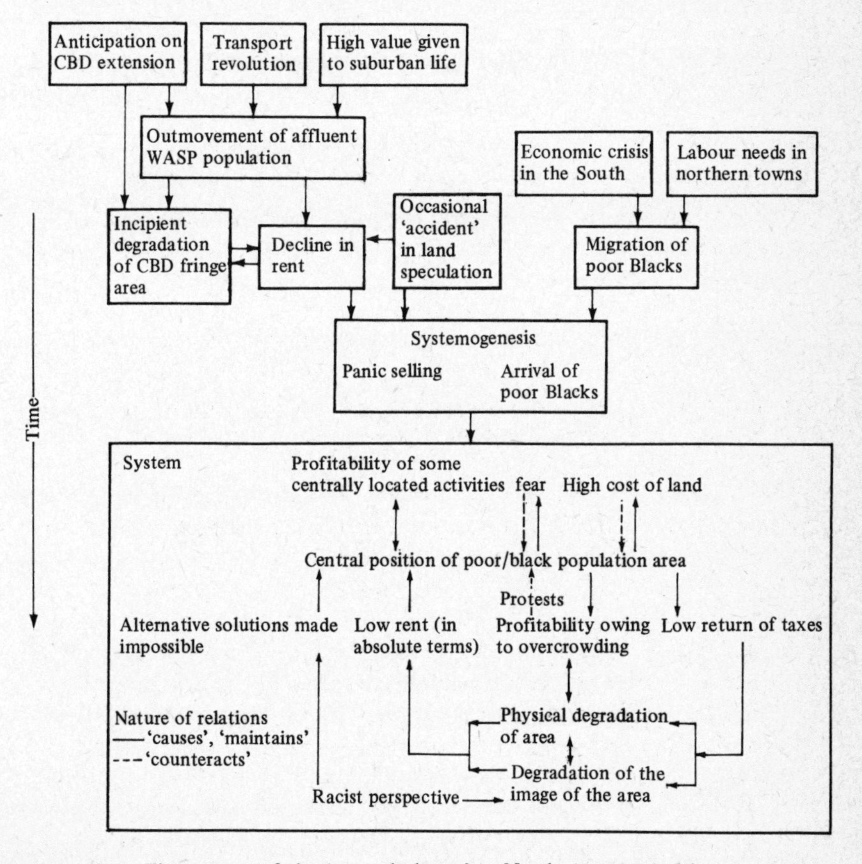

Figure 5.2. The system of the 'central ghetto' in North American cities.

cooperatives have stimulated a redistribution of the social categories involved in viticulture. Against the reduction in capital investment, which leads to the survival of the patrimony of only the larger holdings, the profusion of spatial multiactivity movement ensures the continuation of a vine-cultivating region, which is growing in extent but is becoming less and less agricultural. In this way, the cooperatives constitute a structural hierarchical element with a privileged function in the system. For this reason, vine cultivation has sufficient internal resources to put a stop to external attempts at change, to turn to its own advantage the costly irrigation schemes carried out by local authorities in the Languedoc region, to achieve a qualitative self-improvement (higher-quality vines), and both to concentrate and to redeploy its activities over the land. All of these modifications have been achieved without the existence of the system itself being questioned; they remain secondary in relation to its fundamental homeostatic properties.

The system behaves as a separate identity with respect to the outside world, a French social and economic entity existing under a certain set of circumstances. Through the actions of various economic agents, the system constantly demands sufficiently high prices for its maintenance. This demand has consequent effects on the system itself. Some of these effects are positive, in the sense that the vine-growing system receives a degree of protection from legislative measures which prevent extreme market oscillations; others are negative, as the demand gives rise to reconversion assistance (central government measures) which tend to limit the extension of vine cultivation. But these 'negative effects' have not yet been able to put an end to the functioning of the system, because, as we have already seen, of its stability and solidity.

The structure of the diagram relating to the central black ghettoes is comparable to that for the Languedoc system; despite the major differences between the realities, the logic of the grouping seems to us to be the same; and this we are citing as our second example, although in less detail than the first.

Here, also, the establishment of the elements of the system has been progressive, and has preceeded its formation. These elements have arisen out of phenomena on a totally different scale. On the one hand, certain of them have arisen from an economic change extending at least over the eastern half of the United States (crisis in the South, and the need for labour force in the North); on the other hand, some have arisen from interurban situations, with the beginnings of emigration of the wealthier classes to the suburbs and the anticipation of extension of the CBD. 'Systemogenesis' in this case occurred fairly rapidly. In at least one case, that of Harlem, it began and was accelerated by a speculative crisis: counting on sales of houses at the termini of the recently constructed underground system, investors built numerous buildings for rent.

The capacity thus created exceeded demand, notably because of an economic crisis, and speculators attempted to realise a profit by acquiring the surplus properties and by housing poor Blacks with very high densities of occupation in these buildings. These speculators thrive on panic-selling which enables them to house the Blacks, thus making substantial profits. In this way, at least in this case, we find an element similar to that in the 1907 crisis in the French vine industry, at the beginning of 'systemogenesis'.

The system in the ghettoes is shown here as being different to that in Languedoc. Here, the elements of the system are processes (namely the progressive deterioration of a district) or qualitative aspects of the urban framework (the central position of poor Blacks). The interrelationships are not movements as in the previous case, but are interrelationships of cause and effect. We are able to identify two types of 'retroaction ring'. Certain positive rings tend to further the continuation of the system: for example, a black population, because it is poor, leads to a decrease in the amount of local taxes due, this in turn leads the local authority to reduce maintenance of the district and hence physical degradation is heightened, which tends to keep rents low, or at least relatively so (as they constitute a large proportion of the low incomes of the population). Evidently, these fairly low rents maintain a population with a low level of income, and things turn full circle.

We can identify many other such 'rings'. But there also exist negative rings that hinder the functioning of the system and thus tend to limit its spatial extension. Hence, the degradation of a district tends to give rise to protests, sometimes extremely violent, such as took place during the 'hot summers' of the 1960s. For the present, these negative retroaction rings have not in themselves been able to place the existence of the system and its spatial framework in question, but have in fact combined to stabilise it, or at least to limit its expansion.

The ghetto system contains one fundamental aspect shown on the diagram—the racist feelings of a large part of the white population; it is because of these that the panic sales, so important to the systemogenesis, were able to take place, and that the possible alternative solutions, such as the redistribution of the black population into the rest of the low-income population, were impossible. The less acute and more newly established racist feelings in Western Europe explain why segregatory systems are less common in European cities, where, however, they are not totally absent, as we see in Paris, London, or Stuttgart.

Some general notes on these systems

On the one hand, these notes concern the place of systems analysis in geographical investigation and, on the other hand, general problems of logic and ideology.

In the first instance, systems analysis appears to be an explanatory localisation technique—if you prefer, a part of locational analysis. In effect, we can say that if a spatial phenomenon exists, this is because it corresponds to a system which is relatively stable due to its homeostatic properties. The paradoxical continuation of vine cultivation in Languedoc, and the continuance of central black ghettoes is precisely because of the existence of such systems. Taking this further, we can say that locational analysis demands that we answer the question "why is such a phenomenon found here and not elsewhere?", and that systems analysis helps us to formulate a response. In effect, systems link elements that are themselves localised, establishing relationships between different localities.

These relationships can grow up at different levels, and can cover a variety of aspects. In certain cases it is at the input level that the process of localisation will occur. Some simply make the functioning of the system possible, without playing a fundamental active role: hence the existence of a Mediterranean climate and a wide plain figure among the components of the Languedoc system, where their role is only passive; but in a very different and much more active way, racial prejudice in North America is a fundamental aspect of the ghetto system. One of the authors has tried to show elsewhere, and in a totally different context, how different inputs in tropical regions lead to two different systems of atmospheric circulation, that of the Hadley cell and that of the monsoon (see Durand-Dastès, 1979).

In the case where the inputs only create the *possibility* of a system, the existence of the system derives from other operational causes, notably the juxtaposition of localised elements, often established at different times, and included *within* the system at the time of systemogenesis. Hence, the existence of a dense urban network on a Mediterranean plain is an important element in the Languedoc system. These juxtaposed elements often enter into system interrelationships only during a unique and nonreproducible historic phase, where chance elements play an intermittently important role. The systemogenesis described in the case of Languedoc came into being during an economic fluctuation which took on a fundamental importance. We can imagine that, if this had not taken place, or if it had occurred at a different time or in different circumstances, the system would not have appeared[(2)].

Whatever the case, at one time or another, the proximities (the positions of the elements) play an important role in the appearance of the system.

[(2)] The importance that should be accorded to the presence of an element, and to chance fluctuations, has been the subject of discussion. Hence, Thom believes that the role of chance fluctuations has been exaggerated by certain analysts; he writes, "A relatively comprehensive examination of the substratum enables us to predict *a priori* the possible results of bifurcation, that preexist the causal fluctuation. The role of the latter is in part to set the process in motion and—eventually—to determine, by an apparently arbitrary choice, the subsequent evolution from all the possible outcomes. But it does not create it." (Thom, 1980, page 126).

The Languedoc vine-growing situation can only be understood in terms of the movement between the socioeconomic elements of the system (that is, the different types of holdings, the urban labour markets), but it was the *proximities* that made the establishment and operation of these movements possible within the framework of interrelationships that make up the system. In a slightly different way, social processes create spatial forms that then play the role of 'memory' in a system, as is the case with the physical degradation of districts within the ghettoes. More generally, we can say that spatial inscription often serves as the 'memory' for social and economic processes; this role functions particularly well within the logic of systems.

The above discussion clearly explains how the systems we have presented could be said to be 'geographical' or 'spatial'. More exactly, we have been concerned with the spatialisation of systems of phenomena which remain fundamentally economic and social. It is useful to take as a working hypothesis the manner in which the spatialisation of systems occurs. In the sense that space is an indispensable condition for the existence of a socioeconomic system, we can say that this is 'spatialised'. Certainly, space cannot be seen as a principal factor; for it is social and economic interrelationships that, in their system combination, lead to its establishment. However, the spatialisation of a system may lead to radical and decisive change.

These various reflections pose both logical and ideological problems. Broadly speaking, in the ideological and theoretical debate in France, certain analysts venture to suggest that systems analysis and dialectical logic are compatible. Among the dialecticians, those who believe that formal logic and dialectical logic are irreducibly separated cry heresy. Must we be confined by conceptual exclusions? Whatever the state of the question today, it seems that geographers, as well as others, have to take full part in this debate; it would be most unfortunate if the considerable knowledge amassed from their research did not contribute to it.

The studies described here raise certain questions. Do all fundamental socioeconomic contradictions necessarily arise only from the social or the economic, as the Marxist claims? Can they be fundamentally spatial, or at least of spatial form? Cannot space, which is seen here as the ultimate homeostatic regulator, have other decisive functions (and what are these?) in other systems? In any event, it certainly seems that classic analysis of production methods does not satisfactorily answer the theoretical question of social relationships within production systems, even if they are their essential dynamic foundation. Without doubt the spatialisation of processes will give rise to its own social and economic contradictions.

References

Auriac F, 1979 *Système économique et espace. Un exemple en Languedoc* unpublished PhD thesis, University of Montpellier

Barel Y, 1973 *La Reproduction Sociale* (Anthropos, Paris)

Brunet R, 1973 "Un schéma de l'espace français" *L'Espace Géographique* **2** 249-255
Bertrand G, 1978 "Le paysage entre la nature et la société" *Revue de Géographie des Pyrénées et du Sud-Ouest* **49** 239-258
Dauphine A, 1979 *Espace, Région, Système* (Economica, Paris)
Dollfus O, Durand-Dastès F, 1975 "Some remarks on the notions of system and structure in geography" *Geoforum* **6** 83-94
Durand-Dastès F, 1978 "Sur le concept de combinaison" *Géopoint* **2** 101-106
Durand-Dastès F, 1979 "La notion de système et la circulation atmosphérique" *Bulletin de l'Association des Géographes Français* **56** 391-398
Rose H M, 1972 "The spatial development of Black residential systems" *Economic Geography* **48** 43-65
Thom R, 1980 "Halte au hasard, silence au bruit" *Le Débat* **1** 119-132
Ward D, 1968 "The emergence of central immigrant Ghettoes in American cities, 1840-1920" *Annals of the Association of American Geographers* **58** 343-359

The development and present state of research into 'quantitative geography' in the German-speaking countries

E Giese

Introduction

I should like to preface this chapter with two introductory comments, the first of which relates to my own research activities. Since I am a social and economic geographer, I see the development of 'quantitative geography' from that point of view, and I ask the forbearance of the physical geographers if their contribution to its development is inadequately or incompletely presented.

The second comment relates to the papers by Bartels (1968a), Kilchenmann (1975), and Lichtenberger (1978a; 1978b), which have already examined elsewhere the development of quantitative geography in the German-speaking countries. So as to add something new to these studies, I shall try and give my account a somewhat different emphasis, concentrating particularly on the major current research foci, especially as regards the formulation of theory and model building.

Methodological reorientation

When Burton published his paper on the 'Quantitative Revolution' in geography in 1963, there was no trace of any such revolution in any of the German-speaking countries (GSC). Indeed, not until the mid-1960s was there a beginning to the development of quantitative geography in these countries. Mathematical and statistical techniques began to be applied in geography and were used to try and solve some of the traditional types of questions and problems, in particular those relating to classification and regionalization. Typical of this phase of development were the papers by Steiner (1965a; 1965b) on the application of factor analysis in geography, by Werner (1966) on the geometry of traffic networks, where graph theory was used to assess the economic viability of traffic networks, and that by Ahnert (1966) in which he pointed to the importance of computers and mathematical models for the future of geomorphology. The first techniques to be applied widely were regression and correlation analysis. There was also particular interest in factor analysis and principal components analysis, which was not really surprising since with their help it was possible to describe and analyse complex interrelationships much more clearly than had been the case using conventional methods. In effect it meant that a technique was now available which enabled a traditional and classic field of geographical enquiry, regional analysis (or the analysis of spatial structure), to be investigated much more fully than had previously been possible.

Factor analysis was used primarily as a basis for complex techniques of classification and regionalization, the latter being effected with the help of simple cluster-analysis techniques (the centroid method, the Ward algorithm).

Factor analysis introduced many geographers in the GSC to a 'quantitative phase in their development'. There was an unwritten rule in the geographical world that anyone who was anyone must have completed at least one factor analysis, and by the end of the 1960s, and especially in the early 1970s, the results of these efforts were appearing one after another[(1)]. It was only realized later that they had stumbled on a particularly difficult mathematical–statistical technique, giving rise to heated discussions subsequently (see Giese, 1978c; Kemper and Schmiedecken, 1977); these difficulties played no part in the first innovative phase of quantitative geography.

A survey of the application of factor analysis, and of the closely allied technique of cluster analysis, shows clearly how the use of mathematical–statistical techniques slowly spread throughout the GSC. In this regard the North American influence should not be overlooked: the same process of adaptation occurred as in the United States, but with a time-lag of almost ten years. Berry published his first factor analysis in 1960; the first factor analyses in the GSC date from the late 1960s (Kilchenmann, 1968), apart from the work of Steiner, who was then living in North America.

Without going into the causes of its very belated start, the delay had very considerable consequences for the development of quantitative geography in the GSC. In the late 1960s and early 1970s a large number of new universities were founded in the Federal Republic of Germany, and there was a massive expansion in the number of professorships of which geography took its full share. When the time came to fill the newly created posts, there were few qualified geographers skilled in the use of quantitative methods available for consideration. The result, which has both personal and political ramifications, is that no more than twenty out of around two hundred and twenty professors of geography in the FRG can be counted as quantitative geographers or can be said to have any sympathy for this branch of the subject. In many Geographical Institutes —there are more than forty altogether in the FRG—quantitative geography is either ignored or pursued only half-heartedly. The teaching of the applications of mathematical–statistical techniques is all too often no more than lip service, rather than being encouraged as a good thing and recognized as useful. The situation is much the same in Austria and Switzerland and goes some way towards explaining why quantitative geography has not caught on in the GSC in the way in which it clearly did in the English-speaking world.

[(1)] See, for example, Kilchenmann, 1968; 1970; 1971; Kilchenmann and Moergeli, 1970; Bähr, 1971a; 1971b; Bähr and Golte, 1974; Braun, 1972; 1975; Sauberer and Cserjan, 1972; Sauberer, 1973a; 1973b; 1974; Stäblein, 1972; 1975; Stäblein and Stäblein-Fiedler, 1973; Herrmann, 1973.

Reassessment of scientific theory

In the first instance there was no revision of the content of geography in the GSC to accompany the adoption of the mathematical-statistical techniques, even though such a change should have been an integral part of any such development. The new methodological toy was embraced without any attempt being made to change either the terms of reference, the aims and objectives, or even the fundamental scientific approach. The first decisive step in this direction was taken by Bartels (1968b) with the publication of his paper "On the reassessment of the scientific underpinnings of human geography", together with the translations he initiated of various studies in English [see Bartels (1970a) and Haggett's textbook (1965; 1978) *Locational Analysis in Human Geography*].

These writings of Bartels are important in two respects for the development of quantitative geography. First, human geography had previously stuck doggedly and somewhat one-sidedly to the classical concept of landscape studies and regional geography, but Bartels tried to provide it with a new or at least an alternative theoretical framework. He developed an approach to human geography derived from analytic and pragmatic scientific theory and orientated towards regional science. It is worth noting in this context that, even at this early stage, Bartels's approach incorporated behavioural variables.

A second development in Bartels's writings stated clearly that the goal of this approach to human geography ultimately ought to be the development of new theories and models, as well as the giving of greater precision to existing ones. It meant that the purpose of quantitative geography could not simply be defining, describing, and analysing observations more precisely through the application of statistical techniques. Neither could it be the development of quantitative techniques for their own sake, nor the uncritical translation and application of mathematical-statistical techniques to geographical data. Rather the aim had to be to develop more realistic spatial theories and models, much in the same way that regional science had done, or at least to expand on existing theories. For example, location theories, like those of von Thünen, Christaller, and Lösch, were originally derived to explain the primary, secondary, and tertiary sectors, but more recently regional scientists, such as Isard, von Büventer, and others, have developed them more in the direction of general economic theory.

This task was tackled enthusiastically by the quantitative geographers and at the same time attempts were made to extend the range of techniques available, and to test their applicability. Various different theories and models were adopted and attempts made to make them more precise and to develop them further, as well as to test them empirically. There is a whole series of models dating from the beginning of this phase, all built with the help of regression analysis, factor analysis, or discriminant analysis, but with varying specifications with regard to content.

In physical geography, for example, in the field of hydrology, Herrmann (1972; 1973; 1974), Herrmann and Schrimpf (1976), and Streit (1973) produced models for predicting runoff as well as models of water pollution in the FRG. In the field of geomorphology, Ahnert (1971; 1976) and Rohdenburg et al (1976) produced models of slope degradation.

In human geography there were three main areas of development: first, models to analyse and describe spatial diffusion processes, Bartels (1968c; 1970b), Bahrenberg and Loboda (1973); second, models for analysing central place systems, Deiters (1975), Güssefeld (1975); and third, factor analytic models for analysing inner-city population movements as well as the social segregation of the urban population (the fields of factorial ecology and social ecology), Braun (1972; 1975; 1976). A further focus of research in the GSC 'Quantitative Working Group' was in the field of time-series analysis. Primarily using Markov chains, various types of time-series models were built and used to analyse and predict a number of processes, among them inter-regional migration and regional population growth in the FRG (Gatzweiler, 1975a; 1975b; Gatzweiler and Koch, 1976). In addition, they were used to predict the runoff patterns of rivers (Streit, 1975a). Time-series models were also used to generate synthetically more complete data sets, when only short series of observations of the processes in question were available (Streit, 1975b; 1976). The application of time-series models to spatial variations was a logical extension, but one which only followed later.

Network analysis was a further area of study that emerged fairly early in quantitative geography in the GSC, but it received little sustained attention. Reference should nevertheless be made once again to the work of Werner (1966) on the geometry of traffic networks, which examined the economic feasibility of traffic networks using graph theory. Vetter (1970) extended the work of Werner in his studies of the rail network in Lower Saxony by using graph theory. He developed a simulation model for the construction of railway networks, using techniques similar to those adopted by Kansky (1963) in his simulation models. The graph-theory and matrix studies by Schickhoff (1974; 1977), using as an example the Dutch rail network, provided a critical appraisal of the simulation techniques used by Kansky and Vetter, although to some extent her models are constructed somewhat differently. An interesting study which should stimulate some further work on network theory and its methodological base is that by Güssefeldt (1978). This study used graph theory as a means of checking planning concepts, notably development axes and growth poles, both of which have been the subject of much recent discussion and have already been used in various different aspects of planning in the FRG.

Advances in the formulation of theory and model building

This section examines, in rather more detail, four areas of research into the formulation of theory and model building which are current centres of

interest in quantitative geography in the GSC:
1. Spatial diffusion and spatiotemporal stochastic processes.
2. Partial location theories.
3. Theories of spatial growth and development.
4. Model building for planning purposes.

1. Spatial diffusion research, and analysis of spatiotemporal stochastic processes

The logical starting point for any research into the theory of spatial diffusion is the work of Hägerstrand in the early 1950s. In the GSC Borcherdt (1958; 1961) with his work in agricultural geography was the first to try to come to terms with Hägerstrand's ideas and to introduce his concepts into German geography. He looked at the diffusion of a number of crops introduced into Bavaria at the beginning of the 19th century and then went on to develop an inductive model of the diffusion of innovations in agricultural geography. Unfortunately Borcherdt has not yet taken up the challenge presented by Hägerstrand's inspired idea of simulating the diffusion process, under controlled conditions by using Monte-Carlo techniques, so as to try and understand the mechanism behind the diffusion process by comparing the results with what happened in practice. This was only done later by Hard (1972) in his study of the history of the Rhenish language, in which he successfully built and used a simulation model of the Monte-Carlo type to analyse the diffusion of dialects in the Rhineland. The diffusion studies of Bartels (1968c; 1970b) were also strongly influenced by Hägerstrand's ideas, although more in terms of the explanations they produced, than in terms of the methods used. Bartels studied the emigration of guest workers from the districts in the Izmir region of Turkey and interpreted the spread of the decision to move as the diffusion of a new way of life (adopting the way of life of a migrant worker). He constructed a gravity type regression model to analyse this process in which the space–time difference in the 'adopter quotient' (the number of applicants per 100000 inhabitants) in the Izmir region were categorized as information and communication phenomena.

Bahrenberg and Loboda (1973) have used a multiple regression model to study the spread of television in Poland, but they were influenced less by Hägerstrand than by Berry (1971), who looked at the same phenomenon in the United States. Their analysis was more concerned with the formal space–time characteristics of the diffusion process.

There are three significant features which characterize the more recent development of spatial diffusion research in the GSC:
(a) the extension of research into spatial diffusion into the more comprehensive field of spatiotemporal stochastic process research;
(b) a growing interest in behavioural aspects of diffusion research;
(c) the application of the results of diffusion research to regional planning.

With reference to the first of these features, inspired by the work of Bartels (1968c; 1970b), Giese (1978a), and Giese and Nipper (1979) have examined the complementary issue of the emigration of guest workers from their country of origin, that is the immigration of guest workers into their destination country, in this case the FRG. In accordance with Bartels, they start by assuming that the immigration of guest workers into the destination countries is guided by a spatial diffusion process. They then assume that the spatial spread process can be considered as a stochastic process which varies in time and space and which will be influenced both by exogenous and by endogenous forces. The latter will take the form of spatial and temporal persistence effects. This concept, of a process which evolves and varies in space and time, led to the design of models which can take account both of the causal relationships between the process and the exogenous factors (economic factors, etc), as well as relationships internal to the process (endogenous components, the internal dynamics of the process). Examples of this type of model are the integrated space–time models of the STARIMAR class (Space–Time-AutoRegressive-Integrated-Moving-Average-Regression). This study was a first step towards identifying the characteristics of the diffusion process by using methods based on stochastic processes which vary both in space and time.

In the GSC, Streit (1978; 1979) and Nipper and Streit (1977; 1978) have been particularly interested in stochastic processes and models that vary both in space and time. Streit became interested in them through his work on the runoff characteristics of small rivers. His first analysis was of the temporal variations in runoff data using Markov-chain models, or more accurately various permutations of the ARIMA models described by Box and Jenkins (1970) (Streit, 1975a; 1975b). With the incorporation of the spatial dimension, work began to be done on the spatial persistence of processes, as well as on the statistical modelling of phenomena with spatial autocorrelations. The aim was to extend the ARIMA model, which to that point had been entirely time dependent, so that it could take account both of space and time as does the STARIMAR class of models. It was this problem that initiated the fruitful cooperation, in methodological terms, between Streit and Nipper. The initial interest was the spatial distribution of points, in particular the question of the connection between spatial form and the producing process. The works of Dacey (1968) and Harvey (1968) gave an initial stimulus to his work and led to a greater appreciation of the need for a more strongly process-orientated way of looking at phenomena.

The joint work of Nipper and Streit (1977; 1978) is concerned primarily with the formal, theoretical underpinnings of this model. Their thinking was influenced significantly by the studies of Cliff and Ord (1973) and Cliff et al (1975) on spatial autocorrelation coefficients and their possible applications in inferential statistics and in the modelling of stochastic relationships in phenomena which vary in space. Both these studies were true innovations in the field. As well as these studies by geographers, there was also work on the Kriging technique (variogram technique) (Matheron, 1971; Olea, 1975; David, 1977), which is better known in geology (Nipper, 1981). Work has so far not been done on the structural functions derived by the Russian meteorologist Gandin (1965) or on Agterberg's (1974) two-dimensional autocorrelation function.

Important, though as yet unrealized, links with a more integrated space-time perspective were provided by the studies of Tinline (1970) and Rees (1970), as well as those of Martin and Oeppen (1975), Bennett (1975), and Cliff and Ord (1975). The books by Bennett and Chorley (1978) and by Bennett (1979), both of which adhere strongly to a systems theoretic framework and have an admirably broad compass, ought to stimulate further initiatives. These two books show that both the systems theoretic framework and the mathematical-statistical techniques of stochastic processes that vary both in space and time are already well developed. On the other hand, the practical application, especially to socioeconomic processes, has so far been much less successful and, insofar as it has been realized at all, has produced unsatisfactory results (see Streit, 1979). There are various possible explanations as to why the practical results have so far been unsatisfactory (the quality and quantity of the data, the level of aggregation of the data, the definition of neighbourhood weight matrices, the way the parameters have been interpreted, etc), but there remains a fundamental doubt about the explanatory value of such highly complex models, an issue that is discussed separately at a later stage in this chapter.

With reference to the behavioural aspects of diffusion research, rapid progress can also be observed. Concepts and ideas from the behavioural sciences were adopted and adapted in such a way that there was no attempt to conceptualize the developing diffusion process abstractly in terms of different indicators and indices etc. The aim was rather to concentrate on the potential adopters and then to analyse their reactions to the innovation (Windhorst, 1979).

With reference to the use of the results of spatial diffusion research in applied research (Brugger, 1980), the importance of these studies has been recognized particularly in regional planning and regional economics. Various different concepts based on the theory of polarization have been used, all derived from work on diffusion theory. An initial critical analysis and overview was undertaken by Schilling-Kaletsch (1976).

2. Partial location theories

From the outset, the refinement and implementation of partial location theories has been a major theme in quantitative geography in the GSC. The starting point for this was Christaller's central place theory, which in the 1960s and 1970s stimulated a wealth of empirical and theoretical studies and became one of the most important instruments in regional planning in the FRG. Deiters and Güssefeldt have been particularly active in refining and implementing Christaller's theory. Spurred on by the writings of Berry, Curry, and Dacey in the mid-1960s, they have used various methods and theoretical approaches to test and refine central place theory. Their aim was to produce a more realistic version of the theory on the basis of changes in the pattern of consumer behaviour and the use of spatial characteristics. Both have adopted a probabilistic approach to the theory and trace a behaviouristic approach.

Güssefeldt (1975a; 1975b; 1976; 1980) built a simulation model which linked the action radius, or the spatial demand functions of various social groups, with the accessibility and attractiveness of the service centres (the nature, quality, and quantity of the services available). He used the general logistic function to define the demand function and used nonlinear regression to estimate the parameters. In a similar vein, Deiters (1975a; 1975b; 1976; 1978) also attempted an explanation of central places that was basically probabilistic and behavioural. He also started by assuming that the development of a system of central places was the result of a large number of individual locational decisions on the part of traders, but significantly modified by the prevailing pattern of retail consumer behaviour. In contrast to Christaller, and in accordance with Golledge, Deiters believes that the state of equilibrium in central place systems is the result of a learning process. From this starting point, Deiters developed a probabilistic location model, which explained the spatial pattern of central places as the realization of stochastic random processes.

The formal basis of the model is the 'combined negative binomial distribution' used by Dacey, where the 'extreme cases' are determined by the binomial distribution when describing agglomeration, and by the Poisson distribution when describing dispersal. The parameters of the model encompass the variability of the sphere of influence of a central good. Deiters used quadrat analysis in the empirical test of the model. The preceding determination of centrality of the settlements was conducted by factor analysis.

The refinement of partial location theories in the tertiary sector was by no means restricted only to Christaller's central place theory. Giese (1978b), for example, used work by Alonso (1960; 1964) to examine inner-city land-use theory in order to study the locational requirements of inner-city businesses.

Bahrenberg (1974; 1976; 1978) has looked in great detail at the location–allocation problems of centrally located public services and, as an extension of this work, at spatial optimization models. Stimulated by the work of Revelle, Swain, and Toregas in the early 1970s, interest in spatial optimization models has grown considerably, not least because questions concerning the choice of location and the catchment areas for centralized services, especially centralized public services (hospitals, schools, old peoples' homes, fire stations, emergency centres, etc) are unresolved problems and of great practical importance for planning. Bahrenberg has

tried to develop a general static discrete optimization model for determining the choice of location and the catchment-area boundaries of centralized public services by using the following criteria:
(a) accessibility, measured in various ways;
(b) financial costs; and
(c) size or capacity of the public utilities.

Bahrenberg has come to the conclusion that optimization models ought not to be used to arrive at an optimal, best solution to a problem, which is in keeping with agreed planning goals, as has usually happened in the past. "In the construction of optimization models the more value that is attached to deriving the most precise and elegant solution, the less is its usefulness for spatial planning" (Bahrenberg, 1978, page 17). He believes that a much more important task for applied research is to identify several mutually exclusive feasible solutions. There will always be more than one feasible solution, depending on how the constraints are defined. Bahrenberg et al (1979) are now trying to develop a practical optimization model for planning purposes, and for this reason he and his collaborators have been trying to derive heuristic algorithms for solving static discrete spatial allocation problems with disjunct catchment areas. Deiters and Wäldin (1977) have also investigated the spatial allocation problem for centralized public services, though they have looked at the problem more from the point of view of the school curriculum and other general pedagogic criteria.

Unlike the tertiary sector, relatively little attention has so far been paid by geographers to developing partial location theories in the secondary sector. The studies by Gaebe (1976; 1978) and Wittenberg (1978) are empirical analyses of the extraordinarily complex sets of factors which determine the locational decision of industrial firms. Both are still in the forefront of the attempts to derive a general theory of location. Wittenberg's work is of a spatioeconomic nature, but Gaebe tries to explain the locational decisions of industrial firms not only in terms of the legalities of economic rationale and technical constraints, but also in terms of the noneconomically motivated behaviour of the management.

The trend towards behavioural research and theory formulation in quantitative geography is particularly apparent in the work of Höllhuber (1975; 1978), although so far he has mainly been concerned with the choice of residential location by households. To this end he has examined explanations taken from perception and behavioural research and he has then tried to refine them, or rather reformulate them in operational terms. Taking as his starting point the various explanations of how people choose where they live, Höllhuber developed the concept of marginal improvement in place of residence, which takes into account the social prestige associated with such an improvement. Certainly this ought to explain a significant part of the decisions of householders about where they live. There would be great value in many more empirical studies of how householders go about choosing where they live, but as yet the fundamental theoretical basis for such explanations is still inadequate.

3. Theories of spatial growth and development

Under the banner of 'relevant geography' attention has turned increasingly in recent years towards trying to understand regional disparities and their underlying causes, as well as trying to contribute to a reduction of the conflict over the aims and objectives of programmes for equalizing living standards on a more rational basis (Bartels, 1980, page 53). As a result there has been renewed interest in theories of spatial growth and development. Quantitative geography has not actually done much to add to the wide variety of theories of regional growth and development already available, being more concerned with putting into practice and testing empirically the very generalized explanations and hypotheses. There has been particular interest in the ideas contained in polarization theory (Schätzl, 1973; 1978; Schilling-Kaletsch, 1976; Taubmann 1979a; 1979b; Brugger, 1980).

4. Research orientated towards planning

Interest in the development of models which can be used in regional and area planning, and not merely in pure academic research, is limited to a relatively small group of geographers working with quantitative methods. In physical geography the work has so far concentrated mainly on models for the prediction of river runoff rates (Herrmann, 1974; Herrmann and Schrimpf, 1976) and on models of slope degradation (Ahnert, 1976). In human geography the focus has been primarily either on models for forecasting medium-term population growth and changes in population distribution, especially models for forecasting interregional migration (Gatzweiler, 1975b; Gatzweiler and Koch, 1976; Koch, 1977; 1978), or on urban development models (in particular of the Lowry type—Fischer, 1976).

Currently research is being concentrated on discrete, deterministic regional demographic models. Existing models, which were investigated and tried out as possible starting points, were at the macroscale, such as Rogers's multiregional cohort survival model, or the multiregional accounting system model devised by Rees and Wilson [see Fischer and Sauberer (1979) for Austria]. However, some new macroscale models were also developed, like the one by Koch (1977; 1978), which was used as a basis for the population forecast in the FRG for 1990. In comparison with the complicated models of Rogers and of Rees and Wilson, the one devised by Koch was relatively simple, at least for a forecast model dealing with the hard facts of planning in the FRG. There is a great interest in the GSC in producing models that can be applied as easily and as widely as possible for forecasting regional population growth and changes in population distribution.

In the field of urban planning the situation is rather different, in that the main focus of interest has been the building of microscale regional demographic models. In this context, reference should be made to the small-scale forecast models, or more precisely forecast strategies produced by Dehler (1976; 1978) and Kreibich (1981); Kreibich and Reich (1979). In the field of small-scale population forecasting, there is a growing conviction that the most pressing need is not for more accurate forecasts about the future course of events, but rather for techniques which highlight possible alternative strategies. Any pretence that it is possible to forecast the precise course of future events is being abandoned in favour of projections of population growth that may be used to make clear the effect of land-use development planning. As a direct result of this approach a series of different criteria, which have been used for making forecasts, are now much more important than those which are stressed in methodological–statistical research.

As well as population forecasts, quantitative geography has examined various other planning concepts (uniform functional areas, growth poles, and development axes), which today play an integral part both in the formulation and in the implementation of regional planning and policy. For example, Bartels (1975) undertook an empirical verification of the concept of 'uniform functional areas' as used in the regional policy of the FRG. Güssefeldt (1978) tried to find out whether the policy of point-axial settlement development, which forms the basis of the Land Development Programme for Bavaria and Baden-Württemberg, had reduced regional disparities in the level of development, or whether it had made them even more severe. Ganser (1972), Sauberer (1976), and others have looked at the reorganization of local government boundaries, and used efficient analysis to evaluate the viability of local government units. The simplicity and clarity of its mathematics makes efficiency analysis a useful and easily-applied decisionmaking model in planning.

These summaries of the different studies undertaken by the 'Quantitative Working Group' show that there is a real interest in developing models which can be employed in land-use and regional planning and not only in pure academic research. Nevertheless the bulk of the work is still dedicated to formal theory building and to pure systems theory. Too few people, at least in the German-speaking world, have tried to adapt the discoveries made in the course of formal model building and theory formulation, or in the course of applying new operational techniques, to the needs either of practical planning or of education. If the impact of quantitative geography in the GSC is relatively small, part of the reason is that the Group of quantitative geographers itself has not bothered to explain clearly enough what it is doing and has not sought to foster contacts with its roots.

Advances in methodology and in the analysis of geographical data
In addition to the developments in theory and model building, there has also been significant progress in the various different spheres of quantitative methodology and in the analysis of geographical data. Without the accurate and critical application of the complicated mathematical-statistical techniques, it would be hard to produce sufficiently precise research results in quantitative geography. It will therefore always be necessary to try and develop new quantitative techniques, or to 'discover' them in allied disciplines and then to test them for suitability, as has happened recently with catastrophe theory. Three areas may be picked out as those which have been developed particularly by geographers in the GSC. The first relates to the extension of multivariate techniques to noncontinuous data, especially to nominal data (Kilchenmann, 1973; Kemper, 1978a); the second covers the various different permutations on metric and ordinal path analysis (Kemper, 1978b; Leitner and Wohlschlägl, 1980); and the third includes the whole range of cluster analytic techniques (Fischer, 1978; 1980).

Hence, at the same time as the range of available methods was being expanded, a highly critical discussion broke out about the circumstances under which the statistical techniques then in use ought to be applied. Detailed discussions took place about the problems of applying factor and principal components analysis; for example, assumptions about the distribution of the variables, the transformation of the variables, the use of ordinal and nominal data, and rotation problems (Bahrenberg and Giese, 1975b; Kemper, 1975; Kemper and Schmiedecken, 1977; Giese 1978c). There was also discussion of the problems involved in using regression analysis (problems of ecological falsification, problems of 'outliers'—Bahrenberg and Giese, 1975a; Giese, 1978c).

Trends in development and their associated problems
The recent development of quantitative geography in the GSC is well documented in the past symposia and meetings of the Working Group[(2)], the proceedings of which have all been published. An initial review of developments was undertaken at the symposium in Giessen in 1974 (see Giese, 1975). There followed in quick succession the meetings in Innsbruck in 1975 (see Bartels and Giese, 1976), in Bremen in 1977 (see Bahrenberg and Taubmann, 1978), in Strassburg in 1977 (see Kilchenmann, 1978-1979), in Göttingen in 1979 (Bahrenberg and Giese, 1980) and in Zürich in 1980 (Ostheider and Steiner, 1981).

In conclusion I would like to compare the papers presented at the different symposia since 1974. Clear development trends stand out from such a comparison and these lead to six main questions which need to be

[(2)] Working group on 'Theory and quantitative methods in geography', founded in Bremen in 1977.

answered by future developments of quantitative geography:
(1) If the symposium at Giessen revealed an overemphasis on methodology and technically orientated studies, the emphasis of research has now shifted towards the formulation of theory and formal model building. However, this should not be taken as a sign that the process of adopting and further developing mathematical–statistical techniques is at an end.
(2) The interests of the Working Group have branched out into many subareas. Themes stemming from a great variety of different questions and based on a wide range of theoretical assumptions have been discussed. There is, however, a lack of any central unifying theme; there is also no coherent research programme. The lack of internal coherence within the Working Group is a source of complaint.
(3) Formal theoretical research seems to be somewhat overemphasized and this trend has recently intensified with the growing interest in systems theory (see the studies by Steiner, 1978; 1979a; 1979b).
(4) Now, as in the past, there is a neglect of applied and practical research. Although some attempts are being made to develop simple models and algorithms, which can be applied in practical planning, the schism between applied and practically useful research and formal theoretical research cannot be ignored. Within the Working Group it poses a so far unresolved problem. There is also almost no contact with teachers in schools.
(5) Simple causal models have been overshadowed by complex systems theories with dynamic processes and selfregulating mechanisms. The hypothetical pictures of reality being produced today are becoming ever more complicated. The models are built in the hope that they will represent reality much more accurately than simple causal models, such as were produced when quantitative geography first appeared. This reveals a fundamental problem in quantitative geography, and one that so far has been given little thought. The increase in the empirical explanatory power of models declines as the models become increasingly complex, much as

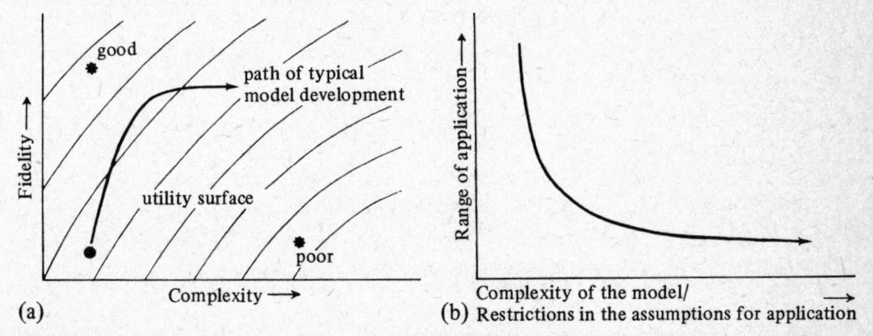

Figure 6.1. The interplay of (a) model complexity and fidelity—a typical time path moves from 'simple and dirty' to complex (after Haggett, 1978; Bennett, 1979); (b) model complexity and restrictions in the assumptions for application, respectively, with the range of application.

predicted by the law of diminishing returns [see figure 6.1(a)]. At the same time the range (area) of application of the models often becomes smaller as the models become more complex, because of the limiting constraints governing their use. The closer the models come to reality, the more complex they become and the more complicated and restricted are the mathematical techniques required to operate them. Hence, the areas in which these kind of models and techniques may be applied are more limited, for their assumptions cannot be fulfilled in reality [see figure 6.1(b)].

(6) This problem is closely related to another. Quantitative geography cannot quite escape the accusation that, because of the difficulties surrounding the development of models, only those problems have been investigated which can easily be expressed in formal terms and solved. The decisive problems facing society are still solved in other ways. In other words the sights of quantitative geography may be firmly set on the construction of formal models, but so far this has only led to a very limited perception of reality.

References

Agterberg F P, 1974 *Geomathematics. Developments in Geomathematics 1* (Academic Press, Amsterdam)

Ahnert F, 1966 "Zur Rolle der elektronischen Rechenmaschine und des mathematischen Modells in der Geomorphologie" *Geographische Zeitschrift* **54** 118-133

Ahnert F, 1971 "A general and comprehensive theoretical model of slope profile development" Occasional Papers in Geography, number 1, University of Maryland, College Park

Ahnert F, (Ed.), 1976 "Quantitative slope models" *Zeitschrift für Geomorphologie* **25** supplement

Alonso W, 1960 "A theory of the urban market" *Papers and Proceedings of the Regional Science Association* **6** 149-157

Alonso W, 1964 *Location and Land Use: Toward a General Theory of Land Rent* (Harvard University Press, Cambridge, Mass)

Bähr J, 1971a "Gemeindetypisierung mit Hilfe quantitativer statistischer Verfahren" *Erdkunde* **25** 249-264

Bähr J, 1971b "Eine Faktorenanalyse zur Bevölkerungsstruktur—dargestellt am Beispiel Südwestafrikas" *Die Erde* **102** 262-285

Bähr J, Golte W, 1974 "Eine bevölkerungs- und wirtschaftsgeographische Gliederung Chiles" *Geoforum* **17** 25-42

Bahrenberg G, 1974 "Zur Frage optimaler Standorte von Gesamthochschulen in Nordrhein-Westfalen. Eine Lösung mit Hilfe der linearen Programmierung" *Erdkunde* **28** 101-114

Bahrenberg G, 1976 "Ein sozial gerechtes Optimierungsmodell für die Standortwahl von öffentlichen Einrichtungen" *Tagungsbericht und wissenschaftliche Abhandlungen des 40. Deutschen Geographentages, Innsbruck, 1975*; (Steiner, Wiesbaden) pp 443-452

Bahrenberg G, 1978 *Ein allgemeines statisch-diskretes Optimierungsmodell für Standort-Zuordnungsprobleme.* Karlsruher Manuskripte zur Mathematischen und Theoretischen Wirtschafts- und Sozialgeographie, Heft 31, Geographisches Institut der Universität, Karlsruhe

Bahrenberg G, Giese E, 1975a *Statistische Methoden und ihre Anwendung in der Geographie* (Teubner Studienbücher Geographie; Teubner, Stuttgart)

Bahrenberg G, Giese E, 1975b "Zum Problem der Normalität und der Transformation bei der Faktoren—bzw. Hauptkomponentenanalyse" *Giessener Geographische Schriften*, Heft 32, Geographisches Institut der Universität, Giessen, pp 9-29

Bahrenberg G, Giese E, 1980 "Quantitative Modelle und ihre Anwendung in der Anthropogeographie und Raumplanung. Einleitung zur Vortragssitzung" *Tagungsbericht und wissenschaftliche Abhandlungen des 41. Deutschen Geographentages, Göttingen, 1979* (Steiner, Wiesbaden)

Bahrenberg G, Loboda J, 1973 "Einige raum-zeitliche Aspekte der Diffusion von Innovationen—Am Beispiel der Ausbreitung des Fernsehens in Polen" *Geographische Zeitschrift* **61** 165-194

Bahrenberg G, Matthiesen G, Steingrube W, 1979 *STAL Heuristische Algorithmen zur Lösung statisch-diskreter Standort-Allokationsprobleme mit disjunkten Einzugsbereichen*, Materialen und Manuskripte, Heft 2, Universität Bremen, Schwerpunkt Geographie, Bremen

Bahrenberg G, Taubmann W, (Eds), 1978 *Quantitative Modelle in der Geographie und Raumplanung* Bremer Beiträge zur Geographie und Raumplanung, Heft 1, Universität Bremen, Schwerpunkt Geographie, Bremen

Bartels D, 1968a "Die Zukunft der Geographie als Problem ihrer Standortbestimmung" *Geographische Zeitschrift* **56** 124-142

Bartels D, 1968b "Zur wissenschaftstheoretischen Grundlegung einer Geographie des Menschen. Erdkundliches Wissen *Beihefte zur Geographischen Zeitschrift 19* (Steiner, Wiesbaden)

Bartels D, 1968c "Türkische Gastarbeiter aus der Region Izmir. Zur raumzeitlichen Differenzierung der Bestimmungsgründe ihrer Aufbruchsentschlüsse" *Erdkunde* **22** 313-324

Bartels D, (Ed.), 1970a *Wirtschafts- und Sozialgeographie* (Kiepenheuer und Witsch, Köln)

Bartels D, 1970b "Geographische Aspekte sozialwissenschaftlicher Innovationsforschung" *Tagungsbericht und wissenschaftliche Abhandlungen des 37.* Deutschen Geographentages, Kiel, 1969 (Steiner, Wiesbaden) pp 283-296

Bartels D, 1975 "Die Abgrenzung von Planungsregionen in der Bundesrepublik Deutschland—eine Operationalisierungsaufgabe" in *Ausgeglichene Funktionsräume-Grundlagen für eine Regionalpolitik des mittleren Weges* Veröffentlichungen der Akademie für Raumforschung und Landesplanung. Forschungs- und Sitzungsberichte, Band 94 (Schroedel, Hannover) pp 93-115

Bartels D, 1980 "Wirtschafts- und Sozialgeographie" in *Handwörterbuch der Wirtschaftswissenschaft 23, Lieferung* (Fischer, Stuttgart) pp 44-55

Bartels D, Giese E, 1976 "Quantitative Methoden und Beiträge zur Theoriebildung. Einleitung zur Vortragssitzung" *Tagungsbericht und wissenschaftliche Abhandlungen des 40. Deutschen Geographentages, Innsbruck, 1975* (Steiner, Wiesbaden) pp 423 ff

Bennett R J, 1975 "The representation and identification of spatio-temporal systems: An example of population diffusion in North-West England" *Institute of British Geographers Transactions* number 66 pp 73-94

Bennett R J, 1979 *Spatial Time Series: Analysis, Forecasting, Control* (Pion, London)

Bennett R J, Chorley R J, 1978 *Environmental Systems: Philosophy, Analysis and Control* (Methuen, London)

Berry B J L, 1971 "Hierarchical diffusion: the basis of development, filtering and spread in a system of growth centers" in *Growth Centers and Regional Economic Development* Ed. N M Hansen (Holt, Rinehart and Winston, New York) pp 108-138

Borcherdt C, 1958 "Über die Vergrünlandung in Bayern" *Berichte zur deutschen Landeskunde* **21** 125-129

Borcherdt C, 1961 "Die Innovation als agrargeographische Regelerscheinung" Arbeiten aus dem Geographischen Institut der Universität des Saarlandes, 6 (Geographisches Institut der Universität, Saarbrücken) pp 13-50

Box G E P, Jenkins G M, 1970 *Time Series Analysis. Forecasting and Control* (Holden-Day, San Francisco)

Braun G, 1972 "Komplexes Faktorensystem räumlicher und zeitlicher Bewegungen" in *Räumliche und zeitliche Bewegungen* Ed. G Braun, Würzburger Geographische Arbeiten, Heft 37, Geographisches Institut der Universität, Würzburg

Braun G, 1975 "Methoden und Modelle zum Schichtungsaufbau und zur räumlichen Mobilität" Giessener Geographische Schriften, Heft 32, Giessen, Geographisches Institut der Universität, Giessen, pp 143-151

Braun G, 1976 "Modelle zur Analyse der sozialen Segregation" *Tagungsbericht und wissenschaftliche Abhandlungen des 40. Deutschen Geographentages, Innsbruck, 1975* (Steiner, Wiesbaden) pp 455-472

Brugger E A, 1980 "Innovationsorientierte Regionalpolitik. Notizen zur einer neuen Strategie" *Geographische Zeitschrift* **68** 173-198

Burton I, 1963 "The quantitative revolution and theoretical geography" *The Canadian Geographer* **7** 151-162

Cliff A D, Haggett P, Ord J K, Bassett K, Davies R, 1975 *Elements of Spatial Structure: A Quantitative Approach* (Cambridge University Press, Cambridge)

Cliff A D, Ord J K, 1973 *Spatial Autocorrelation* (Pion, London)

Cliff A D, Ord J K, 1975 "Space-time modelling with an application to regional forecasting" *Transactions Institute of British Geographers* **66** 119-128

Dacey M F, 1968 "An empirical study of the areal distribution of houses in Puerto Rico" *Transactions Institute of British Geographers* **45** 51-69

David M, 1977 "Geostatistical ore reserve estimation" in *Developments in Geomathematics 2* (Academic Press, Amsterdam)

Dehler K H, 1976 *Zielprognosen der Stadtentwicklung* Schriftenreihe des Bundesinstituts für Bevölkerungsforschung, Band 3 (Boldt, Boppard)

Dehler K H, 1978 *Stadt-Entwicklungsplanung Hanau. Planungsbericht 1: Städtische Bevölkerungsentwicklung; Planungsbericht 2: Perspektiven der Bevölkerungsentwicklung* Magistrat der Stadt Hanau, Hanau

Deiters J, 1975a "Räumliche Muster und stochastische Prozesse—Lokalisationsanalyse zentraler Orte" Giessener Geographische Schriften, Heft 32, Geographisches Institut der Universität, Giessen, pp 122-140

Deiters J, 1975b *Stochastische Analyse der räumlichen Verteilung zentraler Orte.* Karlsruher Manuskripte zur Mathematischen und Theoretischen Wirtschafts- und Sozialgeographie, Heft 8, Geographisches Institut der Universität, Karlsruhe

Deiters J, 1976 "Stochastische Elemente in der Theorie zentraler Orte" *Tagungsbericht und wissenschaftliche Abhandlungen des 40. Deutschen Geographentages, Innsbruck, 1975* (Steiner, Wiesbaden) pp 425-431

Deiters J, 1978 "Zur empirischen Überprüfbarkeit der Theorie zentraler Orte" *Fallstudie Westerwald. Arbeiten zur Rheinischen Landeskunde* Heft 44 (Dümmler, Bonn)

Deiters J, Wäldin E, 1977 "Brand in Tannenweiler. Zur Frage nach dem besten Standort von Feuerwehrstationen" *Raumwissenschaftliches Curriculum-Forschungsprojekt, Materialen zu einer neuen Didaktik der Geographie 9* (München)

Fischer M M, 1976 "Eine theoretische und methodische Analyse mathematischer Stadtentwicklungsmodelle vom Lowry-Tap" Ein methodischer Beitrag zur Regionalforschung. Rhein-Mainische Forschungen, Heft 83 (Kramer, Frankfurt)

Fischer M M, 1978 "Zur Lösung funktionaler regionaltaxonomischer Probleme auf der Basis von Interaktionsmatrizen: Ein neuer graphentheoretischer Ansatz" Karlsruher Manuskripte zur Mathematischen und Theoretischen Wirtschafts- und Sozialgeographie, Heft 25, Geographisches Institut der Universität, Karlsruhe

Fischer M M, 1980 "Regional taxonomy: a comparison of some hierarchic and non-hierarchic strategies" *Regional Science and Urban Economics* **10** 503-537

Fischer M M, Sauberer M, 1979 "Neuere Entwicklungen in der regionaldemographischen Modellbildung. Modellstruktur und Anwendungsmöglichkeiten am Beispiel Österreichs" *Österreichisches Institut für Raumplanung, Materialien,* Nummer 11, Wien

Gaebe W, 1976 "Analyse mehrkerniger Verdichtungsräume. Das Beispiel des Rhein-Ruhr-Raumes" Karlsruher Geographische Schriften, Heft 7, Geographisches Institut der Universität, Karlsruhe

Gaebe W, 1978 "Erklärungsversuche industrieller Standortentscheidungen" in *Seminarberichte der Gesellschaft für Regionalforschung* **13** 161-184

Gandin L S, 1965 *Objective Analysis of Meteorological Fields* Israel Program for Scientific Translations, Jerusalem

Ganser K, 1972 *Nutzwertanalyse zur Bestimmung des Standortes der Kreisverwaltung. Fallstudie für den neuen Kreis Lippstadt-Soest* (Bundesforschungsanstalt für Landeskunde und Raumordnung, Bonn)

Gatzweiler H P, 1975a "Die Anwendung von regulären, homogenen Markov-Ketten-Modellen erster Ordnung zur Deskription und Analyse von Wanderungen" Giessener Geographische Schriften, Heft 32, Geographisches Institut der Universität, Giessen, pp 156-164

Gatzweiler H P, 1975b *Zur Selektivität interregionaler Wanderungen. Ein theoretisch-empirischer Beitrag zur Analyse und Prognose altersspezifischer interregionaler Wanderungen* Forschungen zur Raumentwicklung, Band 1, Bonn

Gatzweiler H P, Koch R, 1976 "Makroanalytisches Simulations-modell der regionalen Bevölkerungsentwicklung und -verteilung in der Bundesrepublik" *Tagungsbericht und wissenschaftliche Abhandlungen des 40.* Deutschen Geographentages, Innsbruck, 1975 (Steiner, Wiesbaden), pp 489-501

Giese E, (Ed.), 1975 "Symposium 'Quantitative Geographie' Giessen 1974. Möglichkeiten und Grenzen der Anwendung mathematisch-statistischer Methoden in der Geographie" Giessener Geographische Schriften, Heft 32, Geographisches Institut der Universität, Giessen

Giese E, 1978a "Räumliche Diffusion ausländischer Arbeitnehmer in der Bundesrepublik Deutschland" *Die Erde* **109** 92-110

Giese E, 1978b "Weiterentwicklung und Operationalisierung der Standort- und Landnutzungstheorie von Alonso für städtische Unternehmen" Bremer Beiträge zur Geographie und Raumplanung, Heft 1, Universität Bremen, Schwerpunkt Geographie, Bremen, pp 63-79

Giese E, 1978c "Kritische Anmerkungen zur Anwendung faktorenanalytischer Verfahren in der Geographie" *Geographische Zeitschrift* **66** 161-182

Giese E, Nipper J, 1979 "Zeitliche und räumliche Persistenzeffekte bei räumlichen Ausbreitungsprozessen—analysiert am Beispiel der Ausbreitung ausländischer Arbeitnehmer in der Bundesrepublik Deutschland" Karlsruher Manuskripte zur Mathematischen und Theoretischen Wirtschafts- und Sozialgeographie, Heft 34, Geographisches Institut der Universität, Karlsruhe

Güssefeldt J, 1975a "Zu einer operationalisierten Theorie des räumlichen Versorgungsverhaltens von Konsumenten" Giessener Geographische Schriften, Heft 34, Geographisches Institut der Universität, Giessen

Güssefeldt J, 1975b "Über ein probabilistisches Simulationsmodell versorgungswirtschaftlicher Interaktionsmuster" Giessener Geographisches Schriften, Heft 32, Geographisches Institut der Universität, Giessen, pp 141-156

Güssefeldt J, 1976 "Der Einfluß raumdifferenzierender Strukturen auf die Ausprägung menschlicher Interaktionssysteme" *Tagungsbericht und wissenschaftliche Abhandlungen des 40.* Deutschen Geographentages, Innsbruck, 1975 (Steiner, Wiesbaden) pp 432-442

Güssefeldt J, 1978 "Die Graphentheorie als Instrument zur Beurteilung raumordnungspolitischer Konzepte, dargestellt am Beispiel der Entwicklungsachsen von Baden-Württemberg und Bayern" *Geographische Zeitschrift* **66** 81-105

Güssefeldt J, 1980 "Konsumentenverhalten und die Verteilung zentraler Orte" *Geographische Zeitschrift* **68** 33-53

Haggett P, 1965 *Locational Analysis in Human Geography* (Edward Arnold, London). Deutsche Übersetzung durch D Bartels, B und V Kreibich, 1973 *Einführung in die kultur- und sozialgeographische Regionalanalyse* (W de Gruyter, Berlin)

Haggett P, 1978 "Spatial forecasting—a view from the touchline" in *Towards the Dynamic Analysis of Spatial Systems* Eds R L Martin, N J Thrift, R J Bennett (Pion, London) pp 205-210

Hard G, 1972 "Ein geographisches Simulationsmodell für die rheinische Sprachgeschichte" in *Festschrift Matthias Zender* Eds E Ennen, G Wiegelmann (Röhrscheid-Verlag, Bonn) pp 25-58

Harvey D W, 1968 "Some methodological problems in the use of the Neyman type A and the negative binomial probability distributions for the analysis of spatial point patterns" *Transactions Institute of British Geographers* number 44, 85-95

Herrmann R, 1972 "Ein multivariates Modell der Schwebstoffbelastung eines hessischen Mittelgebirgsflusses" *Biogeographica* **1** 87-95

Herrmann R, 1973 "Eine multivariate statistische Klimagliederung Nordhessens und angrenzender Gebiete" Beiträge zur Landeskunde von Nordhessen. Marburger Geographische Schriften, Heft 60, Geographisches Institut der Universität Marburg, pp 37-55

Herrmann R, 1974 "Ein Anwendungsversuch der mehrdimensionalen Diskriminanzanalyse auf die Abflussvorhersage" *Catena* **1** 367-385

Herrmann R, Schrimpf E, 1976 "Zur Vorhersage des Abflußverhaltens in tropischen Hochgebirgen West- und Zentralkolumbiens" *Tagungsbericht und wissenschaftliche Abhandlungen des 40.* Deutschen Geographentages, Innsbruck, 1975 (Steiner, Wiesbaden) pp 750-770

Höllhuber D, 1975 "Die Mental Maps von Karlsruhe—Wohnstandortpräferenzen und Standortcharakteristika" Karlsruher Manuskripte zur Mathematischen und Theoretischen Wirtschafts- und Sozialgeographie, Heft 11, Geographisches Institut der Universität, Karlsruhe

Höllhuber D, 1978 "Sozialgruppentypische Wohnstandortpräferenzen und innerstädtische Wohnstandortwahl" Bremer Beiträge zur Geographie und Raumplanung, Heft 1, Universität Bremen, Schwerpunkt Geographie, Bremen, pp 95-105

Kansky K J, 1963 *Structure of Transportation Networks. Relationships Between Network Geometry and Regional Characteristics* RP-84, Department of Geography, Chicago University

Kemper F J, 1975 "Die Anwendung faktorenanalytischer Rotationsverfahren in der Geographie des Menschen" Giessener Geographische Schriften, Heft 32, Geographisches Institut der Universität, Giessen, pp 34-47

Kemper F J, 1978a "Multivariate Analyse nominalskalierte Daten" *Mitteilungen des Arbeitskreises für neue Methoden in der Regionalforschung* 8 3-36

Kemper F J, 1978b "Über einige multivariate Verfahren zur statistischen Varianzaufklärung und ihre Anwendung in der Geographie" Karlsruher Manuskripte zur Mathematischen und Theoretischen Wirtschafts- und Sozialgeographie, Heft 28, Geographisches Institut der Universität, Karlsruhe

Kemper F J, Schmiedecken W, 1977 "Faktorenanalyse zum Klima Mitteleuropas. Ein Beitrag zum Problem der Kontinentalität sowie zur Aussagefähigkeit von Faktorenwerten" *Erdkunde* **31** 255-272

Kilchenmann A, 1968 *Untersuchungen mit quantitativen Methoden über die fremdenverkehrs- und wirtschaftsgeographische Struktur der Gemeinden im Kanton Graubünden* (Schweiz) (Juris, Zürich)

Kilchenmann A, 1970 *Statistisch-analytische Arbeitsmethoden in der regionalgeographischen Forschung. Untersuchungen zur Wirtschaftsentwicklung von Kenya und Versuch einer Regionalisierung des Landes auf Grund von thematischen Karten* (Ann Arbor, Michigan)

Kilchenmann A, 1971 "Statistisch-analytische Landschaftsforschung" *Geoforum* 7 39-53

Kilchenmann A, 1973 "Die Merkmalsanalyse für Nominaldaten. Eine Methode zur Analyse von qualitativen geographischen Daten basierend auf einem informationstheoretischen Modell" *Geoforum* **15** 33-45

Kilchenmann A, 1975 "Zum gegenwärtigen Stand der 'Quantitativen und Theoretischen Geographie" Giessener Geographische Schriften, Heft 32, Geographisches Institut der Universität, Giessen, pp 194-208

Kilchenmann A, (Ed.), 1978-1979 Karlsruher Manuskripte zur Mathematischen und Theoretischen Wirtschafts- und Sozialgeographie, Heft 24, 1978-Heft 34, 1979, Geographisches Institut der Universität, Karlsruhe

Kilchenmann A, Moergeli W, 1970 "Typisierung der Gemeinden im Kanton Zürich mit multivariaten statistischen Methoden aufgrund ihrer wirtschaftsgeographischen Struktur" *Vierteljahrsschrift der Naturforschenden Gesellschaft, Zürich* **115** 369-394

Koch R, 1977 "Wanderungen und Bevölkerungsentwicklung in der Raumordnungsprognose 1990" *Informationen zur Raumentwicklung* Heft 1/2, 13-25

Koch R, 1978 "Ein Beitrag zur Weiterentwicklung regionaler Bevölkerungsprognoseansätze" Bremer Beiträge zur Geographie und Raumplanung, Heft 1, Universität Bremen, Schwerpunkt Geographie, Bremen, pp 107-115

Kreibich V, 1981 "Wohnungsbelegmuster als Grundlage kleinräumiger Bevölkerungsprognosen" in *Theorie und Quantitative Methodik in der Geographie* Eds M Ostheider, D Steiner (Geographisches Institut der ETH Zürich, Zürich)

Kreibich V, Reich D, 1979 "Analyse und Projektion der kleinräumigen Bevölkerungsentwicklung auf der Grundlage des Gebäudebestandes und der Bautätigkeit" in *Demographische Planungsinformationen* Ed. E Elsner (Berlin) pp 168-182

Leitner H, Wohlschlägl H, 1980 "Metrische und ordinale Pfadanalyse: Ein Verfahren zur Testung komplexer Kausalmodelle in der Geographie" *Geographische Zeitschrift* **68** 81-106

Lichtenberger E, 1978a "Klassische und theoretisch-quantitative Geographie im deutschen Sprachraum" *Berichte zur Raumforschung und Raumplanung* **1** 9-20

Lichtenberger E, 1978b "Quantitative geography in the German-speaking countries" *Tijdschrift voor Economische en Sociale Geografie* **69** 362-373

Martin R J, Oeppen J E, 1975 "The identification of regional forecasting models using space-time correlation functions" *Transactions, Institute of British Geographers* number 66, 95-118

Matheron G, 1971 "The theory of regionalized variables and its applications" *Les Cahiers du Centre de Morphologie. Mathematique de Fontainebleau* number 5 (Centre de Morphologie, Fontainebleau)

Nipper J, 1981 "Autoregressiv- und Kriging-Modelle. Zwei Ansätze zur Erfassung raumvarianter Strukturen" in *Theorie und Quantitative Methodik in der Geographie* Eds M Ostheider, D Steiner (Geographisches Institut der ETH Zürich, Zürich)

Nipper J, Streit U, 1977 "Zum Problem räumlicher Erhaltensneigung in räumlichen Strukturen und raumvarianten Prozessen" *Geographische Zeitschrift* **65** 241-263

Nipper J, Streit U, 1978 "Modellkonzepte zur Analyse, Simulation und Prognose raum-zeit-varianter stochastischer Prozesse" Bremer Beiträge zur Geographie und Raumplanung, Heft 1, Universität Bremen, Schwerpunkt Geographie, Bremen, pp 1-17

Olea R A, 1975 *Optimum Mapping Techniques Using Regionalized Variable Theory* Kansas Geological Survey, Series on Spatial Analysis, number 2, Lawrence, Kansas

Ostheider M, Steiner D, (Eds), 1981 "Theorie und Quantitative Methodik in der Geographie" Geographisches Institut der ETH Zürich, Zürich

Rees H J B, 1970 "Time series analysis and regional forecasting" in *Regional Forecasting* Eds M Chisholm, A E Frey, P Haggett (Butterworths, London)

Rohdenburg H, Sabelberg U, Wagner H, 1976 "Sind konkave und konvexe Hänge prozesspezifische Formen?—Ergebnisse von Hangentwicklungssimulationen mittels EDV" *Catena* 3 113-136

Sauberer M, 1973a "Anwendungsversuche der Faktorenanalyse in der Stadtforschung—Sozialräumliche Gliederung Wien" *Seminarberichte der Gesellschaft für Regionalforschung* 7 (Heidelberg)

Sauberer M, 1973b *Quantitative Methoden in der Geographie und Raumforschung. Überblick und Arbeitsbeispiele* Diss Geographisches Institut der Universität, Wien

Sauberer M, 1974 "Zur Planungsrelevanz der Faktorenanalyse" *AMR INTERN, Mitteilungen des Arbeitskreises für neue Methoden in der Regionalforschung* Wien nummer 15, 15-19

Sauberer M, 1976 "Bewertung von Gemeindezusammenlegungen mit Hilfe von Nutzertanalysen" *Tagungsbericht und wissenschaftliche Abhandlungen des 40.* Deutschen Geographentages, Innsbruck, 1975 (Steiner, Wiesbaden) pp 455-472

Sauberer M, Cserjan C, 1972 "Sozialräumliche Gliederung Wiens 1961. Ergebnis einer Faktorenanalyse" *Der Aufbau* 284-306

Schätzl L, 1973 "Räumliche Industrialisierungsprozesse in Nigeria" Giessener Geographische Schriften, Heft 31, Geographisches Institut der Universität, Giessen

Schätzl L, 1978 *Wirtschaftsgeographie* 1 (UTB Schöningh, Paderborn)

Schickhoff I, 1974 *Graphentheoretische Untersuchungen am Beispiel des Schienennetzes der Niederlande. Ein Beitrag zur Verkehrsgeographie.* Dissertation, Münster 1974 sowie Duisburger Geographische Arbeiten, Heft 1, Seminar für Geographie der Gesamthochschule Duisburg 1978, Duisburg

Schickhoff I, 1977 "Matrizentheoretische Verfahren zur Bestimmung der Zugänglichkeit von Knotenpunkten eines Verkehrsnetzes, aufgezeigt am Beispiel 'Eisenbahnnetz Randstad Holland'" *Tijdschrift voor Economische en Sociale Geografie* **68** 152-167

Schilling-Kaletsch I, 1976 "Wachstumspole und Wachstumszentren. Untersuchung zu einer Theorie sektoral und regional polarisierter Entwicklung" *Arbeitsberichte und Ergebnisse zur wirtschafts- und sozialgeographischen Regionalforschung*, Heft 1, Geographisches Institut der Universität, Hamburg

Stäblein G, 1972 "Modellbildung als Verfahren zur komplexen Raumerfassung" *Würzberger Geographische Arbeiten*, Heft 37, Gerling-Festschrift: Geographisches Institut der Universität, Würzburg, pp 67-93

Stäblein G, 1975 "Der Einsatz der Faktorenanalyse und Printermap für die Regionalanalyse und die Regionalisierung" Giessener Geographische Schriften, Heft 34, Geographisches Institut der Universität, Giessen, pp 143-151

Stäblein G, Stäblein-Fiedler G, 1973 "Faktorenanalytische Untersuchungen zur fremdenverkehrsgeographischen Struktur der Provinzen Spaniens" Marburger Geographische Schriften, Band 59, Geographisches Institut der Universität, Marburg, pp 145-161

Steiner D, 1965a "Die Faktorenanalyse; ein modernes statistisches Hilfsmittel des Geographen für die objektive Raumgliederung und Typenbildung" *Geographica Helvetica* **20** 20-34

Steiner D, 1965b "A multivariate statistical approach to climatic regionalization and classification" *Tijdschrift van het Koninklijk Nederlandsch Aardrijkskundig Genootschap* 82 329-347

Steiner D, 1978 "Modelle zur Darstellung geographischer Systeme" *Veröffentlichungen der Geographischen Kommission der Schweizerischen Naturforschenden Gesellschaft* Nummer 5, 79-116, Lausanne

Steiner D, 1979a "Systemtheorie/Systemanalyse und Geographie" zum Buch von G P Chapman "Human and environmental systems—a geographers appraisal" *Geographische Zeitschrift* **67** 185-210

Steiner D, 1979b "A minicomputer-based geographical data processing system" *Proceedings of the NATO Advanced Study Institute on Map Data Processing* (Marata, Italy)

Streit U, 1973 "Ein mathematisches Modell zur Simulation von Abflußganglinien (Am Beispiel von Flüssen des Rechtsrheinischen Schiefergebirges)" Giessener Geographische Schriften, Heft 27, Geographisches Institut der Universität, Giessen

Streit U, 1975a "Zeitreihensimulation mit Markov-Modellen, dargestellt an Beispielen aus der Hydrologie" Giessener Geographische Schriften, Heft 32, Geographisches Institut der Universität, Giessen, pp 165-180

Streit U, 1975b "Erzeugung synthetischer Abflussdaten mit Hilfe eines zeit- und raumvarianten Modells im Einzugsgebiet der Lahn" *Erdkunde* **29** 92-105

Streit U, 1976 "Über eine Anwendungsmöglichkeit von Simulationsmodellen bei der wasserwirtschaftlichen Planung" *Tagungsbericht und wissenschaftliche Abhandlungen des 40.* Deutschen Geographentages, Innsbruck, 1975 (Steiner, Wiesbaden) pp 503-514

Streit U, 1978 "Räumliche und zeitliche Erhaltensneigung als Ansatzpunkt stochastischer Modellbildung in der Physischen Geographie" Karlsruher Manuskripte zur Mathematischen und Theoretischen Wirtschafts- und Sozialgeographie, Heft 30, Geographisches Institut der Universität, Karlsruhe

Streit U, 1979 "Raumvariante Erweiterung von Zeitreihenmodellen. Ein Konzept zur Synthetisierung monatlicher Abflußdaten von Fliessgewässern unter Berücksichtigung von Erfordernissen der wasserwirtschaftlichen Planung" Giessener Geographische Schriften, Heft 46, Geographisches Institut der Universität, Giessen

Taubmann W, 1979a "Erscheinungsformen und Ursachen sozioökonomischer Disparitäten am Beispiel von Jütland/Dänemark" *Münstersche Geographische Arbeiten*, Heft 4, Geographisches Institut der Universität, Münster, pp 149-183

Taubmann W, 1979b "Zur Abgrenzung von sogenannten Problemgebieten—Bemerkungen zu Indikatoren und Verfahrensweisen" Karlsruher Manuskripte zur Mathematischen und Theoretischen Wirtschafts- und Sozialgeographie, Heft 32, Geographisches Institut der Universität, Karlsruhe

Tinline R, 1970 "Linear operators in diffusion research" in *Regional Forecasting* Eds M Chisholm, A E Frey, P Haggett (Butterworths, London) pp 71-91

Vetter F, 1970 "Netztheoretische Studien zum Niedersächsischen Eisenbahnnetz. Ein Beitrag zur angewandten Verkehrsgeographie" *Abhandlungen des 1. Geographischen Instituts der Freien Universität Berlin*, Band 15 (Reimer, Berlin)

Werner C, 1966 "Zur Geometrie von Verkehrsnetzen. Die Beziehung zwischen räumlicher Netzgestaltung und Wirtschaftlichkeit" *Abhandlungen des 1. Geographischen Instituts der Freien Universität Berlin*, Band 10 (Reimer, Berlin)

Windhorst H-W, 1979 "Die sozialgeographische Analyse raum-zeitlicher Diffusionsprozesse auf der Basis der Adoptorkategorien von Innovationen. Die Ausbreitung der Käfighaltung von Hühnern in Südoldenburg" *Zeitschrift für Agrargeschichte und Agrarsoziologie* **27** 244-266

Wittenberg W, 1978 "Neuerrichtete Industriebetriebe in der Bundesrepublik Deutschland, 1955-1971" Giessener Geographische Schriften, Heft 44, Geographisches Institut der Universität, Giessen

7

The impact of institutional forces on the state of university geography in the Federal Republic of Germany in comparison with Britain

E Lichtenberger

The lack of critical study by institutions

The efforts made to give geography a new image and also to redefine it as a science have engendered a lively discussion of basic principles both in English-speaking and in German-speaking countries. In both sets of countries the shape of geography is intimately interlinked with the nature of the institutions in which it is taught or where research is undertaken. However, despite its importance the interrelationship between academic and institutional geography, as in the case of many other subjects, has been little investigated[(1)].

In this chapter a first attempt is made to discuss the institutional growth and present situation of university geography. For this purpose a comparison between Britain and the FRG seems to be of special interest in order

(1) to explain the continuing research advantage of British geography against the institutional background, and

(2) to answer the question of whether changes, similar to the fundamental changes in the training goals on the part of university geography in Britain, decided upon during the past two decades because of the reform in secondary education, are to be expected in the FRG.

The information available has rendered a twofold approach necessary. Whereas a comparison between Britain and the FRG is presented in the first part of the chapter, by making use of simple indicators such as the number of institutes, the number of students, and the composition and age structure of the teaching staff, the second part, dealing with the sociology of organisation and institutional management, is confined to the example of the FRG. Finally problem-solving strategies are briefly suggested.

The growth and present state of university geography in the FRG compared with that in Britain

In both countries, the development of university geography during the past two decades can be described by means of a growth model, with growing demand (the number of students) bringing about an increase in supply (the members of academic staff).

[(1)] Baker (1970); Clarke (1976); Cooke and Robson (1976); Editorial comment (1978); Haggett (1977); Hartke (1960); Lawton (1978); Lichtenberger (1979); Rawstron (1975); Steel and Watson (1972); Stoddart (1967).

As a result of institutional growth, contrasts between Great Britain and the FRG have increased, both in research and teaching, and in the labour market for geography graduates.

Research

In Britain, geography started from a much more favourable position than it did in the FRG, and growth proceeded more slowly and evenly. The traditional contrast between those departments where the main emphasis was either on teaching or on research remained unaffected, although their status was in no way prescribed and merely reflected the particular interests of individual academics. Furthermore, in Great Britain the innovation of quantitative geography coincided with the era of growth in the 1960s and, as a result, surplus research capacity was virtually all invested in a single paradigm (Steel and Watson, 1972; Stoddart, 1967; Tricart, 1969; Whitehand, 1970).

In the FRG, in contrast to Britain, the innovations associated with the 'new geography' and the take-off of 'applied geography' only really appeared in the last phase of the era of growth (Hard, 1979; Leser, 1977; Lichtenberger, 1978; von Rohr, 1975).

In my opinion this time lag in the adoption of the analytical paradigm has been the main reason for the fundamental differences in the distribution patterns of theoretical and quantitative geography in the two countries. In Great Britain analytical geography spread outwards from the Cambridge–Bristol 'quantitative axis', and was thus initiated in the traditional centres of academic excellence in the subject. In the FRG the diffusion of analytical geography was determined by two quite different factors. First, as shown in table 7.1, it took root in the technical universities and has become more firmly established in the new universities than in the already existing ones. As a result it can be viewed as a fringe element in the process of institutional growth. Second, everywhere else the distribution pattern of analytical geography in the FRG is random, depending on whether a department has one or two, invariably young, members of staff with sufficient intellectual sympathy for mathematics, or rather who have studied mathematics to a high enough level to be able to understand the analytical paradigm.

Table 7.1. Distribution of quantitative geography among old and new universities. (Source: Kilchenmann, 1978.)

	With quantitative research programmes	Without quantitative research programmes
Old universities	9	16
New universities (founded since 1945)	10	6

Public financial assistance for research in both countries has enhanced existing tendencies. An extensive support of geography by the German research foundation, Deutsche Forschungsgemeinschaft strengthened, with a feedback from the career norms, a trend towards research abroad, thus encouraging individual aspirants to find yet other *terrae incognitae* somewhere on the globe (Far-Hollender and Ehlers, 1977). On the other hand in Britain research grants doubtlessly furthered a concentration of geographical research on more local research questions.

A very unfavourable development took place in the FRG in the field of publications. German publications by far surpass the British output as to quantity, but there are no 'central' journals at all. Hence "Each institute must have a journal of its own" and institutional particularism is at full height.

In addition there is no 'central' institute comparable to the Institute of British Geographers. As a result an enormous amount of information is not organised methodically but seems to ooze away in the small-groups structure of research. There is no sort of institutionalisation comparable to the British IBG Study Groups. Consequently there is often a 'destruction of production'—research is being carried out, but no cognizance whatsoever is taken of its results.

British universities have also so far been spared self-government through a curia system, as has happened in some German Länder in the wake of the new university laws. These laws have become the instruments of the power struggle between the various interest groups (professors, assistants, students), and the ensuing administrative activity takes up research capacity.

Moreover the process of growth in the FRG has unbalanced, to a great extent, the original relation between academic teachers and students, and thus the research function has turned into a residual function or even into a hobby to be carried out almost as a leisure activity.

Teaching

As can be seen from table 7.2, a university teacher in the FRG has to cope with about three times as many geography students as his British counterpart. If one also takes into account that the vast majority of British geography students take a three-year course, whereas in the FRG nearly twice as many students have a course lasting four-and-a-half to five years provided for them, it is obvious that the balance of teaching loads is even more heavily weighted against academics in the FRG. Overall, teaching commitments in Great Britain are barely half those in the FRG. More serious, however, is the way in which department size in the FRG, amongst other things, has forced teaching to be orientated towards a standardisation of lectures and examinations. The whole of the degree structure is in a straitjacket dictated by the needs of the teaching profession. Equally the sheer volume of work means that the main occupation of university teachers is reading and correcting student exercises.

The very different ways in which teaching is organised in Great Britain and the FRG seems to me to be the key to the secret of the differences in academic potential. Apart from the first year of study, when a more or less standard introductory programme is offered by most departments, the majority of British students are free to choose from a range of special courses those which interest them most for study in the following two years. This system of free choice creates a healthy climate of competition, both in terms of teaching and research, and in terms of the extent to which research results are incorporated directly into teaching, and finally in terms of academics using the reactions of those being taught as a stimulus for their own research (Gould, 1973; Gregory and House, 1973). In departments in the FRG such possibilities are open only to a few selected groups and to a small number of academic staff.

The basic similarities between Great Britain and the FRG stem from the fact that both have a state-run admissions system for new students. However, whereas there is virtually no restriction on entry to geography courses in the FRG, with supply and demand more or less in balance, the admissions ratio in Great Britain is about 0·5, that is, only one out of two candidates for a place on a geography course can actually obtain one.

Table 7.2. Summary of geography-department teaching staff and students in the FRG and Britain. [Sources: *Geographisches Taschenbuch* (1956/57; 1966/67; 1979/80); *Statistisches Jahrbuch für die Bundesrepublik Deutschland* (1958; 1979); A Matter of Degree, *A Directory of Geography Courses* (1976).]

	FRG					Great Britain	
	1956–57	1966–69	1979–80			1956–57	1976–77
Number of departments	30	41	56			32	46
Teaching staff	116	341	668			243	720
Staff per department	4·2	8·3	12·2			7·6	16·0
	1964–65	1977–78				1976–77	
	main subject	total	teacher training	main subject	Master's Degree[a]	total	single honours
Number of students	1887	30398	20604	10082	2522	10915	5565
Students per department	50	540	365	180	45	237	121
Student/staff ratio	6	46	31	15	4	15	8

[a] "Pure" Geography.

Nevertheless entry requirements in Britain vary from department to department, in marked contrast to the FRG. In general, those departments with well-developed graduate programmes have the stiffer requirements. As a result of this, however, the overall drop-out rate is exceptionally low —roughly between 5% and 10%—and the teaching staff are not faced with the built-in problem of their German colleagues of weeding out students below a certain standard (Geipel, 1975). It goes without saying that this is an enormous benefit to the general climate both for teaching and for learning.

Staff structure

British universities have so far been spared experiments in the organisational structures, and hence an increase in the number of positions did not bring about any substantial changes either in the traditional hierarchy of Professors–Readers–Senior Lecturers–Lecturers or in their respective ratios. Things are different in the FRG. In the mid-fifties there used to be a threefold structure consisting of (1) Full Professors, (2) Associate Professors, (3) Assistants, with a 1 : 1 : 1 relationship between the three groups (cf table 7.3). In other words, on average there was only one Assistant per Professor. By the 1960s, the relationship had altered to the extent that it had become 1 : 1 : 2. In effect the increased burden on the teaching staff produced by the rapid increase in student numbers was mainly dealt with by creating a larger number of posts for Assistants. There was also another change which first became apparent in the period 1968 to 1978, namely the hiving-off of civil service posts to deal with specific functions in the realm of teaching and in the administration of resources. Further reorganisation was to follow, partially as a result of

Table 7.3. The increase of university geographers in the FRG and Great Britain. (Sources: *Geographisches Taschenbuch 1956/57, 1979/80.* A Matter of Degree, *A Directory of Geography Courses 1976.*)

Great Britain			FRG		
Title	1956	1976	1956	1979	Title
Head and Professor	27	40	31	177	Full Professor
Professor	1	46			
			-	33	Wissenschaftlicher 'Rat' and Professor (tenured post)
Reader	6	47	27	8	Associate Professor
			28	35	'Honorar-Professor', 'Dozent'
Senior Lecturer	14	128	-	117	'Akademischer Rat' (tenured post)
Lecturer	132	438	41	175	Assistant
Others	63	21	2	123	Scientific coworker
	243	720	116	668	

the new university laws:
(1) In many places the new laws put Associate Professors on the same level as Full Professors, thus replacing the original threefold division by a twofold one.
(2) Simultaneously the number of Professors has trebled, so that they amounted to 30% of the university geographers.
(3) There were also various new posts created with security of tenure[(2)], thus removing from the recipients the pressure and risk associated with the traditional career structure.
In spite of the new laws the problem of the middle ranks could not be solved, and the number of cases of hardship was by no means reduced. It remains an open question, both now and in the future, as to the number of tenured (civil service) posts a university subject should have before the changeover and mobility of personnel, necessary for the continuing academic development of the discipline, is stymied.

As a consequence of these changes there has been a very significant impact of institutional forces upon the evolution of the subject, a feature which is frequently ignored. To clarify the present situation in university geography in the FRG, figure 7.1 compares two pyramids of the academic population, showing the situations before and after the era of growth. The initial situation in the early 1950s, showing an even spread according to age among the various academic levels, is somewhat surprising. At that time nearly a generation divided Professors from Assistants. Nor has the era of growth produced quite the changes one might have expected. Rather than the teaching staff as a whole having a generally younger look, the greater youthfulness is only to be found amongst Professors, both Full and Associate Professors being on average nearly ten years younger than in the mid-1950s. In contrast, the average age of Assistants has gone up by nearly four years. In terms of age the newly formed group 'Wissenschaftliche Räte and Professors' is almost identical with the group 'Ordentliche Professoren' etc, while the tenured posts of 'Akademischer Rat' and the extremely heterogeneous group 'Scientific Coworkers' occupy places between Professors and Assistants. Overall, therefore, a clearly differentiated structure based on age and status, which prevailed in the 1950s, has been replaced by an interlocking structure in terms both of age and status. The 'wages of growth' will be paid by those born between 1935 and 1946, who in terms of numbers make up about half of all the teaching personnel.

I do not intend to consider in any detail the consequences of the present policy of limiting the number of academic posts and the 'closing of the ranks' that has occurred. Suffice it to say that the reduction in mobility, the lack of opportunities for the coming generation whose talents are being

[(2)] 'Wissenschaftlicher Rat' and Professor, 'Akademischer Rat', 'Universitäts-Rat', 'Wissenschaftlicher Mitarbeiter'.

absorbed by nonuniversity research institutes, the lack of new ideas in academic work, etc, are real problems. Equally it is worth remembering that the end of the era of growth affects graduates just as much as the academic staff of departments (Dörr, 1975; Giese, 1979; Kroger, 1979).

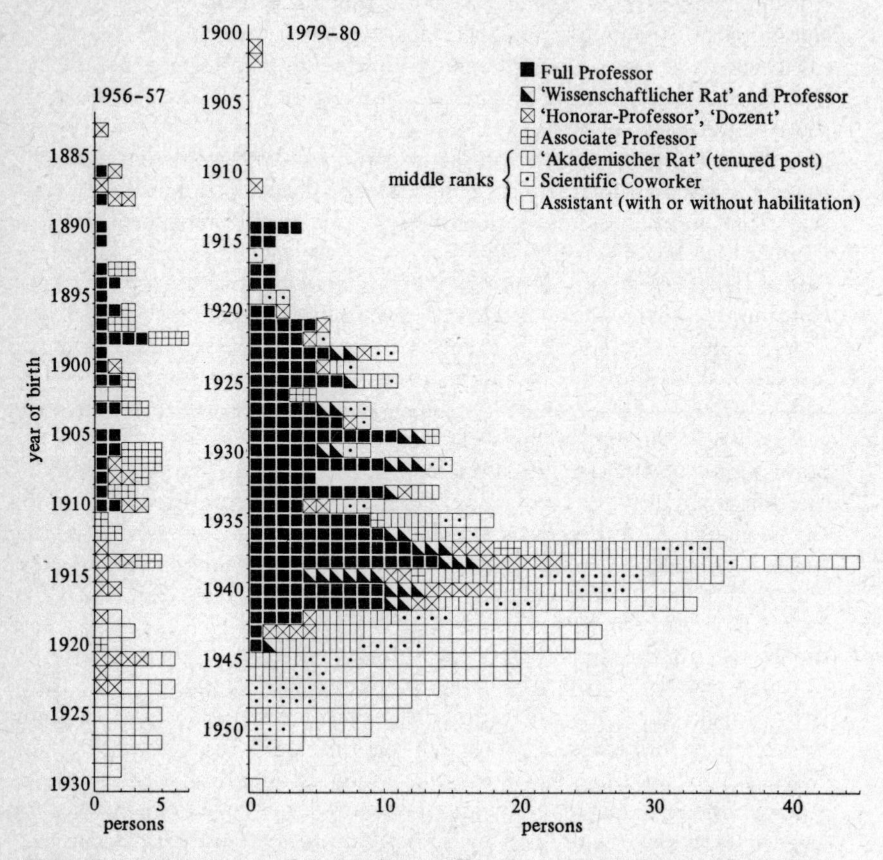

Figure 7.1. Age structure of the teaching staff in departments of geography in the FRG 1956-1957 and 1979-1980. (Sources: *Geographisches Taschenbuch 1956/57; 1979/80*; Lichtenberger, 1978.)

Labour market and course structure

The three-stage education system at British universities with academic degrees (for example, BA/BSc, MA/MSc, PhD) focused on different labour markets has no equivalent in the German-speaking countries (Rich, 1976). In the FRG the new laws regulating university studies have cancelled the possibility of acquiring a master's degree (for grammar school teachers) or a PhD, and introduced a postgraduate course for the latter. At the same time a 'pure geography' (that is, single-honours) master's degree was created, with the express purpose of cutting the number of semesters required, as already practised in other disciplines. So far this aim has largely not been achieved, as the assigned duration of nine semesters tends to be prolonged to an average of fourteen semesters, mainly owing to the 'freedom of learning'. Moreover there is no degree comparable to a BA with a course of three years, with the exception of those at the newly created 'Pädagogische Hochschulen' founded for training teachers. The majority of geography departments are still dedicated to providing interdisciplinary MA courses for grammar school teachers. At present in the FRG no fundamental changes in the goals of education for university geography can be expected comparable to those that have taken place in Britain during the past two decades, owing to a reform of the secondary school organisation and, thus, a drastically reduced demand for teachers for secondary schools (see table 7.4).

The new, pure geography master's degree courses are, to a large extent, aligned with the fields of urban and regional planning[3]. The necessity of opening up the private labour market has been strongly advocated by many members of the 'Verband der Berufsgeographen' (equivalent to the Association of Professional Geographers), but a full incorporation of these concepts in a curriculum has not yet taken place. Hence, the present trend in the FRG towards a master's degree in pure geography must be seen against the background of a general tendency encountered in many disciplines, namely the wish to acquire a professional qualification in order to secure access to the labour market.

Now it is very hard to guess how far the present strong trend towards professional recognition will continue. It is not clear whether it will result in more academic professional careers, or whether the German-speaking countries will follow the USA and accept professional and spatial mobility as a normal part of a person's career. It may be that the present tendency to overevaluate intellectual investment in specific subjects will disappear, an essential development given the much reduced half-life of much knowledge. Whatever happens it is clear that the welfare state, tied up with a consumer society, does not provide the ideal setting for speeding up such developments.

The present dilemma of the new type of courses in geography—as in other fields—consists of the fact that control has been lost of the number of students admitted, especially when the demand for a teachers' training fell rapidly in the second half of the seventies (Winter Semester of 1974–1975, 36000 teacher training students; 1977–1978, 21000), and at the same time the number of students of pure geography courses rose from 1500 to 3000.

[3] Monheim (1976); Partzsch (1976); Rautenstrauch (1975); von Rohr and Soker (1979); Schlick (1979); Seidel and Tiggemann (1979); Verband deutscher Berufsgeographen (Hsg) (1978a; 1978b); Zentralverband der deutschen Geographen (Hsg) (1978).

Large universities like those of Munich, Frankfurt, Münster, Bonn, and the Free University of Berlin all had more than 200 'pure' geography students registered. How could this happen with an entry system governed by *numerus clausus*? Why was there not simply a reduction of the number of students per university teacher? The answer is easy: it was inconceivable that these quotas be brought down to the levels that had existed before the influx; indeed there was little inclination even to press for this in case teaching posts were withdrawn and allocated to other popular subjects with even worse staff/students ratios. There followed an extremely complex chain of events, inspired partly by the Ministry and

Table 7.4. Course structure and labour markets in Britain and in the FRG (1977). [Sources: unpublished data provided by University College London; Verband deutscher Berufsgeographen (Hsg), 1979. Arbeitslosigkeit. Auch bei Geographen?; Walford, 1977.]

Academic degree	Great Britain		FRG	
BA/BSc	**Graduation total 1946**		no equivalent degree	
	in employment 1664			
	further academic study	13·7%		
	teacher training	26·1%		
	other full-time training	5·9%		
	planning	1·3%		
	rest of public sector	9·3%		
	private sector	30·3%		
	temporary employment	8·0%		
	employment overseas	2·2%		
	seeking employment	6·9%		
Interdisciplinary MA			**graduation total (approx.)**	**5000**
			teachers (grammar school) (approx.)	65%
			other teachers (approx.)	32%
			seeking employment	2%
MA and PhD	**Graduation total (approx.)**	**200**[a]	**MA 'pure' geography (approx.)**	**100**
	in employment	136	employment sample	100
	university lecturer	32%	university lecturer	28%
	other teaching	37%	other teaching	8%
	planning	8%	planning	48%
	other public sector	16%	other public sector	
	private sector	7%	private sector	16%
			PhD graduation total (approx.)	**80**
			(no data available)	

[a] Annual average 1972–76.

partly by the departments themselves, which produced a switch of undergraduate places for teachers to places for students studying pure geography.

It may be interesting in this context to explain the solution that has been adopted in Vienna. The Geography Department at the University of Vienna is in the fortunate position of having established two new degrees (MA, PhD) before the crisis in the market for teachers and all the new demand pressures manifested themselves. These two new degrees are in 'Urban and regional research and policy' (course statute: 19-7-1974 and syllabus: 4-12-1975) and 'Cartography' (Arnberger, 1973). Both the courses are based in the geography department, but in terms of content are interdisciplinary, as may be illustrated from the course outline for the degree in 'Urban and regional research and policy'. In the first section about 40% of the syllabus is identical to that for intending grammar school teachers, a further 40% is taken up by practical classes and lectures in formal sciences, and about 20% is made up of interdisciplinary electives. The second section is project orientated but offers in addition courses with outside speakers, mainly experts in planning. Around 60% of the syllabus is explicitly defined, the rest can be selected from the very wide range of courses offered by the other universities in Vienna, depending on the nature of the project work being undertaken, the interests of the individual students, and the issues that are current at the time. The small number of graduates means that so far there has been no difficulty finding jobs for them in research institutes, and in the public and private sectors.

The present institutional dilemma of university geography in the FRG

In the course of growth an extremely complex institutional pluralism has emerged; its range extends from residual structures deriving from the traditional way of life of the scholar on the one hand, to institutions with a complete division of labour on the other hand. Despite a fundamental change in the legal organisation of the universities in various Länder of the FRG, and a participation by the representatives of the 'middle ranks' and the students in various managerial boards of the universities, the traditional concept of autonomy of the German university Professor has remained basically intact. Also the regulations for postgraduate studies and habilitation as well as career norms still rest on the idea of the individual scholar. Teamwork has not succeeded in being acknowledged as a legitimate form of organising scientific work, adequate to the technological state of analytical research. Moreover, radical legal changes in the university organisation did not bring about any modifications as to the fact that the career of university geographers represents, on the whole, a secluded internal process of reproduction. However, despite an adherence to this view of the academic role, the nature of universities as institutions has been greatly modified. If we make a comparison of university management with that in private industry six features stand out.

First, one may take it as read that with university departments one is dealing with organisations that are part of the public sector. This means that many of those things that are essential for the continued existence of a firm in the private sector, such as advertising and market research into the interplay between producer and consumer, are so weakly developed that market response is often poor (Geipel, 1974; Mayr, 1970).

A second feature also derives from analogy with the private sector. It is usually assumed that as a firm grows a functional division of labour will occur, producing the well-known economies of scale that give the big firm the advantage over the small one. However, universities today are in a schizophrenic situation. Academics, who are highly qualified in their own specialisms, have also to undertake tasks such as administration and research management and, because they are only self-taught in the techniques of group politics, are condemned to being little better than dilettantes.

Third, if one understands by rational corporate management the systematic organisation of overall potential output and of the factors of production, and if one views planning as necessary, then any economist would be forced to conclude that any such facet is totally absent from university geography and most other university departments.

A fourth feature, however, is that planning in the economic sense is not an option that is open. An important part of the production process, namely ordinary and extraordinary expenditure, is outside the scope of departmental decisionmaking, because of the lack of central budgetary control in German universities. In practice, planning consists largely of a series of random wishes directed into official channels, and framed in the expectation that the democratic decisionmaking process will produce a result. It is this that gives rise to the general lack of comprehension among many outside the system. The most important factors of production, the workforce, are built into the budget as service sector employees. Decisionmaking, which takes place at the level of head of department or university committee, can therefore only be equated with most middle levels of management. One important difference between the two, however, is that there is no way of controlling the efficiency with which resources are allocated as far as yields, in whatever way they are defined, are concerned. In Germany this is a result in large part of an inflexible quota system, full of bureaucratic procedures and usually organised by the state, which regulates inputs and records outputs.

Fifth, because of a shortage of nonacademic personnel, academic staff are confronted with a multitude of tasks for which they are overqualified. Moreover, a division of labour and a substitution of human labour by machines is rendered practically impossible by the small size of the departments.

Sixth, any rationally run enterprise presupposes hierarchies in all areas of activity, and this expectation highlights further anomalies in German geography departments. In the main these are organised only as partially self-governing teaching establishments, which means that many of the jobs that staff undertake are not included under any institutional structure, for example, public relations, the management of research, and also their own research work. Moreover, even for teaching and internal government the hierarchical system that was originally planned has not been set up, largely as a result of general growth and the enactment of new university laws, as discussed below.

Thus there has been an emerging conflict between the traditional role of the scholar and the increasing demand for more rational decisionmaking within universities. This problem has been made more complicated by interaction with changes in job specifications, an increased burden of teaching, and the increasing pressures of contract research. Each of these factors is discussed in turn below.

The extended range of academic tasks

At the same time as the increase in the size and composition of the academic staff was occurring, new responsibilities were also being loaded onto departments. There is now only a very slim chance for the classical ideals of freedom of research and the unity of research and teaching to be realised, at least in large departments, despite the fact that both are embodied in most state constitutions. As has already been mentioned, most of the new university state laws required departments to become self-governing bodies and, as they emerged from their academic ivory tower, they had to assume new responsibilities.

For the most part these new tasks have been managed in two ways. First, the variety of interests amongst the teaching staff has been retained so that an individual's functions vary over time. A consequence of this has been job mobility within the university, as well as actual changes in the location both of internal and of external information and activity systems. However, any chance of covering the whole range of possible tasks occurs only in very small departments, with very small numbers of students, where there is a first-class provision both of equipment and of nonacademic personnel. In the general run of university life the taking-on of the responsibilities of self-government, preparing specialist reports, managing research projects, etc, means neglecting personal research. After a certain time, all that is possible for a person to do is yet more work in his established research field—now someway from the research frontier—or to take on still more administrative jobs within the university. The impossibility of dealing with so many different jobs is not limited only to Professors, it affects the middle ranks as well. They devote time to self-government, because they believe in group politics, and to contract research, because they see it as one of the jobs associated with the subject and/or because they expect participation either to improve their own job satisfaction, or their career prospects, or even just their private income.

A second method of organising a new task has been to attempt to make the various aspects of the job mutually exclusive. However, this option is not open given the way in which universities are currently organised. When people are asked to evaluate their career, research still occupies first place, even though now it can frequently be viewed as little more than a leisure-time activity. Administrative jobs associated with self-government have been accorded lower academic status, and much the same is also happening with didactics, causing it to be hived off into departmental ghettos.

Teaching

Fundamental changes have occurred in the way in which teaching is organised in the past two decades. If one first of all takes account only of the changes brought about by the growing numbers of students and the increases in academic staff, the following somewhat contradictory pressures emerge: a demand for the standardisation of departments, or rather course content; a demand for more specialist courses; a demand for standardised examination syllabuses, etc. As the structure of teaching establishments becomes ever more complex, so the number of examinations increases. A reduction in the number of final examinations results in a complicated system of examinations, forcing students into an 'acquisitive economy' based on certificates. The whole system becomes a bureaucratic nightmare, spelling an end to the student's freedom to learn and the university teacher's right to teach what he wishes. The new student laws mean that these forms of teaching are prescribed and 'petrified' in syllabuses. It goes without saying that this system can take little account of other external influences such as changes in the labour market.

The emergence of more formal syllabuses in teaching has led directly, in the name of equal opportunity and public accountability, to the introduction of general regulations, but these have ignored a number of significant factors, amongst which are:

(1) The coordination of course content.

(2) A strategy for incorporating new members of the department into the teaching.

(3) The traditional demand for links between teaching and research.

(4) The fact that so far almost the only educational goal has been to produce geography teachers for schools.

One can say without hesitation that all the problems raised, on the one hand, by the particularism of the different Länder and the individual departments referred to above, and on the other, by the innate handicaps stemming from the staff structure and the aspirations of individual academics, will be solved in different ways in each department. In what follows only the implications of the first two points will be considered.

With reference to the first point, there are fundamental difficulties to be overcome before academic content can be properly structured and teaching coordinated. In German university geography there are no

generally accepted criteria for judging the relevance of problems. Equally important is the susceptibility to changing 'fashions', most of which reflect nonacademic pressures and the prevailing general objectives of society. In my opinion it is only possible to establish any connection between the various subdisciplines in geography at two levels:
(a) In terms of practical issues relevant to society. By constructing 'artificial environments' with the help of technical and normative principles, one finds solutions to problems thrown up by human society.
(b) In terms of theory formulation as an abstract system of hypotheses and relationships.

It has not proved possible to use empirical and analytical research to establish a connecting link between the geographical sciences. In those instances where a so-called 'synthesis' is considered educationally desirable, there must of necessity be a touch of amateurism in the end product. This statement applies equally to comprehensive attempts at synthesis both in physical and in human geography.

With reference to the second point, a vertical division of teaching duties which corresponds to the various levels in the academic hierarchy, is still the distinguishing characteristic of the US university system. There the traditional division of responsibility has often been for Assistant Professors to deal with the mass of undergraduates and for Full Professors to formulate the graduate programme. The hierarchical structure of command in the traditional German university fitted into a similar pattern. As part of the general reorganising of academic staff structure, the whole question of the hierarchical division of jobs is being reconsidered with a view to finding some alternative. In fact the constitutional changes are encouraging the emergence of a horizontal allocation of jobs, with the setting up of such transitional forms as oligarchic hierarchies and specialist areas, etc. People at different levels in the academic hierarchy and with different degrees of experience may teach anywhere in the teaching programme as appropriate. There are a number of very different influences pulling in this direction. The fixed content of teaching programmes, as set out in syllabuses, produces a situation where 'second hand' teaching programmes exist side by side with those of scholars, who have their own unique programmes. Equally, increasing specialisation, combined with growing student numbers, means that specialists try to introduce their own particular specialism into the teaching programme. On the other hand, there are Full Professors of high standing who feel obliged to offer basic lectures in order to reach all of the students.

Problems of contract research

Contract research has come to German geography departments as a result of the demand that geographical research be socially relevant. The administrative and academic problems that this has created are discussed below.

(1) If contract research is to be built into the teaching programme it must contain a didactic element. Marrying this teaching requirement and the other terms of a contract, especially the need to keep to a deadline for the end product, is extremely difficult.
(2) For any contract to be successfully completed it is necessary to employ experienced coworkers. If these people are paid from outside funds, conflicts can arise over the use of departmental resources, and over access to facilities, for example, the university computer.
(3) As an academic career is not possible within the framework of contract research, most of those engaged in it only use the appointment as a stopgap before entering a more attractive post.
(4) In general, academics lack the necessary managerial experience, especially as regards job scheduling and financial planning.
(5) Contracts are imposed on departments from outside often with predetermined goals. Indeed they are often only parts of much larger projects and come to universities because their small scope or, more often, the difficulties in costing them make them of little interest to semiofficial or private research agencies (Sieverts and Kossak, 1977).
(6) The danger exists of being asked to do research simply to try and substantiate the preconceived ideas of a person or an organisation. There is also the impediment that the publication of the results of contract work will be prevented or delayed by the person or institution issuing the contract (Harvey, 1974).

Using the matrix of problems and solutions set out in figure 7.2, one can sketch the sort of research that should be of interest to geography departments and also the kind of contracts they are receiving in Germany.

In my opinion there are three types of research contract that are suited to the research work carried on in universities. First, without doubt the most interesting contract is that in which the problem is not clearly defined and where an unknown solution is being sought, but hardly any contracts are issued for such projects. Second, there is a good chance

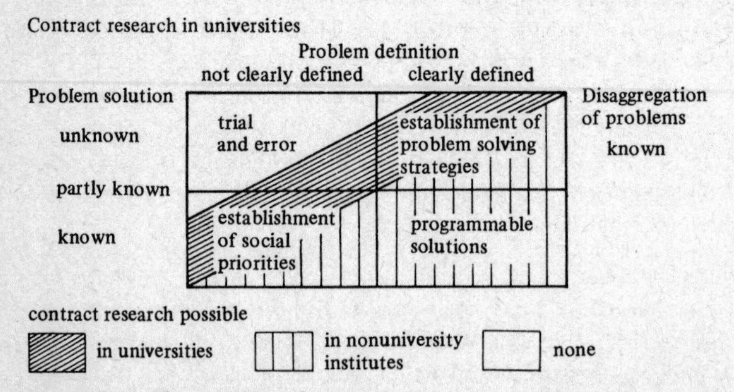

Figure 7.2. The position of contract research in university geography departments, based on an idea of Haggett (1977).

that a contract will prove interesting when the problem, at least initially, is unclear and when the first job is to formulate it in precise terms and to establish a set of social priorities (Coppock, 1974). This sort of work can be stimulating, even though the solutions to the problems associated with each of the alternatives are known. Third, there is the category where the problem itself is clearly defined, but a solution has to be found.

Research contracts involving the routine solution of clearly defined problems are much better suited to nonuniversity research institutes because of the way in which they are organised. As mentioned above, however, residual contracts of this type are often awarded to geography departments in practice.

Solving the problems

In conclusion, the problems described above ought to be presented once more, arranged according to the levels of decisionmaking at which a solution seems feasible.

The two main problems which have to be addressed by university policymakers, both in Great Britain and in the FRG, are as follows:
First. The impact of an imbalance in the age structure of academic staff weighs not only on geography departments, but on all departments that have gone through a period of rapid growth. The danger of stagnation threatens many disciplines as a result of the ending of the era of university growth. The problem is that the talented rising generation may be syphoned off into nonuniversity research establishments, or even into other professions entirely.
Second. What should be the distribution of scarce resources? This distribution can be used as an instrument for promoting a policy of equality and to reduce disparities between old and new universities, incidentally acting also as an instrument of regional policy in deprived areas. Equally it can form part of a demand-oriented locational policy, favouring universities in large centres of population.

It is also very important to remember that university organisation in the FRG differs fundamentally from that in the English-speaking countries. At German universities there is no three-year course leading to a BA or BSc degree, yet it was the existence of such courses that made it easier for the British to change from teacher training to the provision of candidates for the private labour market.

The discussion at the beginning of this essay of the different lengths of student study found in Britain and in the FRG convince me that the time is ripe for orientating teaching in the Pädagogische Hochschulen more strongly towards the private sector. The research function of universities has a strong tradition in the Anglo-Saxon world, as have geography departments specialising in research. This is missing in the FRG. It would be highly desirable, as part of the switch to 'pure' geography master's degrees, that a similar training should be offered, primarily in those departments with adequate resources, even though such a development would fly in the face of state autonomy and departmental independence.

It has been shown that the reduction in the demand for teachers is by no means exclusively a German problem, but also that British geography was able to adapt to the changing situation much earlier. In the FRG the future of teacher training will have to be decided as a part of general educational policy, for it is integral with all that universities are doing.

The FRG faces one additional problem to Great Britain in that the goal of equal opportunity has been satisfied by a reduction in the quality of education. At a time when the numbers of school children are falling and a smaller number of students were being trained as teachers, it would have been appropriate to make a priority of improving the quality of education by reducing class sizes and by improving the staff/student ratios in universities. Another related policy issue for state education is whether teachers in universities will be required to take on additional duties, such as providing postgraduate training for grammar school teachers, as they have in the GDR.

University geography must retain its links with the school subject, especially at a time when other university disciplines, such as sociology and psychology, are trying hard to get a foothold in the teaching in secondary schools. Indeed, geography may take a certain amount of comfort from a comparison with the other subjects traditionally heavily involved with teacher training, such as history and languages. These subjects do not generally have the option of switching to the much underrated alternative of applied research.

In summary, then, a stage has now been reached where further initiatives must come from those actually within university geography. In line with the familiar dictum that "geography is what geographers do" (Bahrenberg, 1979), I believe that separate subunits ought to and could be set up within existing departments for regional research, urban studies, spatial research, etc. In the era of growth that has now ended, the established geographical fraternity succeeded marvellously in building up the resources and personnel of departments. It is now up to the new generation, who owe their jobs to the era of growth in universities and who must now pay the returns for that growth, to introduce new and viable research fields; to formulate new educational goals; and to improve information and communication networks between academics and the field of applied research.

References

Arnberger E, 1973 "Der Studiengang zum wissenschaftlichen Kartographen (akademische Ausbildung)" *Kartographische Nachrichten* **23**(2) 73-79

Bahrenberg G, 1979 "Von der Anthropogeographie zur Regionalforschung—eine Zwischenbilanz" *Osnabrücker Studien zur Geographie 2: Zur Situation der deutschen Geographie 10 Jahre nach Kiel,* Eds G Hard, H J Wenzel (Universität Osnabrück, Neuer Graben/Schloss, Osnabrück) pp 59-60

Baker A R H, 1970 "Some observations on geography in British universities 1968-70" *Area* 2 42-46
Clarke J I, 1976 "The organization of geographical research in the United Kingdom" in *Human Geography in France and Britain* Eds J I Clarke, P Pinchemel (Social Science Research Council, London) pp 8-15
Cooke R U, Robson B T, 1976 "Geography in the United Kingdom 1972-76" *Geographical Journal* **142**(1) 81-100
Coppock J T, 1974 "Geography and public policy: challenges, opportunities and implications" *Transactions, Institute of British Geographers* **63** 1-16
Dörr H, 1975 "Berufsgeographie im Spannungsfeld zwischen Hochschulforschung und planerischer Praxis" *Geographische Zeitschrift* **63**(3) 177-194
Editorial Comment, 1978 "An institute for geographical research?" *Area* **10**(5) 321-322
Far-Hollender U, Ehlers E, 1977 "Geographische Forschung in der BRD und die Deutsche Forschungsgemeinschaft 1960-1975", in *Geographisches Taschenbuch 1977/78*, Eds E Ehlers, E Meynen (Steiner, Wiesbaden) pp 241-253
Geipel R, 1974 "Der bildungs- und regionalpolitische Effekt von Universitätsneugründungen (Beispiel Kassel)" *Deutscher Geographentag Kassel 1973, Tagungsbericht und Wissenschaftliche Abhandlungen* (Steiner, Wiesbaden) pp 53-65
Geipel R, 1975 "Hochschulgründungen und Regionalpolitik" in *Der doppelte Flaschenhals—Die deutsche Hochschule zwischen Numerus clausus und Akademikerarbeitslosigkeit* Hsg. U Lohmar, G E Ortner (Jaenecke Verlag, Hannover) pp 185-199
Giese E, (undated), 1979 "Berufschancen des wissenschaftlichen Nachwuchses in der BRD im Fach Geographie" Als MS vervielfältigt (mimeo) (Geographisches Institut, Universität Gießen, Senckenbergstraße 1, D-6300 Gießen)
Gould P R, 1973 "The open geographic curriculum" in *Directions in Geography* Ed. R J Chorley (Methuen, London) pp 253-284
Gregory S, House J W, 1973 "Preliminary report on high-degree graduates in geography, 1967-71" *Area* **5**(2) 88-95
Haggett P, 1977 "Geography in a steady-state environment" *Geography* **62**(3) 159-167
Hard G, 1979 "Die Disziplin der Weißwäscher. Über Genese und Funktion des Opportunismus in der Geographie" *Osnabrücker Studien zur Geographie 2: Zur Situation der deutschen Geographie 10 Jahre nach Kiel,* Eds G Hard, H-J Wenzel (Universität Osnabrück Neuer Graben/Schloss, Osnabrück) pp 11-58
Hartke W, 1960 *Denkschrift zur Lage der Geographie* Im Auftrag der Deutschen Forschungsgemeinschaft, Wiesbaden
Harvey D, 1974 "What kind of geography for what kind of public policy?" *Transactions, Institute of British Geographers* **63** 18-24
Kilchenmann A, 1978 "Dokumentation über Forschungsprojekte aus dem Bereich Theorie und quantitative Methodik in der Geographie" *Karlsruher Manuskripte zur mathematischen und theoretischen Wirtschafts- und Sozialgeographie* **24** 32 pp
Kroger R, 1979 "Die Zukunft des Geographen? Thesenpapier zur Abschlußveranstaltung des Deutschen Geographentages 1979" *Geographische Rundschau* **31** 418-419
Lawton R, 1978 "Changes in University Geography" *Geography* **63** (number 278) 1-13
Leser H, 1977 "Aufgaben und Möglichkeiten der Geographie in der Schweiz heute" in *Geographisches Taschenbuch 1977/78,* Eds E Ehlers, E Meynen (Steiner, Wiesbaden) pp 179-200
Lichtenberger E, 1978 "Klassische und theoretisch-quantitative Geographie im deutschen Sprachraum" *Berichte zur Raumforschung und Raumplanung* **22**(1) 9-21
Lichtenberger E, 1979 "The impact of political systems upon geography: the case of the FRG and the GDR" *Professional Geographer* **31**(2) 201-211
Mayr A, 1970 "Standort und Einzugsbereich von Hochschulen. Allgemeine Forschungsergebnisse unter besonderer Berücksichtigung der Untersuchungen in der BRD" *Berichte zur Deutschen Landeskunde (Bad Godesberg)* **44**(1) 83-110

Monheim H, 1976 "Der Geograph in der räumlichen Planung. Berufsfelder, Qualifikationen und Berufschancen" *Geographische Rundschau* **28**(5) 200-210

Partzsch D, 1976 "Zur Situation der Raumplanerausbildung in der Bundesrepublik Deutschland" *Berichte zur Raumforschung und Raumplanung* **20**(1) 15-20

Rautenstrauch L, 1975 "Berufsinteressen und Reform der Planerausbildung" *Stadtbauwelt* **47** 155-159

Rawstron E M, 1975 "Geography as an institution" *International Social Science Journal* **27**(2) 259-274

Rich D C, 1976 *Geography Degree Course Guide 1976/77.* Careers Research and Advisory Centre, Hobson's Press, Cambridge

Rohr H G von, 1975 "Geographen zwischen Hochschulausbildung und Berufswirklichkeit. Replik auf einen Beitrag von H. Dörr" *Geographische Zeitschrift* **63** 291-298

Rohr H G von, Soker E, 1979 "Geographiestudenten WS 1978/79. Oder: Der unaufhaltsame Vormarsch des Diploms?" *Standort* **2** 21-24

Schlick W, 1979 "GEO und Geographen–Ergebnisse einer Diskussion" *Standort* **2** 28-29

Seidel W, Tiggemann R, 1979 "Noch einmal: Zur berufsständischen Lage der Geographen in Landes-, Regional- und Stadtplanung in Nordrhein-Westfalen" *Standort* **2** 25-27

Sieverts Th, Kossak E, 1977 "Hochschulforschung im Bereich Städtebau und Landesplanung heute" *Stadtbauwelt* **55** 241-248

Steel R W, Watson J W, 1972 "Geography in the United Kingdom 1968-72" *Geographical Journal* **138** 139-153

Stoddart D R, 1967 "Growth and structure of geography" *Transactions, Institute of British Geographers* **41** 1-20

Tricart J, 1969 *The Teaching of Geography at University Level* (Harrap, London)

Verband deutscher Berufsgeographen, (Hsg), 1978a "Der Beruf des Geographen. Qualifikation, Anforderungen, Arbeitsmarkt, Arbeitsbedingungen" *Material zum Beruf des Geographen 1*, PO Box 106325, D-2000 Hamburg

Verband deutscher Berufsgeographen, (Hsg), 1978b "Arbeitsmarksituation der Diplom-Geographen" *Standort* **3** 20-23

Verband deutscher Berufsgeographen, (Hsg), 1979 "Arbeitslosigkeit. Auch bei Geographen?" *Material zum Beruf des Geographen 2*, PO Box 106325, D-2000 Hamburg

Walford R, 1977 "Geographical education in Britain" *Progress in Human Geography* **1**(3) 503-509

Whitehand J W R, 1970 "Innovation diffusion in an academic discipline: the case of the 'new geography'" *Area* **3** 19-30

Zentralverband der deutschen Geographen, (Hsg), 1978 *Rahmenordnung für den Diplomstudiengang Geographie*, Geographisches Institut der Universität, Am Hubland, D-8700 Würzburg

Sources:

A Matter of Degree *A Directory of Geography Courses 1976.* Sonderpublikation des Geographical Magazine in Zusammenarbeit mit dem Institute of British Geographers und der Geographical Association. o. J. (1976). o. O, 31 pp

Geographisches Taschenbuch 1956/57. Ed. E Meynen (Steiner, Wiesbaden)

Geographisches Taschenbuch 1966/67. Ed. E Meynen (Steiner, Wiesbaden)

Geographisches Taschenbuch 1979/80. Eds E Ehlers und E Meynen (Steiner, Wiesbaden)

Statistisches Jahrbuch für die Bundesrepublik Deutschland 1958. Ed. Statistisches Bundesamtsbericht (Kohlhammer, Stuttgart)

Statistisches Jahrbuch für die Bundesrepublik Deutschland 1979. Ed. Statistisches Bundesamtsbericht (Kohlhammer, Stuttgart)

Quantitative and theoretical geography in Italy

M P Pagnini, A Turco

This essay sets out to provide an account of research into Italian theoretical and quantitative geography, though the authors realise that an analysis of the evolution of the discipline rests on particular judgement of values and choice of field. They consider that Italian geography, though showing linkages to international experience, also presents strong specific features which divorce it considerably from the scientific contexts in other countries (Turco, 1980a; 1980b; Celant, 1980). They agree with the idea that "whatever the problems a scientist is working on, the current explanatory principles have strongly influenced the type of solution at which he might arrive" (Kuhn, 1964, page 14). In this report, therefore, they certainly intend to carry out an information giving function, but they would also like to stimulate interest in a deeper and more direct approach to Italian geographic thought.

At the outset it must be emphasised that there is a distinct division between physical and human geography in Italy. It is true that there is no lack of 'integrating' trends, but we believe that these, mainly wearing themselves out in ritualistic declarations of principle, have no appreciable influence on research practice. Hence, it is argued that Italian geographers are developing their subject within the sphere of two scientific communities. Here we shall speak only about the more numerous of the two communities, the geoanthropic sphere, in which the authors have worked and are still operating.

Paradigmatic conflicts

Ideological-disciplinary ferment cannot always be entirely explained by philosophical theories alone; however, T S Kuhn's (Kuhn, 1978) conceptual categories and interpretative schemes offer an adequate base for the analysis of contemporary Italian geography. The paradigm with which the major part of the community identifies itself is called 'idiographic'. It is difficult to give both a synthetic and an effective explanation of this term, as it evokes almost the entire tradition of academic geography: its epistemological status, the investigative practices and research inheritance, the code of communication, and lastly the mental attitude towards selecting problems and resolving 'puzzles'.

Two lines of thought have formed during the seventies against the idiographic monopoly. The first could be called marxist, even if only in a very wide sense. Taking original inspiration and, within certain limits, intellectual support from the works of Gambi (1964; 1973), this emerging potential paradigm has been enriched with some significant contributions of its own (AA.VV., 1975; Quaini, 1974; 1975; 1978), which are only partially related to the theory maturing within the framework of studies of other European and American marxist geographers.

The second line of thought could be called, again in a very wide sense, 'theoretic-quantitative'. Its pioneers were mainly Toschi (1941; 1954; 1959), Nice (1953), and above all Bonetti (1961; 1964; 1967); Dematteis, who today constitutes a reference point in the marxist line, brought out the first empirical 'new-geography' study of significance (Dematteis, 1966) and also the first global criticism on the epistemological meaning of the 'quantitative revolution' (Dematteis, 1970).

Thus, in Italy one can observe a triangular contest which could dominate community life during the next decade. Its results are of major importance, since they will determine the cultural vitality of the discipline and its capacity for including and resolving the many serious problems of space which strangle Italian society.

Obviously, the strategy of the idiographic paradigm is to conserve its dominant status. Its basis seems to be to avoid the disciplinary crises arising from conflict with rival paradigms. This strategic aim is pursued with different tactics, according to which methodology is considered—marxist or theoretic-quantitative. In the first case, the attitude is one of substantial indifference; the contribution of theory and methods from the marxist criticisms is ignored. The conviction is also rather widespread, though seldom clearly expressed, that the scientific value of the marxist contribution is very slight as it is polluted by the value judgements of an ideological-political nature on which its studies are based.

The theoretic-quantitative trend, on the contrary, is formally accepted by the dominant paradigm. In fact, judging from certain topical manifestations of community life [for example, in the National Geographical Congresses (AGEI, 1980)], it would seem that the 'old geography' considers the 'new geography', if not a legitimate child, at least an adopted child. However, the true reality of these disciplinary changes does not permit any great illusions. The dominant paradigm, remaining faithful to its own character, cannot create, and has not created, a favourable climate for the development of the 'new-geographical' thought and research. By not stimulating adherence to the theoretic-quantitative scientific plan, it has in practice prevented the emergence of new 'schools' or sound methodological trends. Having provided neither a financial policy, nor an academic career policy intended to preserve space for those who have commited themselves to the theoretic-quantitative field; it has rendered impossible the cumulative development of the 'new geography'.

The few exceptions have always been individual cases and have mostly favoured secondary level research workers, though they have only rarely maintained the quantitative-study appointment. In short, the presence of the dominant paradigm has led, as far as the 'new geography' is concerned, to a heavy (but certainly not at all surprising) paralysing effect.

The relationship between the two trends, marxist and theoretic-quantitative, is no better. In fact, the two lines of thought communicate very little with each other, both being anchored to positions of mutual—not always sufficiently motivated—aversion, not only of a scientific, but also often of a political or personality based nature. The priority should be to bring together marxists and new geographers in an intense, even aggressive, but open, constructive, and unreserved intellectual confrontation free from extreme protagonistic ambition. Indeed, the experience of other countries (France in particular) clearly shows that the reciprocal interest between the two trends is not only possible but extremely fruitful. It is therefore desirable that the recent signs of joint approaches becoming evident in Italy (Vallega, 1979a; Coppola, 1980) multiply and spread into wider reaches of scientific debate.

The status of the struggle for theoretic-quantitative geography in Italy
The expression 'theoretic-quantitative geography', with reference to Italy, helps to indicate the activity of a certain number of geographers who, without any coordination, follow more or less critically the 'quantitative revolution' and its conceptual and methodological evolution. It should be realised, however, that probably each of those ready to follow 'theoretic-quantitative geography' has a personal idea of what the expression means and what it may lead to at an ideological-disciplinary level. Research efforts, even though presented in this chapter as homogeneous, are in reality quite heterogeneous.

The 'new geography' accounts for about forty—mostly young—devotees who in the main hold subordinate academic appointments or positions of limited influence. This number of research workers is obviously rather low, not only in comparison with the entire community of geographers (about ten times greater), but also taking into consideration other circumstances. In fact, for many authors the new geography has resulted in few isolated and occasional pieces of work, although its fundamental orientations have been quite different. Sometimes, perhaps in complex but interrupted—or unimportant—research directions, the new geography has been too burdensome and unsatisfactory at the level of scientific prestige and therefore for the professional career. Thus many have shifted to 'idiographic' research in accordance with the laws of this dominant paradigm. It is not by chance, therefore, that the demand for expertise in the new geography, that has occurred in Italy during the last twenty years, has been satisfied principally by disciplinary areas other than geography, viz urban studies, sociology, and economics (Turco, 1980a).

Theoretic-quantitative research in Italy is atomised. Research workers have not had any institutional point of reference; there is no Commission within the Associazione dei Geografi Italiani (AGEI), the representative body for Italian university geographers; there is no autonomously constituted working group; there is no journal, not even an informative circular. Without any kind of coordination, communication among research workers can only be isolated, although there is evidence of examples of systematic and successful collaboration within limited groups.

Apart from the empirical research, with which we shall deal below, research workers try to support theoretic-quantitative geography along three main lines of development.

The first line of development concerns the technical requirements (mathematical, statistical, and data elaboration) considered a *sine qua non* for a serious, efficient scientific presence. Besides the critical presentations of quantitative instruments already experimented with in the international geographic field (Celant, 1974; Bonetti, 1976; Staluppi, 1976b; Cori, 1976; Buzzetti, 1979), there are also texts which penetrate the internal structure of mathematical models with the aim of testing, and possibly strengthening, their descriptive value (Staluppi, 1976b; Zanetto, 1979; Vlora, 1979). We are, however, far from achieving a satisfactory situation in this field. For example, even today there is no real manual on quantitative geography conceived and produced by Italian research workers. Moreover, it is still necessary to explore the possibilities of elaborating new models and even creatively interpreting existing complex models by means of their structural integration or modification in the Italian context.

The second line of development involves epistemological thought. Many authors have explored the logical basis of the theoretic-quantitative paradigm (Dematteis, 1970; Vallega, 1976), its connections with the contemporary philosophical currents (Pagnini, 1976), and lastly the most appropriate way to lead geography towards a nomothetic emphasis (Vallega, 1979a; 1979b; Massi, 1972; AA.VV., 1973). These same lines of thought and argument have been criticised by the marxist current (Vagaggini and Dematteis, 1976).

The third line of Italian development, more pragmatic but no less important than the previous two lines, concerns what could be called the fertilisation of the community *humus*; that is the fostering of conditions suitable to the activation of circular and cumulative development of theoretic-quantitative geography. This development is perhaps the most critical among those in which the Italian 'new geography' is involved. It should guarantee the diffusion of research, favour contacts between individuals and groups, institute scientific unity, and stimulate unification of cultural and linguistic schools. In fact, all this has been accomplished only minimally. Certainly it is extremely difficult to proceed in such a direction in a rather unreceptive atmosphere. Nor should the significance and range of several initiatives be underestimated, such as the Italian translation of Christaller (Christaller, 1980); or of textbooks such as Carter (1975), Boudeville (1977), Lloyd and Dicken (1979) which, contrary to what happened to an important study by Isard in the early nineteen-sixties (Isard, 1962), have been widely circulated; or of the organisation of international meetings (Racine et al, 1978). But it is evident that more must be done in future years if we do not want theoretic-quantitative geography in Italy to become the temple of a minor sect practising esoteric rites or, even more simply, dying of consumption!

Empirical research

Empirical research (carried out at different levels—nationally, regionally, locally) is spread over a very wide range of topics. The major sources of inspiration, however, have been found in the great changes which have occurred in Italy from the time of the so-called 'economic miracle'. First, emphasis has been placed on analysis of urban growth both by using normative and by using descriptive models (central place theory, the rank-size rule, the economic base). Next, considerable effort has been devoted to surveying and interpreting the spatial effects of the violent and uncontrolled economic growth that has occurred: regional imbalances; backwardness in the South; internal migrations; and industrial localisation. However, little attention has been given to the problems of agriculture and to the localisation of tertiary activities.

In the space available here it is not possible to review all of the empirical developments of Italian quantitative geography. Hence, the reader is referred to the excellent bibliographical reviews, arranged according to subjects and/or to analytical techniques which exist. Besides a review of physical geographical work based on quantitative methods (Belloni et al, 1980) two reviews of a general character are concerned with human geography (Turco, 1980a; and Ruggiero and Skonieczny, 1980). Among the major bibliographic reviews of research into urban geography (Mainardi, 1970; Dematteis, 1977) there are some important reviews specifically devoted to central place theory (Wapler, 1972; Pagnini, 1974; Cori, 1977).

References

AA.VV., 1973 *Studi su: città, sistemi metropolitani, sviluppo regionale* (Pàtron, Bologna)

AA.VV., 1975 *Colloquio sulle basi teoriche della ricerca geografica* Déjoz (Valle d'Aosta) 11-12 Ottobre 1974 (Giappichelli, Turin)

AGEI, 1980 *Convegno Quaderno sullo stato della ricerca geografica in Italia 1960/1980* Associazione dei Geographi Italiani, Varese

Belloni S, Lupia Palmieri E, Pellegrini G B, 1980 "Utilizzazione dei metodi quantitativi in geografia fisica" in AGEI (1980), pp 631-646

Bonetti E, 1961 *La Teoria della Localizzazione* publication 5, Università di Studi—Facoltà di Economia e Commercio, Istituto di Geografia, Trieste

Bonetti E, 1964 *La Teoria delle località Centrali* publication 6, Università di Studi—Facoltà di Economia e Commercio, Istituto di Geografia, Trieste

Bonetti E, 1967 *La Localizzazione delle attività al dettaglio* (Giuffrè, Milan)

Bonetti E, 1976 "L'analisi fattoriale e le sue applicazioni nella ricerca geografica" *Cultura e Scuola* **60** 151-163

Boudeville J-R, 1977 *Lo Spazio e i Poli di sviluppo* (Angeli, Milan)

Buzzetti L, 1979 "Il problema della valutazione della distanza" *Geografia* 2 34-38

Carter H, 1975 *La Geografia Urbana. Teoria e Metodi* (Zanichelli, Bologna)
Celant A, 1974 "La teoria dei grafi: uno strumento di analisi della geografia economica" *Rivista di Politica Economica* **11** 1400-1440
Celant A, 1980 "I paradigmi nella ricerca geografica" in AGEI (1980), pp 713-728
Christaller W, 1980 *Le Località Centrali della Germania Meridionale* translated by M P Pagnini (Angeli, Milan)
Coppola P, 1980 "Geografi al bivio: vecchi e nuovi orizzonti" in AGEI (1980) pp 729-740
Cori B, 1976 "La teoria della 'rank-size rule'" in Commissione di Ricerca del Comitato dei Geografi Italiani: *Studi su citta, sistemi metropolitani, sviluppo regionale* (Giardini, Pisa)
Cori B, 1977 "Gli studi geografici sulle aree di gravitazione urbana in Italia: metodi, risultati e limiti" *Storia Urbana* **2** 169-183
Dematteis G, 1966 *Le Località Centrali nella geografia Urbana di Torino* Università degli Studi—Laboratorio di Geografia Economica "P Gribaudi", Turin
Dematteis, G, 1970 *Rivoluzione Quantitativa e Nuova Geografia* publication 5, Università di Studi—Laboratorio di Geografia Economica "P Gribaudi", Turin
Dematteis G, 1977 "La rete urbana italiana, 1945-1975. Rassegna degli studi" *Storia Urbana* **1** 1-10
Gambi L, 1964 *Questioni di Geografia* (ESI, Naples)
Gambi L, 1973 *Una Geografia per la Storia* (Einaudi, Turin)
Isard W, 1962 *Localizzazione e Spazio Economico* (Cisalpino, Milan)
Kuhn T S, 1974 "La nozione di causalità nello sviluppo della fisica" in *Le Teorie della Causalità* Ed. M Bunge (Einaudi, Turin) pp 5-15
Kuhn T S, 1978 *La Struttura delle Rivoluzioni Scientifiche* (Einaudi, Turin)
Lloyd P E, Dicken P, 1979 *Spazio e Localizzazione* translators M Cost, M P Pagnini (Angeli, Milan)
Mainardi R, 1970 *Metodi di Analisi delle Reti Urbane. Bibliografia annotata* (Centro di Documentazione, Milan)
Massi E, 1972 "Analisi regionale e sviluppo polarizzato" in *Atti della Tavola Rotonda di Geografia Applicata sul Tema: "Poli Assi e Aree di Sviluppo Economico con Particolare Riguardo alle Regioni Sottosviluppate" Bollettino della Società Geografia Italiana,* Supplement to volume 1, Series X
Nice B, 1953 *Geografia e Pianificazione Territoriale* Memorie di Geografia Economica, IX, Istituto di Geografia di Università, Naples
Pagnini M P, 1974 "Scale e gerarchia dal punto di vista geografico" *Dibattito Urbanistico* **39** 22-33
Pagnini M P, 1976 "Teorie della percezione, strutturalismo e geografia" in *Studi sullo Strutturalismo, Volume II* Ed. G Proverbio (Società Editrice Italiana, Turin)
Quaini M, 1974 *Marxismo e geografia* (La Nuova Italia, Florence)
Quaini M, 1975 *La Costruzione della Geografia Umana* (La Nuova Italia, Florence)
Quaini M, 1978 *Dopo la Geografia* (Espresso Strumenti, Milan)
Racine J-B, Raffestin C, Ruffy V, 1978 *Territorialità e paradigma centro-periferia* (Unicopli, Milan)
Ruggiero V, Skonieczny G, 1980 "Le tecniche matematico-statistiche nell'analisi spaziale" in AGEI (1980) pp 831-844
Staluppi G, 1976a "L'analisi della minor distanza" *La Geografia nelle Scuole* **1** 1-10
Staluppi G, 1976b "A methodological approach to distribution of settlements in geographical research" in *Italian Contributions to the 23rd International Geographical Congress 1976* Eds A Pecora, R Pracchi (Consiglio Nazionale delle Ricerche, Rome)
Toschi U, 1941 *La Teoria Economica della Localizzazione delle Industrie Secondo Alfredo Weber* (Macri, Bari)

Toschi U, 1954 "La distribuzione geografica nella teoria economica" *Bollettino della Società Geografica Italiana* 257-264
Toschi U, 1959 *Geografia Economica* (Unione Tipografica Editrice Torinere, Turin)
Turco A, 1980a "L'emploi des modèles dans l'analyse des problèmes territoriaux en Italie" in *Les Modèles comme Source d'inspiration dans la Géographie Contemporaine* Actes du Symposium 8 et 9 Décembre 1978, Institut de Géographie, Université de Lausanne, Lausanne, pp 43-73
Turco A, 1980b "I modelli nei paradigmi della geografia italiana" in AGEI (1980) pp 865-879
Vagaggini V, Dematteis G, 1976 *I Metodi Analitici della Geografia* (La Nuova Italia, Florence)
Vallega A, 1976 *Regione e territorio* (Mursia, Milan)
Vallega A, 1979a "Porti e regionalizzazione: un paradigma sistemico" *Bollettino della Società geografica italiana* 10-12,577-597
Vallega A, 1979b "Neopositivismo e marxismo in geografia: riflessioni su un dibattito" *Rivista Geografica Italiana* 2 129-152
Vlora N R, 1979 *Città e Territorio* (Mursia, Milan)
Wapler G, 1972 *La Ricerca sulle Località Centrali nella Letteratura Geografica Italiana* (Umbria Editrice, Perugia)
Zanetto G, 1979 "Il potenziale da modello a strumento" *Rivista Geografica Italiana* 3 298-320

The limits to geographical research

G Battisti

More than twenty years after the 'quantitative revolution' recognised by Burton (1963), we can see that quantitative methods have spread throughout the geographical literature. In Italy research including these new methods is certainly increasing (see the bibliography of Pagnini and Turco, chapter 8 in this volume). We are, however, far from conceiving a 'regional science' characterised solely by the statistical–mathematical approach, for which *Geographical Analysis*(1) or the *London Papers in Regional Science*(2) represent good examples. On the contrary, at the very moment of publication of what can be considered the first handbook of quantitative geography (Vlora, 1979) a step backward in interests has been registered, towards subjects which seemed to have been abandoned. The study of the landscape, for example, should be one of the major themes of the next Italian Geographical Congress. This is owing to the influence of the studies of perception, which have been strong in Italy (Pagnini, 1975; Brusa, 1978; 1979; Bianchi and Perussia, 1978); however, we are still in a phase of uncertainties. Perhaps this is the right moment for a comprehensive reconsideration of the meaning and the methods of our discipline.

The creation of theoretical models and their verification in practice, in the framework of a deductive restructuring of geography, is the consequence of the kind of problems which geographers started to deal with from Mark Jefferson and Walter Christaller onward. As Racine and Reymond noted (1973, page 109), today "the geographer has to abandon the bells of grazing herds or the steps of peasants to turn to the assessment of commuters' migrations through the calculation of trips made and of urban hierarchies through the statistical data of census-taking". They think that the long-lasting use of the inductive method in France is mainly connected with the delay in the migration of people from the countryside. "Christaller and Lösch worked out their theories in a country where population active in the agricultural sector amounted to 18% only; their theories were accepted in England (primary sector: 8%) and above all in the U.S.A. (primary sector: 21%), not so in France where the primary sector still covered 37% of the active population" (Racine and Reymond, 1973, page 108). This explanation may seem extremely simple but the corresponding datum for Italy was down to 17·2% as late as in 1971!

Apart from considerations on the usefulness of statistical models and indicators in general, it is certain that this new approach has led to a sort

(1) Published by Ohio State University Press.

(2) Published by Pion Limited, London.

of inurement to reality. By concentrating efforts on the phase of elaboration, in many cases the geographer tends to use available statistics, rather than undertake field work. Such a method is justified by the saving of time and by the fundamental consideration that the use of models with a mathematical basis may be profitable only if the model (or the calculation scheme) can be applied to a space that is different from the one in which it had originally been worked out. It is clear that this is possible only by resorting to numerical data collected with standardised methods, otherwise we would fall back to a sort of 'subjective quantification'. With this term we usually mean (Corna Pellegrini, 1974, page 168) the procedure of "attributing a weight to any phenomenon or specific typology, which is conventionally fixed by the researcher and varies from zero to a maximum value according to the intensity of manifestation of the phenomenon itself".

In any case, whether the modern geographer should undertake the task of collecting information himself or not, it is certain that the amount of knowledge on the world around us has reached such levels that it strongly affects the method with which the researcher carries out his work. When the researcher had to observe and analyse the 'unknown' (in the times of exploration and of the voyages of discovery), research on the spot (and, in addition, the inductive method) was a typical and characteristic aspect of geography. Now, in contrast, in a time when science is increasingly based on the steady foundations of acquired knowledge and on a great amount of available data (collected day after day), the use of written sources and of the deductive method becomes more necessary. As a result there is an urgent need to have an organic structuring of the bulk of data acquired by the expert (this is the very task of synthesis-sciences).

Jaja himself had reached such a conclusion in what was the first work of geographical methodology published in Italy. In 1910 he wrote: "as to the so-called 'general geography', I tend to believe that it is still a new-born science. We shall have a general geography, that is the collection and knowledge of fundamental laws governing the interdependence of earth's phenomena, only when geography reaches its second phase, that is—as I said—the deductive phase of potentiality explication of existing laws—if existing—or of their application to the specific cases. So far geography has been an inductive science, in its first phase" (Jaja, 1910, page 55).

The reevaluation of field surveys

As a consequence of the fact that researchers tend to concentrate their attention on instrumental elements, there is the risk of losing contact with reality. The habit of resorting to 'official data' may in fact lead to conformity and intellectual dependence on the power ruling over statistics and 'data banks'. This may represent a real obstacle to research, and not only in geography. In many countries, and Italy is one of them, official information is not up-dated. One example is offered by the Land Register, which dates back to forty years ago. Another example is the evolution of official census techniques, where the municipal data used in urban researches has generated the problem of under' and overbounded cities (Grytzell, 1963). Moreover the structure of society constantly changes. The growing discrepancies between, on the one hand, employment and unemployment statistics, and, on the other, data relating to the 'black manpower' market, recently led in Italy to the codification of the concept of 'submerged economy', a category of activities which is very near to productive forms typical of the Third World. In such situations the resort to the *computer*, as Racine and Reymond did, as a fundamental element of a hypothetical 'global method' able to reproduce a space structure in detail, is a sort of escape from reality.

With humility we have to admit that there is no short-cut in the field of geography and that no formula, though complex or ingenious, will ever be able to mechanise our work. As stated by Bonetti (1969) we must not confuse means with goals. The creation of models in fact should always be connected with field investigation. For example, Pagnini (1971) inspected all twelve thousand shops of the Persian Bazaar, at a time before it had been affected by Western influences (some years later, the Shah ordered its demolition) and, after having reproduced them on an urban map, she tried to analyse their structure with the aid of a computer. This example demonstrates that if the geographer is supposed to carry out accurate bibliographic research, a careful investigation on the spot, and lastly an up-dated statistical elaboration of acquired data, then his work will be quite hard!

However, it is not true that there are contrasting methods for geographic research. The fundamental opportunity which the 'new geography' presents does not lie in its method (in the sense of cognitive process), but rather in the choice of the data to be collected and evaluated. In other words, the alternative is between data which refer directly to the object of research, and data which allow us to get to it through a mediation and thus allow us to assess its general framework and origins. A typical example is afforded by the particular category of functional regions represented by market areas. These may be identified by carrying out surveys of the consumers and/or of the selling units, and through the calculation of gravitational areas created on the basis of centrality indicators. These are two aspects of a single reality which the geographer should tackle simultaneously in his research; and this once again calls for an encyclopaedic kind of man with enormous working capacity.

The problem gets even more complex when we seek to study dynamic elements. In fact, the traditional idiographic or *pedibus calcantibus* form of research calls for the individualisation of the 'geographical facts' inherent in the landscape and therefore somehow characterised by a prolonged stay in the environment. A glacier, a forest, a country landscape, a network of settlements are all sensitive facts, elements of the territory which are conceived as static reality. But when, as Dematteis (1973) said, our interest is focused onto the "latent structures of society", or onto the complex network of relations which support a given environmental reality, then the research has to cover a whole range of values which—at first sight—could be considered as 'nongeographical', since they have no unequivocal manifestation in space. For example, commuters' movements are a dynamic reality which can be 'fixed' in space with pictures of dozens of trains and hundreds of buses underway, but to be wholly understood there must also be a description of the whole region with its dwellings, its work places, its centres of study, its recreational areas, and the complex network of traffic connecting them in different ways and at different times.

The study of all these flows has led to a conceptual revolution in geography, not least because it calls for the use of a mathematical form of expression. This has created a problem for defining what the field of action for the geographer should be. I am referring to the sense of uneasiness expressed almost twenty years ago at the Geographical Congress in Trieste by Caraci. Faced with the wide range of problems raised by Migliorini (1962) with his report on "Internal Migrations and Territorial Movements of the Italian Population", Caraci (see Migliorini, page 415) wondered: "Should we even try to understand the frequent, if not regular movements which occur in towns, mainly in large towns, for recreational purposes (stadia, cinemas, theatres, etc.)?". Only two years later the first work by Hägerstrand (1963) on time-geography was published. Today these subjects have been accepted even by the Italian establishment of geographers, as evidenced by the review by Gentileschi (1978) and the general interest in social geography. Older geographers, however, still consider geography as 'landscape science' (Biasutti, 1947).

The geographer's field of action

Such examples bring us face to face with the first and perhaps the major obstacle hindering the development of geographical research: *self-limitation*. The fact of excluding a given field of phenomena *a priori*, simply because it is not considered to be noteworthy, is a risk which we run more or less continually. After having somehow chosen our field of action, the typical problem arises of how to choose the information on which the study is to be based. A major development in this area is data collection supported by the aid of artificial satellites. This rapid progress has brought sudden changes in the field of theoretical problems with which the researcher has to cope. Available computers, operating in real time and able to process quite complex data at high speed, together with cameras on board spacecraft are able to monitor specified areas twenty-four hours a day. As a result, artificial satellites will be able to carry out a considerable amount of work in the field of geography (see Marino, 1979; Sabins, 1978).

Such a triumph in technology allows us to have the whole earth reproduced on a smaller scale. This will increasingly widen the geographer's field of action. An almost complete availability of data, the possibility of taking pictures—in the broadest sense of the word—of the Earth's surface at regular intervals, if not constantly, will free the geographer from a considerable amount of work in the phase of data collection. In addition, there should be an increase in data quality from this kind of surveying: we may finally be able to have all the data required for some types of quantitative analyses. This means that in the future the task of the geographer can in some cases be much more directed to the interpretation of reality, rather than to its description. He will not analyse the elements of the geographic space and their distribution any longer, but rather their connections, their interrelations, and their changes. Geography will not be characterised as much by the study of the earth's surface, but more by the analysis of the processes occurring on it.

Of course data availability from satellites will not be the same for all the categories of phenomena which are the subject of geographical studies. It will certainly suffice for geomorphology and climatology, for botanical geography and, to a considerable extent, for zoological geography. The study of migration, for example, is already commonly carried out with satellite data (for reindeer, wolves, fish-shoals) both by means of direct surveying and through radio signals emitted by automatic devices somehow attached to the animal in question. This type of example suggests that there are two categories of data limitation which affect geographical research: one spatial, and one temporal. Each is discussed in turn below.

Spatial data limits to geographical research

A first spatial limit is to geographical research cases in which data must be obtained by means of interviews (both individual and group interviews), questionnaires, etc. Here a distinction must be made between methods and goals. The recent trends in psychological geography have sometimes represented a step back to the past, insofar as they eventually aim at identifying material—and visually perceivable—elements in the environment. But if we interview people on their movements, on their incomes, on their favourite political party, on their shopping habits, we collect data on a series of phenomena dynamically related to the territory, but not on the territory itself. In other words we deal with elements affecting the environment in an ecological sense: the behaviour of shoppers makes it possible for the shop-keeper to retain his position in space. The analysis of the latter allows us to determine whether, in that area, there is a surplus of expenditure capacity which is not catered for by the existing retailing facilities, and therefore to foresee the possibility that the points of sale may increase in number within a certain time. This is a survey of an indirect type.

A second spatial limit is met when the choice between direct and indirect survey is not free owing to time or money factors. A major example of what can be considered as such a space limit to research is to be found in the town of Trieste. Here, for about fifteen years, a whole retail-trade sector has developed, which is orientated to Yugoslavian customers (Battisti, 1979). Hundreds of thousands of people come to the town to get supplies of consumer goods which, for the greatest part, are then sold in the Balkan hinterland. For obvious reasons a survey on the points of departure of these people cannot be proposed. The alternative of a direct investigation of dealers in Trieste often results in threats to the physical health of the investigator himself. Moreover, the flow of money fed by this trade is not comparable at all with the number and the outer aspects of the points of sale, which often consist of simple stalls. Similarly the customers, apart from some rare exception, are certainly no more willing to be interviewed. For this reason a systematic field investigation must be limited largely to counting the cars coming from beyond the border and, if possible, recording number-plates.

Temporal data limits to geographical research

Even if data are available, both qualitative and quantitative, there are still problems to be solved. Once the problems related to *space limits* and resulting from the inherent nature or from the localisation of the analysed facts and phenomena are overcome, we are faced with the *temporal data limits* to research. These act in different ways. One problem that arises is the choice of the right time intervals for surveys. Continuous surveying is practicable in a limited number of sectors, for example, in meteorology and in astronomy. For the social sciences, however, this is not usually possible, and creates considerable problems for geographical inference.

An example will make this clear. The survey of urban spread in the area of Milan, carried out by comparing land-use maps with the employment structure of the population (Battisti, 1977), proved that there is a clear connection between the two phenomena. In fact, the localisation of nonagricultural population shown in the 1961 census reflects the distribution of nonagricultural employment in the 1951 census. In this case we have a good example of phase displacement between the space extension of social and demographic phenomena and formal geographical facts. This phase displacement is not a chance phenomenon. Changes in human behaviour, such as finding a job in another municipality, occur first as phenomena of individual mobility, and, in a second phase only, are consolidated on the ground as material facts which will appear in a census. Those who work far from home at first hope to find a job in their neighbourhood, and only after a certain period of time do they change their residence. Similarly, the decline of agriculture in the surrounding areas of fast developing towns is not immediately followed by the urban transformation of the previously agricultural area. The 'way of life' cannot stem out of nothing, a certain 'technical' time is required.

Conclusion

The spatial and temporal limits to geographical research create a number of important dilemmas for quantitative geography. Most important of these is the delay in producing data which undermine the relevance of research. For example the ten-yearly rhythm with which censuses are carried out does not make it possible for us to formulate relationships in precise terms. In the case of the rural exodus which occurred in Lombardy from 1945 to 1951, the first survey that could be carried out was the agricultural census of 1961 (when the process was well under way). To this fifteen-year period we have to add the time required for scientific elaboration. Analysis of rates of change was possible only with the 1971 census, whose temporary results were available in 1973. The survey by Battisti (1977) was concluded in 1975 and the final publication is dated 1977. A delay of 6 years, at best, in addition to the original 25 years since the start of the process may well support the assertion that geography in this case is historical!

Such a conclusion raises unpleasant questions about the validity and the function of scientific research. To the above technical times we have to add the period of time required for the spread of the research among those who need to use it. The knowledge of present facts is the foundation on which our decisions for the future are based. But we have to base our future planning on a perception of the present day which in fact is but a synthesis of the past. Only the past is knowable, but not the future; not even the present is fully knowable. And this principle of 'incomprehensibility of the universe' in real time must set an ultimate limit to the potential of geographical research.

References

Battisti G, 1977 "Contributo alla delimitazione territoriale della regione-città milanese" in *Milano, Megalopoli Padana, Valli Alpine. Studi sulle reti urbane* Ed. G Corna Pellegrini (Pàtron, Bologna), pp 179–216

Battisti G, 1979 *Una regione per Trieste. Studio di geografia politica ed economica* (Istituto di Geografia della Università, Trieste)

Bianchi E, Perussia F, 1978 *Il centro di Milano, percezione e realtà. Una ricerca geografico-psicologica* (Unicopli, Milan)

Biasutti G, 1947 *Il paesaggio terrestre* (Unione Tipografico-Editrice Torinese, Turin)

Bonetti E, 1969 "Modelli e analisi spaziale" *La Geografia nelle scuole* **15**(1) 161–178

Brusa C, 1978 *Geografia e percezione dell'ambiente: Varese vista dagli operatori dell'ente pubblico locale* (Giappichelli, Turin)

Brusa C, 1979 *Evoluzione di un'immagine geografica. Il Varesotto turistico secondo i Baedeker, le Guide del Touring e alcune fonti locali* (Giappichelli, Turin)

Burton I, 1963 "The quantitative revolution and theoretical geography" *The Canadian Geographer* 7 151–162

Corna Pellegrini G, 1974 *Geografia e politica del territorio. Problemi e ricerche* (Vita e Pensiero, Milan)

Dematteis G, 1973 "Metodi moderni per lo studio della geografia urbana e regionale: rassegna critica e proposte" in Comitato dei Geografi Italiani, *Studi su: città, sistemi metropolitani e sviluppo regionale* I Quaderno *Sul metodo della ricerca* (Pàtron, Bologna) pp 1-60

Gentileschi M L, 1978 "Un'approccio per lo studio delle migrazioni in una prospettiva geografico-umana" in *Italiani in movimento* Ed. G Valussi (Grafiche Editoriali Artistiche Pordenonesi, Pordenone) pp 29-47

Grytzell K G, 1963 *The Demarcation of Comparable City Areas by Means of Population Density* Lund Studies in Geography, Series B, Human Geography, 25 (Gleerup, Lund)

Hägerstrand T, 1963 "Geographic measurement of migration: Swedish data" in *Les Déplacements Humaines* Ed. J Sutter (Entriens de Monaco en Sciences Humaines, Monaco) pp 61-83

Jaja G, 1910 *La geografia come ramo della logica dei metodi* (Lapi, Città di Castello)

Marino C M, 1979 "Un fotografo nello spazio" *Geodes* **1**(5) 64-75

Migliorini E, 1962 "Migrazioni interne e spostamenti territoriali della popolazione italiana" in *Atti XVII Congresso Geografico Italiano* (Istituto di Geografia della Università, Trieste) **1** 363-416

Pagnini M P, 1971 *Strutture commerciali di una città di pellegrinaggio: Mashhad (Iran Nord-orientale)* (Istituto di Geografia della Università, Trieste)

Pagnini M P, 1975 "Immagini mentali e documentazione cartografica" *Bollettino dell'Associazione Italiana di Cartografia* **35** 7-16

Racine J B, Reymond A, 1973 *L'analyse quantitative en géographie* (Presses Universitaires de France, Paris)

Sabins F F, Jr, 1978 *Remote Sensing. Principles and Interpretation* (W H Freeman, San Francisco)

Vlora N R, 1979 *Città e territorio* (Pàtron, Bologna)

Quantitative methods in Dutch geography and urban and regional planning in the seventies: geographical curricula and methods applied in 'spatial research'

F M Dieleman, D Op't Veld

Introduction

In this chapter an attempt will be made to give some, occasionally fairly personal, impressions about the development and use of quantitative methods in Dutch geography and urban and regional planning during the 1970s. The chapter consists of three main parts. In the first part we give a number of impressions about the development of human geography in Holland during the last ten years. In the second part we shall describe the gradual development of the training in quantitative methods and research methodology at the five Dutch departments of human geography during the 1970s. Only limited attention is paid to the curricula in related disciplines such as regional economics and urban and rural sociology. Finally, in the third part of the chapter we attempt to throw light on the use of quantitative methods in, what we have called arbitrarily, 'spatial research' or 'spatial sciences' in the Netherlands during the 1970s. Unlike the discussion of the development regarding training facilities, the treatment of this topic is not restricted to the contributions of human geographers only. The main reason for extending attention to other disciplines is that in applied spatial research geography does not function independently, but is part of a complex whole of interlinked disciplines showing considerable overlap. Disciplines like regional economics, transportation research, rural and urban sociology, and physical planning also make contributions to spatial research, especially where quantitative methods and mathematical model building are concerned. Moreover, this broadening of scope offers the opportunity of an interesting comparison of curricula in human geography and spatial research activities in general.

Quantitative techniques and Dutch geography

The introduction of advanced quantitative techniques and social science research methodology into Dutch geography dates back roughly to 1970. In Anglo-American geography the new emphasis on quantitative methods was just one aspect of more general paradigm changes (Johnston, 1979). However, for Dutch human geography this is far less true. It is easy to find Dutch geographers in the late sixties who specify regional geography as the ultimate aim of geography, and view geography as an essentially idiographic science. Keuning (1969, page 135) for example states that:

> "Nomothetic thinking ... has attracted many supporters in recent years, partially due to the influence of the strong application of mathematical

> and statistical-mathematical methods in ... for example economics and sociology. However ... it seems to me that human geography ... in its pursuit of the individual aspect and the unique aspect ... must nevertheless be viewed as a branch of science which is by definition idiographic."

It can safely be assumed that views like this were prevalent among a significant portion of Dutch geographers around that time. Nevertheless it would be incorrect to conclude that the introduction of quantitative methods around 1970 is evidence of a reaction against a mainly idiographic human geography in the Netherlands. Heinemeyer (1977) points out correctly that the involvement of Dutch geography with a type of research along the more general lines of social science research methodology is of a much longer standing. Much of the work long before 1970 was of a more thematic than regional nature, as for example in urban and rural geography; this point is substantiated in publications by Heslinga (1974; 1980). Even Keuning, whose explicit view on the nature of human geography we mentioned above, was partly involved in research of a much more nomothetic character, like his work on the identification of a hierarchy of urban centres in the Netherlands (Heinemeyer, 1977).

Moreover, long before 1970 Dutch geography had established close contacts with urban and regional planning. A considerable number of human geographers were working in 'economisch-technologische' and 'sociografische' institutions in the Netherlands and received a research training oriented towards these studies.

The fact remains, however, that Dutch geography was fairly unaffected for a long time by the turbulent developments in the field in the United States. This is certainly strange because in the international journals these developments were clearly evident, and Dutch geographers had always been actively inspired by ideas from abroad. If Dutch geographers drew any inspiration from work in the United States in the early sixties, it was more from the human ecologists and the work, for example, of Stuart Chapin than from 'spatial geography' or 'quantitative geography' (Heinemeyer, 1977).

It was only at the end of the 1960s that a number of lecturers were appointed at various university departments of geography who were given the task of mastering the quantitative methods and the positivist-oriented scientific methodology. Our impression is that these appointments met with general consent. Even geographers who were not attracted at all by a quantitative type of approach to geographical problems tended to feel that at least some Dutch geographers should pay more attention to these matters. Anyway, it was one important step in a rapid reorientation of Dutch geography in the 1970s towards Anglo-American developments in the discipline. One of the manifestations of this reorientation was a seminar on "Mathematical methods in human geography" in 1971. Hauer (1971) discussed the 'new geography' and advocated abandoning regional geography as the ultimate aim of the discipline. The pursuit of systematic principles, the application of general social science methods of investigation, and the use of quantitative techniques were recommended. At the seminar, a great amount of attention was devoted to factor analysis. In Dutch geography this technique was still quite new, and it shows how limited at that time the knowledge about quantitative methods still was. It is also indicative of the great interest in especially quantitative methods, and less in research methodology, that arose in Dutch geography in the 1970s, a point which will be illustrated below.

It is difficult to answer the question why little attention was paid for so long a time to the paradigm changes in Anglo-American geography. There is no written discussion in Dutch geographic literature on this theme, and so one can rely only on personal impressions. Perhaps the following issues played a role:

(a) Until the seventies, the influence of French geography, where the regional monograph was quite important, remained significant. Lectures on the nature of human geography, for example, often included a review of the opinions of well-known German and French geographers about concepts and methods which were especially meaningful for regional geography.

(b) Many geographers had a humanities background, and their interest was fairly history-oriented. Understandably, they had very little desire to start working with unfamiliar quantitative methods and to devote their attention to a general scientific research methodology.

(c) The computer was introduced into the Netherlands at a relatively late date, and it was not until the seventies that the universities had easy access to it. This is probably one of the reasons why the use of quantitative methods did not develop rapidly.

(d) Changes in the way a field of science is practiced are partly a question of generation differences, the new generation replacing the old [as stated by Taylor (1976), Johnston (1979), and Heslinga (1974)]. Until the seventies, Dutch professors were in a position to shape the academic training in human geography to fit their own personal preferences, but this changed when the staffs increased in size and the staff units ('vakgroepen') began to take over many of the professors' responsibilities. Moreover, a generation of professors retired in the early seventies, and the arrival of a new generation was one of the factors leading to a different approach to the field of geography.

One of the most striking aspects of Dutch human geography in the period from 1970 to 1980 was the great expansion of the university staffs and the increase in the number of students which made this expansion possible. Heslinga (1980) spoke of a "miraculous multiplication". Until 1950, there were only two universities where students could major in human geography, only a few professors, and an extremely small number of staff members. An institute of human geography was then established in each of three other universities, one in 1950, one in 1959, and one in 1961. Between 1959 and 1979, the number of geography professors and lecturers rose from four to twenty (Heslinga, 1980), and in the past decade the number of staff members tripled or quadrupled. At each of the five universities, the human geography departments now employ between forty and sixty staff members.

Without a doubt, this spectacular growth of the field of human geography at the universities can be viewed as the driving force behind the changes in the actual practice of the discipline and in the curriculum for the students. Regional geography and historical geography, with a considerable interest in the study of the landscape, continued to contribute significantly to the activities in the discipline. In the seventies, however, stimulated by the increase in the number of staff members, other specialisations such as urban and rural geography, economic geography, geography of under-development, educational geography, and urban and regional planning expanded more rapidly.

With the growth of these specialisations within Dutch human geography, there was also a clear shift in orientation towards the Anglo-American literature. The Anglo-American approach to geography was incorporated into the Dutch counterpart almost automatically, and certainly without any fuss. Students were amply confronted with quantitative methods and empirical-analytical research methodology. In geographic research and education the thematic orientation became more dominant, and the importance of regional studies declined. The stages which Johnston (1979) distinguished in the development of Anglo-American geography can, to a certain extent, also be found in Dutch geography in the 1970s. For example, in the early 1970s attention was devoted to geostatistical methods (ter Hart, 1968), social physics (Hakkenberg, 1969), factor analysis (Dieleman, 1971; van der Knaap, 1971) and the rank-size rule (Deurloo, 1972). Within less than a decade, similar to recent trends in Anglo-American geography, there were already the first reactions to the partially empirical-analytical—if not positivist—orientation in Dutch geography. Kouwenhoven (1979) felt drawn towards such geographers as Buttimer, Yi-Fu Tuan, Entrikin, etc and chose to follow the lines of 'humanist' geography. Jansen (1980) preferred a phenomenological approach in the spatial sciences and expressed his opposition to geography with a logical-positivist tint.

So, within a decade, a reorientation took place in Dutch geography which bore many similarities to the changes in the Anglo-American counterpart in the last three decades. The great majority of human geographers at the Dutch universities now feel closely involved with the profession as practiced in America and England. Moreover, the younger generation of human geographers—and there are a great many of them—have received an academic education with a clear Anglo-American accent.

Nevertheless, we still have the impression that Dutch human geographers are less intrigued in general by the use of quantitative methods, empirical-analytical research methodology, and model building than their Anglo-American colleagues are or were. There are three main reasons for this:
(1) In urban and economic geography the orientation in the 1970s was mainly towards the Anglo-American practices, and the use of quantitative methods expanded rapidly in these fields. In historical geography, the study of landscape, the geography of underdevelopment, and educational geography—all of which play an important role in the Netherlands—this was much less the case. For example, some geographers continued to devote their attention to regional studies, and a large number of books were written about a wide range of countries. In the field of historical (urban) geography, Deurloo and Hoekveld (1981) are an exception to the rule regarding the use of quantitative methods. Consequently, when we discuss below the use of quantitative methods in spatial research, we can, as far as human geography is concerned, restrict our attention to research on the Western world (primarily in the fields of urban-and-rural and economic geography).
(2) In the past decade, Dutch human geography, particularly urban, rural, and economic geography, has retained close links with urban and regional planning. Many geographers are employed in the field of urban and regional planning. The research done by urban, rural, and economic geographers was and is largely centred around problems which are considered to be significant within the framework of urban and regional planning in the Netherlands. This will probably be even more so in the future, once the new way of financing university research—more dependent on non-university financial sources—becomes more widespread. This is why research into the above mentioned specialisations of geography is rather practically oriented: quantitative methods and mathematical models are used and developed if considered necessary, but the majority of the work remains problem oriented. Extensive experiments with quantitative methods or extremely theoretical investigations are fairly rare and mainly practiced by staff members appointed to teach quantitative methods.
(3) This point is again related to the increase in staff concomitant with the expansion of Dutch human geography in the past ten years. Although many of the recently appointed younger staff members have had an Anglo-American oriented education, this was hardly the case with the professors and lecturers appointed during the period. When the increase in the number of chairs for human geography gave the universities the opportunity to appoint new professors and lecturers, the generation of human geographers who had received a training in the use of quantitative methods was still too young to fill these chairs. Consequently, in the large majority of cases, professors and lecturers were appointed with hardly any inclination to apply new quantitative techniques. In spite of the fact that most of them are familiar with the Anglo-American literature, they have

less affinity with these techniques and model building. This is one of the reasons why, in Dutch human geography, there is still a great diversity of opinions regarding the actual practice of the profession. We have the impression that writing a book about a country or a region has remained just as acceptable as conducting research with essentially theoretical or quantitative aspects [although Kouwenhoven (1979) and Jansen (1980) seem to have a different opinion on this matter]. Hopefully the relative tolerance and mutual respect that exists now will not disappear in the debate on the conduct of geographic research that is developing at the moment.

Just as has been the case with the other specialisations of geography, the field of research methods and techniques in human geography has profited from the increased interest in geography as a university subject. At a rapid pace and, in our view, with widespread approval, the various university departments of geography have introduced courses in statistics, quantitative methods, and research methodology. In the past decade, there has been a great increase in the number of university staff members who devote a large part of their time to quantitative methods, which has led in turn to an increased amount of knowledge about this field. At the departments, three to five staff members are in a position to teach and do research involving quantitative methods. A number of chairs of geography have been reserved for this specialisation. Added to the staff members who are specialised in philosophy of science and the history of the profession, this means that at the five human geography departments alone, there is considerable manpower engaged with the technical, methodological, and theoretical aspects of human geography.

If one analyses the use of quantitative methods in spatial research in general—as we shall attempt to do later in this chapter—one sees that, in addition to the work of human geographers, the contribution of related specialisations is considerable. In this respect, special reference must be made to regional economists (particularly with regard to mathematical model building), the subdepartments of the Universities of Technology in Eindhoven and Delft, the Research Centre for Physical Planning (PSC)-TNO in Delft, and sociologists such as van Doorn and van Vught (1978) and van de Vall (1980).

Quantitative methods in human geography curricula

If we take the curricula at the five departments of human geography at the Dutch universities to be an indicator of the geographers' knowledge of quantitative methods then, as might be expected, the 1970 knowledge level was quite low. Table 10.1 shows when, and for which groups of students, various quantitative methods were introduced into the curricula at the five universities. The dates when certain branches of statistics or the philosophy of science were introduced into the curricula do not give a fully accurate picture of the gradual manner in which the various subjects

Table 10.1. The introduction of statistics, quantitative methods and research methodology into the curricula of the five Dutch University departments of human geography in the early 1970s. Source: Dieleman, 1975.

University	Descriptive statistics, probability distributions, regression and correlation analysis	Factor analysis, geostatistics, simulation, network analysis, Markov chains, etc
Groningen	Has been included for a long time in the curriculum of graduate[a] students, who attended lectures on the subject at the economics department. Given for the first time at the human geography department itself in 1974. In 1974, statistics was also included in the first-year and second-year curricula.	Since c. 1970/71, factor analysis has been included in the curriculum for certain groups of graduate students. An introduction to matrix algebra was provided for them by the department of economics.
Utrecht	Since c. 1959/60, Rijken van Olst (first volume) and Gregory have been studied by undergraduate students. Not in detail, however. Since c. 1966, extensive lectures on statistics have been included in the curriculum for undergraduate students.	Since c. 1970, lectures in most of the above mentioned techniques were included in the curriculum for some of the third-year students, in rather great detail.
Nijmegen	Since 1971/72, lectures have been given on statistics to undergraduate students. Before then Rijken van Olst was studied, although no questions were asked about it on examinations.	Since 1971/72, the use of these methods in geography has been dealt with by a small group of graduate students. No separate lectures on the subject.
Amsterdam (University of Amsterdam)	Since the 1950s seminars have been held for first-year and second-year students where research methodology and statistics are both taught. Since 1968, there has been a separate course in descriptive statistics, and since 1970 there has also been a course on inferential statistics.	Since c. 1971, graduate students have been in a position to become familiar with them. No separate lectures.
Amsterdam (Free University)	Since c. 1968, they have been dealt with superficially for postgraduate students. Since 1969, lectures on statistics have been given to undergraduate students.	They have been taught superficially to approximately half the graduate students since c. 1968, and in greater detail since 1970/71.

[a] The terms undergraduate student and graduate student are used here for the Dutch 'prekandidaatsstudent' and 'doctoraalstudent'. These are only the rough equivalents of the words, since the differences between the Anglo–American university systems and the

Table 10.1 continued

Use of the computer	Philosophy of science	Research methodology
Since c. 1971 for a number of graduate students.	Ever since c. 1970, in addition to a survey of the history of the geography discipline, the philosophy of science (De Groot is dealt with) has also been included in the curriculum for a group of graduate students.	Since c. 1970, attention has been devoted to this by a group of graduate students specialising in research. In the third study year, methodology is also dealt with in a practical course involved with field work.
Since c. 1970, a number of graduate students have been given the opportunity to become familiar with it. From 1974 on, undergraduate students have been offered this opportunity as well.	Since c. 1965, some attention has been devoted to it, although not in great detail. Since c. 1972, it has been dealt with in the first study year.	Since c. 1966, methodology has been dealt with in the framework of the thesis written in the final year. Since 1970, undergraduate students have also devoted an increasing amount of attention to it.
From 1972 on, a small group of graduate students were briefly introduced to it.	Since 1970/71, attention was devoted to it starting in the first study year, and all the students devoted a great amount of attention to it throughout their training.	Since 1970/71 undergraduate as well as graduate students have devoted a great deal of attention to it.
In 1970, lectures were introduced for graduate students.	Ever since c. 1960, it has been dealt with on the reading list for undergraduate students. In 1966, second-year students were more or less confronted with it by means of lectures on 'quantitative methods'.	For a long time, a large amount of attention has been devoted to these methods by undergraduate as well as graduate students.
In 1974 it was introduced for approximately half the graduate students.	Ever since c. 1966, graduate students have had to study a number of books about it. In c. 1972/73 it was introduced to the first-year curriculum.	Ever since c. 1968 extensive attention has been devoted to this topic in the framework of the third-year research-training programme.

Dutch one, where the 'kandidaats' taken midway is not an actual 'graduation', make it impossible to translate them with more precision.

were introduced. Usually, there was a gradual development. First, a suitable-looking book was selected to be read by the students, though no questions were asked about it at examinations (Gregory, *Statistical Methods and the Geographer*, was sometimes used in this way). This situation gradually changed either because a staff member acquired the necessary knowledge or because a new staff member possessing this knowledge was appointed. Consequently, the years given in table 10.1 must be seen in this light, and not as absolute dates.

Table 10.1 makes clear that prior to 1965 there were no serious courses in statistics at the geography departments of the Dutch universities. At most of the departments, basic statistics (column 1) were not introduced as a separate course until 1970. It is no wonder, in view of the late start of courses in statistics being given to large groups of geography students at an early stage of their studies, that the introduction of more advanced quantitative methods (column 2) has been even more recent. Moreover, initially, these courses are often given to a few of the graduate students and the subject is dealt with in a superficial manner. For the large majority of Dutch geography students, the computer (column 3) remained a very vague concept up until the mid-1970s. It is still not something they can utilise to its full extent and use independently.

In columns 4 and 5 of table 10.1, facts are also given with respect to the introduction of the philosophy of science and social science research methodology into the human geography curricula. The introduction of these subjects was an even more gradual process than it was in the case of statistics and quantitative methods. It is obvious that for a long time courses had been given in the methodology of human geography, and students had been taught to conduct research while writing their theses. However, given the interest in regional description, these parts of the curriculum partly consisted of dealing with the concepts and methodology judged as being fitting and proper for regional description. Consequently, the philosophy of science was initially not touched upon in these lectures. Nor were themes dealt with such as the empirical cycle, the operationalisation of concepts, and the testing of hypotheses. Gradually, however, these topics were either added to already existing parts of the curriculum, or certain parts were adapted in such a way as to fit into the basic lines of general research methodology. This is why the dates given in columns 4 and 5 of table 10.1 are only a rough estimation of when this development took place. However, despite these reservations concerning dates, it can be noted that no real attention of any importance was devoted to social science research methodology prior to 1970 (with the exception of the University of Amsterdam).

Roughly speaking, the teaching of research methodology and quantitative methods at the five human geography departments was in a preliminary stage up until 1975. A number of staff members, often a mathematician in conjunction with a geographer, became familiar with the basic material and tried to find a way to integrate it into the human geography curriculum. Since approximately 1975, lectures on the basic principles of descriptive and inferential statistics and research methodology have become fixed and integrated parts of the curricula of all the five departments, as is clear from an inventory drawn up by the "Working-group on Mathematical Geography" in 1978. After 1975, basic statistics and the fundamental principles of the quantitative methods came to demand less attention, so that it became possible to devote more attention to the improvement of graduate courses. Consequently, after 1975 a wider range of subjects could be discussed and dealt with in greater depth. Partially this was also due to the fact that research had provided a greater amount of experience with the subjects involved.

If we compare the 1978 graduate curricula of the five human geography departments, then considerable differences in emphasis can be noted (see also Dieleman and Hauer, 1980). In the first place, these differences pertain to the various specialisations from which graduate students can choose. We have the impression that it is mainly the group of the students majoring in the geography of Western industrialised countries, economic geography, urban geography etc (many of whom are to be employed in the field of urban and regional planning in the Netherlands) who are intensively confronted with quantitative analysis. This seems to be much less the case with students majoring in historical geography, political geography, educational geography, and geography of underdevelopment. At Nijmegen, Utrecht, and Amsterdam (Free University), special chairs in quantitative methods and research methodology make it possible for graduate students to major or minor in these subjects.

A second difference in emphasis between universities pertains to whether it is research methodology or techniques which occupy a more central position in the department curricula. In Nijmegen (at least in 1978), research methodology seems to dominate; a relatively small amount of attention is devoted to quantitative methods, whereas basic analytical tools like correlation analysis, regression analysis, and factor analysis occupy a large part of the curriculum. At departments where the techniques are emphasised more strongly than research methodology, the above mentioned quantitative methods also receive a large amount of attention. In addition, however, input–output modelling, gravity models, entropy models, Markov models, linear programming, network analysis, and so forth also occupy a more or less important position. It is almost self-evident that, as a result of this development of the graduate curricula after about 1975, the use of the computer has also been introduced; many graduate students learn to work independently with standard programmes such as SPSS. Thus, by the end of the 1970s all the human geography departments were offering a diversified training programme on quantitative methods and research methodology. Nevertheless, a number of shortcomings can be noted.

For example, neither cluster analysis nor multidimensional scaling (Everitt, 1977) received significant attention in the curricula. To a certain degree there was also a dearth of attention devoted to spatial interaction models, allocation models, multicriteria analysis, and so forth. In spatial research in the Netherlands these methods do play an important role, as will become clear in the next section, but they are only dealt with very briefly in human geography curricula.

Quantitative methods in Dutch 'spatial research': main accents and prevailing trends during the 1970s

Data collection

In order to acquire an impression of the main accents and trends in the application and development of quantitative methods in Dutch 'spatial research' in the seventies, some quantification is called for. First of all a yardstick to measure scientific activity is needed. Obviously, an attractive, operational measure concerns the number of publications dealing with quantitative methods. As an operationalisation of scientific activity, the number of publications clearly has some shortcomings. For instance, it does not take into account the sometimes considerable differences in scientific scope and quality that exist between publications. Nevertheless, the number of publications has been used here because of its indicative value and ease of measurement. In addition, it is necessary to define the set of titles that are relevant as exponents of quantitative 'spatial science'. Here, 'spatial science' is thought of as the diversified subgroup of social sciences and related applied disciplines, characterised by an essentially locational and/or regional approach and interest in their research, such as human geography, regional economics (including spatial econometrics), urban and rural sociology, transportation sciences, and physical planning. Obviously, this is not exactly a strict and unambiguous scientific definition of the subject, but is an attempt to distinguish relevant from irrelevant titles. The other part concerns the treatment on an advanced quantitative level which was used as a yardstick to select within the field of 'spatial science' the subgroup of publications dealing with quantitative methods.

The general character of these guidelines inevitably introduces some subjectivity into the selection of relevant titles. But, we are fairly confident that it meets the requirements of representativity and reliability that apply here.

Some 457 different titles published within the period from 1971 to 1980 (approximately May) were collected (table 10.2). To arrive at an

Table 10.2. The number of publications per annum dealing with quantitative methods in Dutch spatial science in the period 1971–1980.

Year	'71	'72	'73	'74	'75	'76	'77	'78	'79	'80 (until May)
Number of publications	8	25	16	50	62	63	62	52	97	22

Table 10.3. The number of times quantitative methods have been dealt with in Dutch spatial science in the period 1971–1980, classified according to type of method and thematic category.

Method		(0) Textbooks	(1) Residential structure	(2) Residential location process	(3) Spatial structure of industry and offices	(4) Location process of industry and offices	(5) Spatial structure of service activities	(6) Location process of service activities	(7) Spatial choice of service activities	(8) Private transport	(9) Public transport	(10) Movement of goods	(11) Movement of messages	(12) Spatial structure (generally)	(13) Spatial interaction (generally)	(14) Spatial system (generally)	(15) Strictly method[a]	(99) Not classifiable	Total
Textbook	(0)	4															7	1	12
Analysis of variance	(1)																		
Contingency analysis	(2)																		
Correlation and regression	(3)		4	15	2		2		1	3	1			22	23	2	1	3	79
Discriminant analysis	(4)				1					1				1	2	1			6
Factor analysis	(5)		7		4	1	1		1					9	2	2	2		29
Cluster analysis	(6)		3	3	5		2							7	5	1	5	1	32
Multidimensional scaling	(7)					1			2					4	2		2	4	15
Geostatistical analysis	(8)				1		3							1			1		6
Graph theory	(9)									1					1	1			3
Trend-surface analysis	(10)													1					1
Monte Carlo simulation	(11)			1														1	2
Markov chains	(12)			4			1	1					2			2	1		11
Time-series analysis	(13)													3					3
Gravity and entropy modelling	(14)		2	15			1		15	6	3	3		8	58		2		113
Logit modelling	(15)			2					16	7	1	2			24		1		53
Spatial choice models	(16)								4										4
Spatial preference models	(17)								3										3
Behavioural models	(18)								2										2
Multiregional economic models[b]	(19)				3	4								7	2	9	5	3	33
Location–allocation techniques[c]	(20)													2					2
General urban and regional models[d]	(21)			2	1									3	1	25			32
Multiregional demographic accounting	(22)		4	2										1					7
Mathematical programming	(23)		4	1	2	3		1		1	1	3		4	8	6	11	21	66
Multicriteria analysis	(24)			1		2		1				1		1	1		13	26	46
Assignment techniques	(25)									1	2				9				12
Spatial autocorrelation	(26)			2	1			1						3		1	4	4	16
Shift-and-share analysis	(27)				3									4		1	2		10
Linear structural models[e]	(28)																		
Canonical correlation	(30)													2	1				3
Geocoding methods	(31)														1	1		2	4
Interdependency analysis	(32)													2	1				3
Cost–benefit analysis	(33)		1								1						1	3	6
Total		4	25	49	23	11	10	4	44	20	9	9	2	85	141	52	59	69	614

[a] No thematic relation; [b] input–output, attraction, multiregional growth models; [c] *P*-median; [d] Wilson, Batty-type; [e] Jöreskog's LISREL.

idea of the favoured quantitative methods, the main research themes in quantitative research, and the trends during the period, the titles were classified according to three aspects: method, thematic, and the time dimension.

The categories of the method dimension were developed starting from the basic and generally applicable methods of correlation and regression, analysis of variance, contingency analysis, factor analysis, and cluster analysis; added to this set were other more specific techniques frequently used to analyse past observations and simulate possible future conditions in 'spatial science' (trend-surface analysis, Markov chains, Monte Carlo simulation, gravity and entropy modelling, logit models, interregional economic modelling, etc).

Apart from techniques used for the analysis of empirical phenomena, there are also the decision aids used in economic and physical planning. The techniques concerned are cost–benefit analysis, multicriteria analysis, and to some extent mathematical programming. This resulted in the use of thirty-two specific quantitative methods that, together with a 'textbook' category, make up the method dimension used for classification purposes (see table 10.3). For the sake of brevity the characteristics of and differences between the various methods are not discussed in detail.

The set of research topics that make up the thematic dimension was arbitrarily established by combining the division of activities into the categories of home, work, and services, with the three organisational concepts of structure, interaction, and system. From these, together with three other categories, 'textbook', 'strictly method', and 'not classifiable', we arrived at the sixteen categories shown in table 10.3.

To obtain suitable results the following practical rules were adopted:
(a) Usually each publication was characterised by no more than one thematic category. However, if different techniques were used or dealt with in a publication, it was assigned to the respective method categories.
(b) In the case of a reader or a colloquium report each of the contributions was treated as a separate publication.
(c) Identical publications available at different places were introduced in the classification as only one relevant title. This rule was not applied though in the case of identical publications appearing in different languages, usually Dutch and English.

Only a selection of the collected and classified publications is included in the list of references of this chapter. The full list of publications can be found in Dieleman and Op't Veld (1981).

The annual number of publications

If we now turn to the results of the process of collection and classification of titles on quantitative spatial research, it is hardly surprising that the number of quantitative studies in Dutch spatial science shows a considerable increase throughout the seventies (see table 10.2). The pace of the increase, however, is somewhat exceptional. The number of quantitative studies published in 1979 amounts to six times the number published in 1973, which is not even the year with the smallest number of publications. Besides the important catalysing influence exerted by the "Vervoersplanologisch Colloquium" (Transportation Planning Research Colloquium) organised annually by a group of active researchers since 1974, this swift increase relates to the expansion of a number of existing scientific institutions in terms of manpower and activities as well as to the start of a few new research and training centres. In this respect one can think of the rapid development of regional economics as an independent scientific tradition at the Universities of Groningen and Rotterdam and at the Free University in Amsterdam, and the rapid expansion of the spatial research activities at the Netherlands Economic Institute in Rotterdam. Each of these institutes established its own series of Research Memoranda and clearly regional economists at these institutes contributed considerably to the increase in the number of studies published during the 1970s.

The development of groups of staff members specialised in quantitative methods and research methodology, sometimes causing the creation of an independent section on quantitative methods at the five Dutch geography departments, is another factor behind the fast growth of quantitative spatial studies. The contribution of academic geography to the number of quantitative studies is relatively small as compared with the abundant production in regional economics. Partly this reflects a different attitude towards the publication of research results between both groups of scientists.

In addition to the activities in regional economics and human geography, developments in the field of urban and regional planning were also important for the rapid growth of quantitative spatial studies in the 1970s. At the Universities of Technology in Eindhoven and Delft, and at the University of Utrecht, subdepartments concerned with research and education in urban and regional planning started and/or expanded rapidly in the seventies, paying considerable attention to quantitative methods. In addition the Research Centre for Physical Planning (PSC) came into existence in 1971 as part of the Dutch National Organisation for Applied Scientific Research (TNO). The importance of the latter institution during the seventies is illustrated by the fact that the method oriented periodical *Planning, Methodiek en Toepassing*, sponsored and coedited by PSC-TNO, has rapidly taken over the position of the *Tijdschrift voor Economische en Sociale Geografie* as the leading Dutch journal dealing with quantitative methods in spatial science.

Main accents

Table 10.3 summarises the scientific production in Dutch quantitative spatial research in the 1970s. It is obvious that, according to the number of publications, gravity and entropy modelling received by far the most attention during that decade (113 cases). Other quantitative methods showing high overall frequencies are general correlation and regression (79), mathematical programming (66), logit modelling (53), multicriteria analysis (46), multiregional economic models (33), cluster analysis (32), general urban and regional models (32), and factor analysis (29).

The leading role of gravity and entropy modelling is largely the product of two factors. First, gravity and entropy models can easily be applied to different types of spatial interaction, ranging from migration to the movement of messages. Second, the annual Transportation Planning Research Colloquium, owing to its catalysing influence on the publication of research results in transportation modelling, accounts for a considerable number of publications dealing with gravity and entropy models (for example, Le Clerq, 1977; Nijkamp, 1979; Ruygrok et al, 1979). Apart from transportation research, gravity or entropy type models were frequently used in relation with aggregate residential location processes (Somermeyer, 1971; Drewe, 1972; Klaassen and Drewe, 1974; Drewe and van der Zouwe, 1979) and the spatial choice of service facilities. The latter group of studies usually concerns shopping studies (for example, Timmermans, 1979; Timmermans and Veldhuisen, 1979; van der Linde, 1980). In a few cases gravity models were used on data regarding outdoor recreation (de Kievit, 1976). Although logit models (see Ruygrok and Wieleman, 1975; Verster et al, 1976; Le Clerq, 1979; van Lierop and Nijkamp, 1979; Ruygrok, 1979; van der Zwam et al, 1979) do not differ meaningfully from gravity and entropy models in terms of their applicability to various types of spatial interaction, they have hardly been discussed in connection with migration analysis and have never been applied to the modelling of migration flows. This is even more remarkable because the relative frequencies of both methods, apart from the 'residential location' category, differ only slightly.

The figures relating to correlation and regression analysis also show an accent on spatial interaction, although they are also fairly often discussed in relation to the analysis of spatial structure. They are probably employed frequently because of their general applicability and because of their mathematically simple structure, not only in regional economics and transportation research but also in geography and urban and regional planning (for example, Deurloo, 1972; van Duyn, 1975; Ruygrok and Wieleman, 1975; Bartels and Bertens, 1976; Bartels, 1977; van der Knaap, 1978).

Because the complex decision problems in economic and physical planning, for which mathematical programming and multicriteria analysis are used, can hardly be classified within the thematic categories, the majority of publications dealing with these methods are found in the category 'not classifiable'. The remaining cell-frequencies are too low to allow reliable conclusions. Inspection of the names of the main authors reveals that regional economists contribute most to the development and use of both methods (Nijkamp, 1972; Mastenbroek and Paelinck, 1976; Nijkamp and Rietveld, 1976; van Delft and Nijkamp, 1977; Paelinck, 1979), although, in multicriteria analysis a considerable number of studies are accounted for by physical planners (Voogd, 1975; 1976; 1980).

Unlike the case of the quantitative methods discussed above, cluster and factor analysis show an accent on spatial structure, which in itself is hardly surprising. A second interesting difference concerns the prominence of geographers in the use of both methods (Dieleman, 1971; 1978; van der Knaap, 1971; 1978; Deurloo, 1976; Masser and Scheurwater, 1978; van Ginkel, 1979; Op't Veld, 1979; Op't Veld and Timmermans, 1980; Timmermans, 1980). In cluster analysis this leading role has to be shared with physical planners, although some of these are geographers by education.

In contrast, multiregional economic modelling is evidently a strictly regional economic affair (van Wickeren, 1971; van Leeuwen et al, 1976; Ancot and Paelinck, 1980). In the construction of general urban and regional models physical planners occupy a dominant position (van Est, 1979; Veldhuisen and Kapoen, 1979); and some studies of the Netherlands Economic Institute are also important (Klaassen, 1973; Beumer et al, 1977).

If we look at the less frequently appearing quantitative methods (between ten and sixteen cases), it is striking that problems of spatial autocorrelation have only received attention from regional economists (Bartels and Hordijk, 1977; Hordijk and Nijkamp, 1977; Brandsma and Ketellapper, 1979). The same applies to multidimensional scaling, which is restricted to use by regional economists and physical planners (Nijkamp and Voogd, 1978; 1980; van Setten and Voogd, 1979). Markov chains and shift-and-share analysis, on the other hand, represent quantitative methods receiving considerable attention from Dutch geographers (see for example, van der Knaap, 1978; Dietvorst, 1979; Timmermans and Op't Veld, 1980 on Markov chains; and Wever, 1972; Oosterhaven and van Loon, 1979 on shift-and-share analysis). Assignment techniques have been applied mainly, of course, by traffic engineers (Edelman, 1975; Le Clerq, 1979).

For quantitative methods showing marginal frequencies below twenty, the figures are too low to arrive at reliable conclusions. It is therefore impossible to say anything more conclusive about a large group of quantitative methods. At first glance, one can think of three reasons why this large group has not been applied more: first, the majority of the researchers in quantitative spatial science are not familiar with the particular method because it was not part of the curriculum at the time they attended lectures; second, computer-programs involving the particular method are difficult to obtain; and third, the particular quantitative method is badly suited for the analysis of spatial phenomena.

In our opinion the first reason is the most important. As far as geographers are concerned it applies particularly to such methods as contingency analysis and time-series analysis.

Prevailing trends in the 1970s

To reveal the main trends in quantitative methods in Dutch spatial science in the 1970s it is possible to report on changes observed when the decade is divided into two five-year periods: 1971–1975, and 1976–1980 (see Dieleman and Op't Veld, 1981). One can see that, although in both periods gravity and entropy modelling, and correlation and regression are the most popular quantitative methods, considerable differences exist between the two periods.

During the 1970s, logit modelling, multicriteria analysis, cluster analysis, multidimensional scaling, and Markov chain analysis become increasingly popular, whereas the other quantitative methods become less important. The enormous increase in studies dealing with logit models, and regarding all types of spatial interaction except migration processes, stems from the wish to incorporate more theory on individual behaviour into the models. Disaggregate multinomial logit is considered to be theoretically more attractive than gravity and entropy models because of its firmer underpinning in terms of individual behaviour.

The rapid increase in scientific attention devoted to multicriteria analysis, as an important instrument for the solution of multifacetted decision problems in economic and physical planning, is partly explained by the increased external pressure exerted on decisionmakers and politicians by, for instance, the environmentalist movement asking for a considered judgement on complicated evaluation problems. Without the activities of a few researchers who accepted the challenge to develop the methods asked for, however, multicriteria analysis would not have received the attention that it has.

As opposed to other quantitative methods showing the same marginal frequencies for both halves of the decade, the activities concerning general urban and regional models prove to be clustered around 1976 and to fade away gradually towards the end of the decade. So, at the start of the eighties the attention devoted to general urban and regional models is the same as it was at the start of the seventies.

Two other important changes occurred during the 1970s. The first relates to the use of general correlation and regression. This is applied far more often to the analysis of residential location processes in the second than in the first half of the 1970s, presumably as a result of the increasing value attached in physical planning to information regarding migration behaviour (deriving from the increasing importance of migration processes on future population distribution and in affecting forecasts).

The second, remarkable, difference concerns the surprising increase in the number of quantitative studies dealing with the spatial choice of service activities, which the second half of the decade shows as compared with the first. This rather sudden change in emphasis was caused largely by the fact that the making of a 'retail development plan' became obligatory within the framework of Dutch Physical Planning law.

Consequently municipal and provincial authorities spent a considerable amount of money on the research necessary to prepare these plans. In addition to this direct influence on physical planning research, the obligation affected academic research indirectly by the resulting demand for know-how regarding the development and planning of shopping centres.

Conclusions and speculations

The preceding paragraphs have illustrated the rapid development of quantitative analysis in Dutch geography and urban and regional planning in the 1970s. The emphasis was very clearly on quantitative methods and the construction of mathematical models. When compared with related social sciences, less attention was paid to general research methodology and its translation to specific problems in the discipline. Given the present opinions on this matter, it seems reasonable to expect that geographers, especially those working in the field of quantitative spatial analysis, will turn more to questions of research methodology in the first part of the 1980s than they did during the 1970s.

Despite the fact that a considerable range of quantitative methods were incorporated into the curricula, geographers, as far as spatial research was concerned, contributed mainly to the use of quantitative methods such as factor analysis, cluster analysis, and Markov chain analysis. The development and use of mathematical models within a more deductive research design was left to regional economists and physical planners. We expect though that in the next few years ties between regional economists and human geographers on the research front will become closer again. The reason for this expectation is that a few geographers have started to work on mathematical model building, while some regional economists are directing more attention to the analysis of empirical data of a low level of measurement and to the discussion of geographical problems without the use of a mathematical framework.

With respect to the popularity of various quantitative techniques applied in Dutch spatial research, it will be interesting to see if the increasing interest in the analysis of individual behaviour and motivation will result in a further change in emphasis than that noticed in the last few years. If the shift towards the analysis of individual behaviour persists, it might lead to a further relaxation of the emphasis on gravity and entropy models, with an increasing importance of disaggregated logit-modelling. Quantitative methods suited to the analysis of individual behaviour and to the analysis of data at a low level of measurement (like contingency analysis, linear structural equations, spatial choice and preference models, multidimensional scaling, and multicriteria analysis) might become more popular in the first part of the 1980s.

The question of whether the number of publications in spatial research, on or using quantitative methods, will continue to grow, is more difficult to answer. Certainly the number of geographers and planners better trained in quantitative methods is still increasing and might be a stimulus in this direction. In addition the abundance of problems faced in regional economics, transportation, and physical planning, which can be solved by the use of quantitative methods and models, could exert a second positive influence. On the other hand a further stagnation of the economy could force the government to cut expenses on research and higher education still more. A further uncertainty relates to the projected reorganisation of the financial structure of research at the universities, and the extent to which this may create shifts in research activities in the different disciplines in the next five years.

References

Ancot J P, Paelinck J H P, 1980 "Recent results in spatial econometrics" *NEI, Foundations of Empirical Economic Research, 1980/11* Netherlands Economic Institute, Rotterdam

Bartels C P A, 1977 "The structure of regional unemployment in the Netherlands" *Regional Science and Urban Economics* **7** 103-135

Bartels C P A, Bertens A M, 1976 "A factor and regression analysis of regional differences in income-level and concentration in the Netherlands" *Applied Economics* **8** 178-192

Bartels C P A, Hordijk L, 1977 "On the power of the generalized Moran contiguity coefficient in testing for spatial autocorrelation among regression disturbances" *Regional Science and Urban Economics* **7** 83-101

Beumer L, Gameren A van, Hee B van der, Paelinck J, 1977 "A study of the formal structure of J.W.Forrester's urban-dynamics model" *NEI, Foundations of Empirical Economic Research, 1977/6* Netherlands Economic Institute, Rotterdam

Brandsma A S, Ketellapper R H, 1979 "A biparametric approach to spatial autocorrelation" *Environment and Planning A* **11**(1) 51-58

Clerq F le, 1977 "Een Amsterdams verkeers- en vervoersmodel: combinatie van aggregate en disaggregate technieken" (The Amsterdam Transportation Model: A Combination of Aggregate and Disaggregate Methods), in *Colloquium Vervoersplanologisch Speurwerk* Eds G R M Jansen, P H L Bovy, F Le Clerq, J P J M van Est, Delft, pp 103-131

Clerq F le, 1979 "A model for multimodal trips" in *New Developments in Modelling Travel Demand and Urban Systems* Eds G R M Jansen, P H L Bovy, J P J M van Est, F le Clerq (Saxon House, Teakfield, Farnborough, Hants) pp 93-116

Delft A van, Nijkamp P, 1977 *Multi-criteria Analysis and Regional Decision Making* (Martinus Nijhoff, The Hague)

Deurloo M C, 1972 "De wet der urbane concentratie" (The rank-size rule) *Tijdschrift voor Economische en Sociale Geografie* **63** 306-314

Deurloo M C, 1976 "Clusteranalyse van kwantitatieve gegevens, uitgewerkt voor een agrarische karakteristiek van de Nederlandse landbouwgebieden" (Clustering regions according to agricultural characteristics) *Tijdschrift voor Economische en Sociale Geografie* **67** 213-229

Deurloo M C, Hoekveld G A, 1979 "The population growth of the urban municipalities in the Netherlands between 1849 and 1970, with particular reference to the period 1899-1930" in *Patterns of European Urban Growth since 1500* Ed. H Schmal (Croom Helm, London)

Dieleman F M, 1971 "Factoranalyse en multidimensionale groepering" (Factor analysis and multidimensional classification) *Tijdschrift voor Economische en Sociale Geografie* **62** 217-225

Dieleman F M, 1975 "Methoden en technieken in het geografisch onderwijs en onderzoek" (Methods and techniques of research) *Geografisch Tijdschrift* **9** 214-224

Dieleman F M, 1978 *Een Analyse van Spreidingspatronen van Vestigingen en van Werkgelegenheidsbegieden in Tilburg en Eindhoven; een Methodisch-technische Studie* (An analysis of the location of economic activity and employment areas in Tilburg and Eindhoven; a method-technical study) (Krips, Meppel)

Dieleman F M, Hauer J, 1980 "Inleiding" (Introduction) in *Wegen in het Ruimtelijk Onderzoek* (Directions in spatial research) Eds F M Dieleman, J Hauer, J A van Staalduine (Bohn, Scheltema and Holkema, Utrecht)

Dieleman F M, Op 't Veld D, 1980 *Quantitative Methods in Dutch Geography and Urban and Regional Planning in the Seventies: Geographical Curricula and Methods Applied in 'Spatial Research'* Research Centre for Physical Planning-TNO, Delft

Dietvorst A G J, 1979 "Telefoonverkeer en ekonomische struktuur in Nederland: een toepassing van de veldtheorie" (Telephone calls and economic structure in the Netherlands: An application of graph-theory) Geografisch Instituut der Katholieke Universiteit, Nijmegen

Doorn J van, Vught F van, 1978 *Planning, Methoden en Technieken voor Beleidsondersteuning* (Planning, methods in policy research) (van Gorcum, Assen)

Drewe P, 1972 "Onderzoek naar Vooruitberekeningsmodellen voor de Interregionale Migratie in Nederland" (An analysis of models forecasting interregional migration in the Netherlands) Netherlands Economic Institute, Rotterdam

Drewe P, Zouwe K van der, 1979 "Interregionale Migratie en Spreindingsbeleid: nog meer modellen voor de interregionale migratie in Nederland" (Migration policy and interregional models: more models) Memorandum 20, Instituut voor Stedebouwkundig Onderzoek, Technische Hogeschool, Delft

Duyn J J van, 1975 "The cyclical sensitivity to unemployment of Dutch Provinces, 1950-1972" *Regional Science and Urban Economics* **5** 107-132

Edelman W F, 1976 "Een toedelingsmethode voor openbaar vervoernetwerken met meervoudige routekeuze" (A public transport assignment method using multiple route choice) in *Colloquium Vervoersplanologisch Speurwerk 1975* Eds F le Clerq, J P J M van Est, G R M Jansen, P H L Bovy, The Hague, pp 571-589

Est J van, 1979 "The Lowry model revised to fit a Dutch region" in *New Developments in Modelling Travel Demand and Urban Systems* Eds G R M Jansen, P H L Bovy, J P J M van Est, F le Clerq (Saxon House, Teakfield, Farnborough, Hants) pp 222-252

Everitt B S, 1977 *The Analysis of Contingency Tables* (Chapman and Hall, Andover, Hants)

Ginkel J A van, 1979 "Suburbanisatie en recente woonmilieus" (Suburbanization and characteristics of residential areas) Geografisch Instituut, Rijksuniversiteit, Utrecht

Gregory S, 1963 *Statistical Methods and the Geographer* (Longman, Harlow, Essex)

Hakkenberg A, 1969 "Enkele kritische kanttekeningen bij het verschijnsel 'social physics'" (Some critical remarks regarding social physics) *Tijdschrift voor Economische en Sociale Geografie* **60** 375-379

Hart H W ter, 1968 "De Toepassing van enkele geostatistische methods in het stadsgeografisch onderzoek" (The application of some geostatistical methods in urban geographic research) *Tijdschrift voor Economische en Sociale Geografie* **59** 25-32

Hauer J, 1971 "Een poging tot plaatsbepaling van de nieuwe geografie" (An attempt to characterise the 'new geography') in *Mathematische methoden in de sociale geografie* (SISWO en KNAG, Amsterdam) pp II1-II19

Heinemeyer W F, 1977 "De sociaal-geografische bemoeienis met stad en stedelijk systeem. Aperçu van de Nederlandse stadsgeografie sinds 1950" (Dutch urban geography since 1950) *Geografisch Tijdschrift* **11** 259-272

Heslinga M W, 1974 "Over de Vooys en de geografie" in *Een Sociaal–Geografisch Spectrum* Eds J Hinderink, M de Smidt, Geografisch Instituut, Rijksuniversiteit, Utrecht, pp XI–XLIII (Rijksuniversiteit, Utrecht)

Heslinga M W, 1980 "Sociale geografie in meervoud" (Human geography and plurality) *Geografisch Tijdschrift* **14** 177–181

Hordijk L, Nijkamp P, 1977 "Dynamic models of spatial autocorrelation" *Environment and Planning A* **9** 505–519

Jansen A C M, 1980 "Een fenomenologische oriëntatie in de ruimtelijke wetenschappen?" (A phenomenological orientation in the spatial sciences?) *Geografisch Tijdschrift* **14** 99–111

Johnston R J, 1979 *Geography and Geographers, Anglo–American Human Geography Since 1945* (Edward Arnold, London)

Keuning J H, 1969 *De Denkwijze van de Sociaal-geograaf* (The social-geographical way of thinking) (Het Spectrum, Utrecht)

Kievit J L de, 1976 "Onderzoek Midden-Randstad: model voor de openlucht-rekreatie" (Midden-Randstad study: modelling outdoor recreation) *Stedebouw en Volkshuisvesting* **57** 230–236

Klaassen L H, 1973 "SPAMO II: a spatial model" *NEI, Foundations of Empirical Economic Research, 1973/11*, Netherlands Economic Institute, Rotterdam

Klaassen L H, Drewe P, 1974 *Migration Policy in Europe: A Comparative Study* (Saxon House, Teakfield, Farnborough, Hants)

Knaap G A van der, 1971 "Een indeling van Nederland naar 'ekonomische gezondheid', een verkennend onderzoek" (A subdivision of the Netherlands according to 'economic health': an exploration) *Tijdschrift voor Ekonomische en Sociale Geografie* **63** 332–350

Knaap G A van der, 1978 *A Spatial Analysis of the Evolution of an Urban System: The Case of the Netherlands* (Erasmus University Press, Rotterdam)

Kouwenhoven A O, 1979 *Perspectief op de Cultuur van de Geografie* (Perspective on geographical culture) (Vrije Universiteit, Amsterdam)

Leeuwen I van, Paelinck J, Wagenaar S, 1976 "Towards estimation of the static inter-regional attraction model for the Netherlands: an application of spatial econometrics" *NEI, Foundations of Empirical Economic Research, 1976/12* Netherlands Economic Institute, Rotterdam

Lierop W van, Nijkamp P, 1979 "Ruimtelijke-keuze- en interaktie-modellen: kriteria en aggregatie" (Spatial choice and interaction models: criteria and aggregation) in *Colloquium Vervoersplanologisch Speurwerk* Eds F le Clerq, J P J M van Est, G R M Jansen, P H L Bovy (The Hague) pp 377–405

Linde L van der, 1980 "Modellen, toepassingsmogelijkheden voor onderzoekers en bestuurders" (Models, potentialities in research and policy) in *Wegen in het Ruimtelijk Onderzoek* (Directions in spatial research) Eds F M Dieleman, J Hauer, J A van Staalduine (Bohn, Scheltema and Holkema, Utrecht)

Masser I, Scheurwater J, 1978 "The specification of multi-level systems for spatial analysis" in *Spatial Representation and Spatial Interaction* Eds I Masser, P J B Brown (Martinus Nijhoff, The Hague) pp 151–173

Mastenbroek P, Paelinck J, 1976 "Multiple criteria decision making: information exhaustion, uncertainty, and non-linearities" *NEI, Foundations of Empirical Economic Research, 1976/4*, Netherlands Economic Institute, Rotterdam

Nijkamp P, 1972 *Planning of Industrial Complexes by Means of Geometric Programming* (Erasmus University Press, Rotterdam)

Nijkamp P, 1979 "Gravity and entropy models: the state of the art" in *New Developments in Modelling Travel Demand and Urban Systems* Eds G R M Jansen, P H L Bovy, J P J M van Est, F le Clerq (Saxon House, Teakfield, Farnborough, Hants) pp 281–321

Nijkamp P, Rietveld P, 1976 "Multi-objective programming models: new ways in regional decision making" *Regional Science and Urban Economics* **6** 253-274
Nijkamp P, Voogd J H, 1978 "The use of multidimensional scaling in evaluation procedures: methodology and application to an industrial location problem" RP-13, Research Centre for Physical Planning TNO, Delft
Nijkamp P, Voogd J H, 1980 "New multicriteria methods for physical planning by means of multidimensional scaling techniques" RM-1980-1, Economic Faculty, Free University, Amsterdam
Oosterhaven J, Loon J van, 1979 "Sectoral structure and regional wage differentials: a shift and share analysis on 40 Dutch regions for 1973" *Tijdschrift voor Ekonomische en Sociale Geografie* **70** 3-16
Op 't Veld D, 1979 "Interregionale Migratie in Nederland. De Regio-indeling" (Interregional migration in the Netherlands: the system of regions used) Planologisch Studiecentrum TNO, Delft
Op 't Veld D, Timmermans H, 1980 "De identificatie van ruimtelijke systemen voor de analyse van interregionale migratie: theoretische en methodisch-technische overwegingen" (The identification of spatial systems for the analysis of interregional migration: theoretical and method considerations) in *Wegen in het Ruimtelijk Onderzoek* (Directions in spatial research) Eds F M Dieleman, J Hauer, J A van Staalduine (Bohn, Scheltema and Holkema, Utrecht)
Paelinck J, 1979 "The multi-criteria method qualiflex: past experiences and recent developments" *NEI, Foundations of Empirical Economic Research, 1979/15* Netherlands Economic Institute, Rotterdam
Rijken van Olst H, 1961 *Inleiding tot de Sociale Statistiek Deel I* (van Gorcum, Assen)
Ruygrok C J, 1979 "Disaggregate choice models: an evaluation" in *New Developments in Modelling Travel Demand and Urban Systems* Eds G R M Jansen, P H L Bovy, J P J M van Est, F le Clerq (Saxon House, Teakfield, Farnborough, Hants) pp 13-38
Ruygrok C J, Wieleman J T, 1975 "Enkele aspecten van gedisaggregeerde gedragsmodellen voor discrete keuzesituaties" (Some aspects of disaggregate behavioral models dealing with discrete choice situations) in *Colloquium Vervoersplanologisch Speurwerk 1975* Eds F le Clerq, J P J M van Est, G R M Jansen, P H L Bovy (The Hague) pp 75-99
Ruygrok C J, Essen P G van, Eems C van der, 1979 "The Apeldoorn transportation study" in *Colloquium Vervoersplanologisch Speurwerk 1979* Eds F le Clerq, J P J M van Est, G R M Jansen, P H L Bovy (The Hague) pp 1-73
Setten A van, Voogd J H, 1979 "Interaction modelling under fuzzy circumstances" RP-17, Research Centre for Physical Planning TNO, Delft
Somermeyer W H, 1971 "Multi-polar human flow models" *Papers of the Regional Science Association* **26** 131-145
Taylor P J, 1976 "An interpretation of the quantification debate in British geography" *Transactions, Institute of British Geographers* **1** 129-142
Timmermans H J P, 1979 *Mathematical Shopping Models and Multi-attribute Planning: Some Technical and Operational Observations* Technische Hogeschool, Eindhoven, Afdeling Bouwkunde, Vakgroep Urbanistiek en Ruimtelijke Organisatie
Timmermans H J P, 1980 *Centrale Plaatsen Theorieën en Ruimtelijk Koopgedrag* (Central place theories and spatial shopping behaviour) (Ergonbedrijven, Eindhoven)
Timmermans H J P, Op 't Veld A G G, 1980 "Temporal changes in the retailing component of the Dutch urban system: a Markov approach" RP-19, Research Centre for Physical Planning TNO, Delft
Timmermans H J P, Veldhuisen K J, 1979 "Het ruimtelijk koopgedrag van consumenten in een multi-attribuut planning" (Spatial shopping behaviour of consumers in multi-attribute planning) *Planning, Methodiek en Toepassing* **9** 11-20

Vall M van de, 1980 *Sociaal Beleidsonderzoek, een Professioneel Paradigma* (Social policy research, a professional paradigm) (Samson, Alphen aan den Rijn)

Veldhuisen K J, Kapoen L L, 1979 "A regional allocation model—a regional household allocation model based on preferences measured with questionnaires" in *New Developments in Modelling Travel Demand and Urban Systems* Eds G R M Jansen, P H L Bovy, J P J M van Est, F le Clerq (Saxon House, Teakfield, Farnborough, Hants) pp 252-281

Verster A C P, Leeuwen I L van, Langen M de, 1976 "An application of urban shopping models" in *Colloquium Vervoersplanologisch Speurwerk 1976* Eds J P J M van Est, G R M Jansen, P H L Bovy, F le Clerq (The Hague) pp 369-409

Voogd J H, 1975 "Planevaluatie: een noodzakelijke voorwaarde bij het werken met alternatieven" (Project evaluation: a prerequisite in dealing with alternatives) *Stedebouw en Volkshuisvesting* **56** 287-298

Voogd J H, 1976 "Concordance analysis: some alternative approaches" RP-2, Research Centre for Physical Planning TNO, Delft

Voogd J H, 1980 "Qualitative multicriteria evaluation methods for development planning" *Planologisch Memorandum 1980-5* Department of Urban and Regional Planning, University of Technology, Delft

Wever E, 1972 "De shift- and share-analysis: mogelijkheden en beperkingen" (Shift- and share-analysis: opportunities and restrictions) *Tijdschrift voor Ekonomisch en Sociale Geografie* **61** 295-297

Wickeren A C van, 1971 *Interindustry Relations, Some Attraction Models* (Enschede)

Zwam H H P van, Putten Th H van, Smith S A, Ponsioen A M G J, 1979 "Zuidvleugel study" in *Colloquium Vervoersplanologisch Speurwerk 1979* Eds F le Clerq, J P J M van Est, G R M Jansen, P H L Bovy (The Hague) pp 229-280

The discrimination between alternative and interacting causes in geographical research

A G M van der Smagt

Introduction

This essay is concerned with a misconception which characterises much statistical analysis in geographical research. This misconception is based on the property that geographers (and they are not alone in this) do not distinguish adequately between *the discrimination of alternative and mutually exclusive causes* and *the analysis of causes that are interacting to produce a single result*. In the first case the job of the researcher is to separate as much as possible the effect of one cause from the effects of the others. In the second case such a separation is not possible because the causal operation of one factor assumes by definition the operation of the others. Therefore, "the analysis of interacting causes is fundamentally a different concept from the discrimination of alternative causes" (Lewontin, 1974a, page 401), and the disregarding of this distinction leads directly to a misconception of causality. This distinction between the two types of analysis has implications for geographical research, and in this chapter we shall investigate these implications. We shall apply this distinction to the discrimination of alternative causes in the next section, and to the analysis of interacting causes in the third section. The findings are then evaluated by reference to a particular problem in geography which is used to investigate the extent to which parametric statistical models can be used within the framework of either of the two types of analysis.

The discrimination of alternative causes

Where several alternative factors are influencing a phenomenon our first task is to describe each of the causal relationships separately. That is to say, all the possible factors apart from the one under investigation are treated as disturbance factors, and by means of idealisation we exclude the interference caused by these factors. For example, the point-mass in classical mechanics is a construct which has no direct relation to reality but which is, nevertheless, an effective device. It is by the use of such relatively simple devices that we are able to approach *ceteris paribus* conditions reasonably well in practical research. In experimental research this means randomising or matching; in statistical research it means holding factors constant, for example, by using partial correlation, regression, and crosstabular analysis.

The above mentioned techniques make it possible to eliminate the additive disturbing variables and thus to realise the *ceteris paribus* conditions.

Nevertheless, the usefulness of this sort of approach should not be overestimated. It is useful in the sense that it can separate out manifestations of behaviour which appear to be identical but which differ in meaning. However, at this stage nothing has yet been explained. Recently Harré put this as follows: "The role of statistics ... should be comparable to the role of purification techniques in chemistry, that is, statistics should be a device for selecting the typical member for intensive study. In real science, the work begins by idiographic investigation of the selected typical member" (Harré, 1978, page 55).

These points can be illustrated by considering the three components which contribute to the production of the manifestation (event E): the research object (described by the object theory, T), the measuring instrument (described by the auxiliary theory, A), and the disturbing factors (included in the *ceteris paribus* clause, C). If we wish to make a statement T on the basis of the manifestation E, then the validity of that statement is directly dependent upon how far we can trust our knowledge of A and C. Assumptions must be made about both of these, assumptions which must necessarily be regarded as background knowledge taken as read. Were we to reverse the procedures in order to test T, then we encounter the same problem. It is not theory T *alone* which is tested; in fact the whole conjunction $T \wedge A \wedge C$ is tested. We would never know with certainty which component was to blame if there were to be a falsification (see, for example, Duhem, 1916; Lakatos, 1970; Derksen, 1980, pages 143–144). Of most interest to us in this discussion, however, is the relationship between T and A on the one hand, and between T and C on the other. There is a fundamental difference between these two relationships. The measuring instrument A in interaction with T contributes to the production of the manifestation E. The relationship between T and A is a coproduct relationship. The absence of the measuring instrument makes the existence of E in fact impossible.

The role played by the disturbing variables is completely different. These variables, as we have already seen, influence E in such a way that the joint working of T and A, instead of becoming apparent, is concealed, disturbed, or by a contraworking, neutralised. C, in contrast to A, is not a coproducer but an alternative (additive) producer. This difference is the reason why, when E is deduced from T, *the auxiliary theory is incorporated into the deduction but the ceteris paribus clause is not* (Derksen, 1980, page 144). Hence, there is an important role played by *auxiliary theory*. This mediates the process of measurement and observation.

The analysis of interacting causes

It is noteworthy that we make very many instrumental observations in our daily lives without being conscious of the role of the measuring instrument. Viewed from this naive standpoint, the role of the observation instrument is a neutral one which makes no essential contribution to the manifestation itself. In reality, however, the measuring instrument does play an extremely fundamental and active role. This means that the instruments involved in interaction with the phenomena observed cannot be regarded as anything other than *coproducers*. The main reason that we do not recognise the coproductive character of manifestations when confronted with them in our daily lives is that all human beings are equipped with similarly constructed sense organs. As a result, the role of the measuring instrument usually remains implicit.

The fact that the role of measuring instruments is not made explicit often has the undesirable consequence that knowledge about the object observed is *reified*. As a result of the specific structure of the eye, we ignore all those characteristics of an object to which it is not sensitive. However, as soon as the measuring instrument is replaced or its structure changed, then the ground on which abstraction has been made gives way. Instruments with a different structure bring into being different abstractions and necessitate a different classification. *Reification* arises if an abstraction which is in principle relational (that is, relational to the instrument) is not recognised as such but is regarded as being a description of absolute characteristics of a concrete object, characteristics which the object would possess in all contexts, in 'all possible worlds'. It is incorrect to attribute concreteness and absoluteness to the object in this way.

In summary, it is the specific structuring effect of measuring instruments which give us only a selective and incomplete image of reality. As long as we carefully specify the structure of the measuring instrument which lies behind the selection and abstraction, this need create no problems since the manifestation provides information which, by means of a careful rational reconstruction, can be used to say something further about the object of study. Although this never results in 'pure' knowledge, we still can claim to get more knowledge about the object of study.

An example in geographical research

It is useful to take the argument given above and develop it further by applying it to a problem formulated in geographical research. I shall do this by using a recent article by Massey (1979). The topic to be described and explained is the spatial distribution of employment. Two factors play a part in explaining this spatial manifestation: *the conditions of production* themselves (which are dependent on the relations to the economic system, technological developments, etc), and the *spatial distribution* of these conditions of production. It is usually assumed (according to Massey) that every new investment in economic activity "will respons to geographical inequality in the conditions of production in such a way as to maximise profits". However, she adds, "While this is correct, it is also trivial. What it ignores is the variation in the way which different forms of economic activity incorporate or use the fact of spatial inequality *in order to* maximise profits. This manner of response to geographical uneveness will vary both between sectors and, for any given sector, with changing conditions of production" (Massey, 1979, page 234). Her main criticism is that geographers pay too little attention to these 'changing conditions of production'. The regional distribution of employment is seen not only as a spatial manifestation but also as being the result of a purely spatial process. Thereby, the conditions of production are assumed to be constant (Massey, 1979, page 240). Let us attempt to place this example in the framework developed in the previous section. In the first place we see that Massey's conditions of production serve as an auxiliary theory. It is impossible to understand the relationship between the regional distribution of employment (E) and the regional distribution of the conditions of production (T) without a knowledge of the specific nature of the production process itself (A). Structural changes of an economic or technological nature in these conditions of production necessarily involve a redefinition of the spatial distribution of conditions.

Consider the following possibility. Suppose that at time t_1 the regional supply of unskilled labour is of crucial significance for investment in the region: at time t_2 this supply can have ceased to be important as a result of radical technical innovations in the production process. The conclusion is inescapable: that the spatial manifestation (E) is to be regarded as the coproduct of an existing spatial distribution of the conditions of production (*inter alia* the product of past investments) and the condition of the production-process itself.

Geographers usually do not make the role of auxiliary theories explicit enough. It is therefore not surprising that the coproductive role of the conditions of production is hardly ever recognised. This is more apparent in situations where it is absolutely necessary to establish fundamental differences between such conditions. The answer is as simple as it is pragmatic: a typology. Instead of a theoretical interpretation of economic and technological changes, we make a nominal classification into (in this example) industrial sectors. However, this is a rather unsatisfactory procedure. Instead of defining the auxiliary theory as a problem, the reaction is simply to construct domains, with the aim of being able to neglect the auxiliary theory again *within* each domain. This approach solves nothing and, indeed, denies the role of auxiliary theory.

Just as in the previous analogy about measuring instruments, every failure to explicate the auxiliary theory necessarily leads to reification. If a specific structure of the conditions of production is given, then we can define the spatial distribution and establish regularities between properties of this distribution and the behaviour observed. The realised abstraction can be justified functionally *just as long* as we specify the qualitatively invariant structure of the conditions of production. If that is omitted, however, then a functional abstraction changes to a reification.

Consider the statistical analysis of the research problem posed by Massey. If we investigate the investment behaviour of firms and want to make use of correlation and regression analysis, then we have to establish in advance whether all firms in fact possess that type of behavior within their repertoire of possibilities. In this, the conditions of production play an important role. Firms in the same environment but with divergent production structures do *not possess the same repertoire of possibilities*. This conforms to Harré's remark that "every member of the domain must have the property amongst its repertoire of possible properties" can without difficulty be linked to the qualitative invariance of the auxiliary theory. Massey gives the example of firms which are able to establish production units elsewhere and firms with conditions of production which make this impossible. Firms of the first sort, for example, are able to use regional variations in wage levels in a completely different way.

It has been shown in the previous sections that there could be no objection to establishing empirical relationships between, for example, the spatial distribution of production conditions and the distribution of phenomena we want to explain. This is, however, conditional upon the qualitatively invariant conditions of production being made fully explicit, for the above relationship is valid only in such conditions. Only in this way can reification be avoided. This demands a careful specification of the qualitatively invariant conditions assumed. It is only for units which possess qualitatively identical conditions (of production) that we can guarantee that the empirical relationship observed is among the repertoire of possible properties. Being alert to this assumption helps us to avoid reification.

In conclusion, let us consider how far statistical techniques can help us in the two sorts of analysis: the discrimination of alternative causes and the analysis of interacting causes. In the first sort of analysis we can use techniques of additive analysis such as partial correlation, multiple regression, and path analysis. The only problem with these techniques, however, is that they cannot contribute much of any relevance when applied in geographical research. Partial correlation and multiple regression are merely purification techniques, nothing more.

What is of real interest to geographers is the stage which follows that 'purification': the analysis of interacting causes. Techniques of parametric analysis are, sadly, of little use here. Bivariate correlation, simultaneous equation estimation, and regression constitute the only exceptions to this. However, the parameters thus estimated will quickly lose their relevance because the qualitative invariance of the auxiliary theory is, in reality, strictly limited.

In such cases, there is a great temptation to make the contexts within which the parameters show different values into a new variable in its own right, although the additive versions of multiple correlation and regression analysis can then, of course, no longer be used. However, introducing the context as an element of an interaction variable is a misconception, since it completely *underestimates* the interaction effect. The separate contributions of (in our terms) T and A are estimated before the contribution of the interaction between T and A is examined. In this respect there is no difference between either procedure and techniques of additive analysis. The apparent separation which such analyses of causes brings about is, however, in this case purely illusory (see Lewontin, 1974a; 1974b). Johnston, in his recent introduction to multivariate analysis, desribes partial correlation as follows: "[it] handles the data in such a way ... that we can identify the effect of the variables, as if the other were not there" (Johnston, 1978, page 62). This might be sensible for the discrimination of alternative causes, but for the analysis of interacting causes it is patently absurd. If, in fact, 'the other were not there' then there is *nothing left* to analyse because the manifestation (E) is not produced! It is precisely because the manifestation is purely a product of interaction that every attempt *to separate out* the independent contributions of the various causes must be doomed to failure.

Conclusions

We said in the introduction to this chapter that it was a misapprehension to assume that external conditions can be held constant. Such an approach is, indicative of a common error often made in geographical research which results from not distinguishing between alternative causes and the fundamentally different analysis of interacting causes. Only when there are alternative causes is holding constant identical to elimination. In such cases, the *ceteris paribus* clause is a valid assumption.

When there is a coproduct relationship between factors, elimination is not possible. The most serious solution here would be to describe the coproduct relationship using nonadditive statistical models. However, we have already argued above that such models do not constitute a valid operationalisation of the assumed theoretical relationship.

References

Derksen A A, 1980 *Rationality and Science* (in Dutch) (Van Gorcum, Assen)
Duhem P, 1977 *The Aim and Structure of Physical Theory* (originally published in 1916) (Atheneum Paperback, New York)
Harré R, 1978 "Accounts, actions and meanings—the practice of participatory psychology" in *The Social Context of Method* Eds M Brenner, P Marsh (Croom Helm, London) pp 44-65
Johnston R J, 1978 *Multivariate Statistical Analysis in Geography* (Longman, London)
Lakatos I, 1970 "Methodology of scientific research programs" in *Criticism and the Growth of Knowledge* Eds I Lakatos, A Musgrave (Cambridge University Press, Cambridge) pp 91-196
Lewontin R C, 1974a "The analysis of variance of causes" *American Journal of Human Genetics* **26** 400-411
Lewontin R C, 1974b "Darwin and Mendel—the materialist revolution" in *The Heritage of Copernicus* Ed. J Neyman (MIT Press, Cambridge, Mass) pp 166-183
Massey D, 1979 "In what sense a regional problem?" *Regional Studies* **13** 233-243

Part 2

Applications in physical geography

Applying the theory of spatial stochastic processes in physical geography

U Streit

Introduction

Methods of analysing and modelling stochastic processes are gaining increasing interest in certain areas of physical geography, in particular in hydrology and climatology, where they have been used to supplement and to replace deterministic techniques. An excellent survey covering a remarkable number of relevant papers published in English is given by Unwin (1977).

Although classical statistics rely on stochastically independent variables, frequently it is processes which possess intrinsic interdependencies that are of particular interest in physical geography. Such strong dependence or autocorrelation may be regarded as a consequence of temporal and/or spatial transfer and storage processes of mass and energy. Well-known examples are the slow temporal variations of sea levels with their buffering capacity against stochastically independent precipitations, or the relatively smooth surfaces in the isobaric relief of spatial atmospheric pressure distributions.

To represent stochastic interdependence in time and space, various methods and models have been developed: temporal and spatial autocorrelation coefficients, spectra, and variograms serve to analyse empirical data; in addition Markov chains, ARIMA, STARIMA processes[1], and so-called 'Kriging' are available for modelling the underlying stochastic processes.

As regards applying and developing these kinds of stochastic procedures, German physical geography will have to make up a backlog of several years in research. This is particularly true for pure spatial and combined spatiotemporal stochastic techniques (Nipper and Streit, 1978; Streit, 1979). Possible reasons for the late interest in stochastic processes in physical geography have been summarized by Unwin (1977); they are

(a) widely-acknowledged strongly deterministic position;
(b) model requirements such as stationarity, which are regarded as being too stringent and unrealistic;
(c) insufficient data concerning quality and quantity;
(d) inadequate mathematical training of geographers.

Another aspect, which is specific to the German situation, is the traditionally strong predominance of geomorphology. In contrast to

[1] ST(ARIMA)—Space-Time(AutoRegressive Integrated Moving Average).

climatological and hydrological problems, it is obviously more difficult to submit problems like the genesis of reliefs (especially from a macromorphological point of view) to quantitative analysis; hence, there is still a strong preference for the application of nonnumerical conventional methods of terrain observation and interpretation.

It seems all the more important, therefore, to demonstrate with real problems the applicability and power of stochastic approaches. Thus in the first instance we should dispense with a detailed derivation of the pertinent mathematics. It is interesting to note that this approach has proved to be quite successful with temporal series (judging by the slowly but steadily increasing number of relevant publications). The spatial aspect so characteristic to geographical investigations, however, has been almost completely neglected as far as these methods are concerned.

This chapter presents some examples in which spatial stochastic techniques are applied and, at the same time, tries to outline the present state of affairs in this particular field as far as German geography is concerned.

The question as to whether or not spatial stochastic processes do exist in reality is a philosophical one which will not be tackled here. Instead, attention is concentrated on a number of interesting questions and problems relevant to practical geography. These can be tackled with some promise of success using stochastic approaches. Amongst these problems are not only the analysis of spatial data for persistences, hidden trends, and anisotropic effects, but also the generation of spatial forecasts, that is the spatial interpolation and extrapolation to provide missing data, a procedure which is often so important in practice.

Spatial autocorrelograms

The concept of spatial autocorrelation is based largely on the work of Cliff and Ord (1973). It relies on the assumption that the process is weakly stationary, that is, the mean values and variances are equal in all units of space, and the covariance, although being dependent on the distance and direction of two units of space, is not dependent on absolute position. These assumptions are frequently regarded as being unrealistic for natural processes. However, no proof can ultimately be provided for a stationary or unstationary state on the basis of empirical data (Klemes, 1974). It should be mentioned here that a number of suitable techniques exist for eliminating or implementing assumed nonstationarities, for example difference operators (Bennett, 1979, page 113).

The fundamental problem in constructing spatial autocorrelation coefficients for units of space with a particular shape and arrangement is the necessity of defining neighbourhood relations over various spatial lags.

Formally, this can be solved by defining weights $w_{ij}^{(k)}$, where, for $i, j = 1, ..., n$; and $k = 1, 2, ...$,

$$w_{ij}^{(k)} \begin{cases} > 0\,, & \text{if the units of space } i \text{ and } j \text{ are neighbours with respect to the spatial lag, } k \\ = 0\,, & \text{otherwise,} \end{cases} \qquad (12.1)$$

for each pair of spatial sites (i, j).

The freedom of choice in defining this spatial neighbourhood relation is, at least theoretically, a significant advantage of this concept of spatial autocorrelation. For example, theories or hypotheses prevailing in a specific discipline and concerning endogenous control mechanisms of spatial processes can, and should, be taken into consideration in the weight definition. From a more pragmatic point of view, however, this may occasionally cause some difficulties. There is a considerable risk in using purely geometrical neighbourhood definitions in an uncritical way, and the interpretation of estimated coefficients, especially for higher spatial lags, may become troublesome. Problems of this kind are certainly encountered in the examples given below.

A spatial autocorrelation coefficient of order k for metric data can be estimated as a normalized covariance:

$$I(k) = \sum_{i=1}^{n} z_i z_i(k) \Big/ \sum_{i=1}^{n} z_i^2\,, \qquad (12.2)$$

where

$$z_i = y_i - \bar{y}\,, \qquad \bar{y} \text{ is the mean of the measured values } (y_i \,|\, i = 1, ..., n)$$

$$z_i(k) = \sum_{j=1}^{n} w_{ij}^{(k)} z_j\,, \qquad \sum_{j=1}^{n} w_{ij}^{(k)} = 1 \text{ for all } i \text{ and } k\,.$$

The statistical significance of this statistic can be checked approximately using the ordinates of the normal distribution, as suggested by Cliff and Ord (1973). In the examples below, a slightly modified coefficient is used, namely

$$R(k) = I(k) \left(\frac{\mathrm{var}[z_i]}{\mathrm{var}[z_i(k)]} \right)^{1/2}\,. \qquad (12.3)$$

This modified form satisfies the property that it is always within the interval $(-1, +1)$, (see Haggett et al, 1977, page 375).

Example 1

In the first example to be discussed, the 'Münsterland' (North-West Germany) precipitation rêgime is to be analysed for spatial effects of persistence and their variabilities with time. For this purpose, the monthly precipitation amounts (mm) as measured at sixty-five test points have been analysed for each year of a series covering seventeen years.

A definition of binary neighbourhood weights, $w_{ij}^{(k)}$, is used based on the Thiessen polygon method, frequently used in hydrology and climatology. This assigns to each test point a surrounding reference area: two points i and j will be first-order neighbours if they share a common polygon segment; they are neighbours with a spatial lag k (>1) if i has a neighbour of order $(k-1)$ which is itself a first-order neighbour of j. This recursive definition, according to the criterion of contingency, is based on the experience in climatology that, at least in the plains and by using means obtained over a long period of time, the amounts of precipitation measured at one point can be regarded as representative of a larger surrounding area. The spatial lags represent a roughly-scanned isotropic distance function.

On the basis of this special structure, spatial autocorrelation coefficients up to a lag of k equal to 5 have been calculated. Table 12.1 shows that extreme changes in spatial persistence may occur from one year to another. For example, while in January 1970 a strong spatial clustering of similar values of precipitation may be supposed with a neighbourhood radius of three spatial lags (~22 km), precipitation in the following year shows no intrinsic interdependence pattern whatsoever (that is, it is known as a 'white noise'). General patterns for the temporal variations of the spatial correlograms, for example depending on wet and dry years, could not be detected. A comparison of different months of the same year (1959 being a characteristic example) gives an indication of a tendency towards higher persistences during the winter months; this obviously reflects a smaller percentage of convective precipitation.

Table 12.1. Spatial autocorrelation coefficients, $R(k)$, of precipitation data, based on sixty-five meteorological stations in the 'Münsterland' (NW Germany).

Month, year	$R(1)$	$R(2)$	$R(3)$	$R(4)$	$R(5)$	$\bar{P}$ (%)[a]
January, 1970	0·86*	0·85*	0·87*	0·23	−0·06	78
January, 1971	−0·01	−0·10	−0·05	−0·05	−0·05	101
July, 1959	0·40*	0·30*	−0·24*	−0·31*	−0·53*	65
January, 1959	0·64*	0·45*	0·24*	−0·09	−0·45*	129

* Significant at the 0·05 level.
[a] The actual spatial mean as a percentage of the spatiotemporal mean.

Example 2

The second example demonstrates a different method for defining neighbourhood weights between points in space. Symader (1979) has analysed, at 208 sampling points located in the Northern Eifel (between Aachen and Bonn), the heavy metal content in the soil surface layer. For manganese he found concentrations between 10 ppm and more than 900 ppm, and a mean value of approximately 200 ppm. The high manganese concentrations were found especially in sandstones and in conglomerates of the lower red sandstones. To examine whether the covariation of these data provides evidence for directional differences, as well as differences in distance, sector-specific correlograms were calculated. The value of 1 is assigned to the neighbourhood weight $w_{ij}^{(k)}$ of two points i and j, when j with respect to i, is located in the sector observed, as well as being in the distance ring k; it is zero otherwise. In the east–west direction the exponential decay shows a good approximation to the theoretical autocorrelation function of the first-order autoregression model (figure 12.1). In the north–south direction, however, there is a rapid change from positive autocorrelation in the first three spatial lags to negative values; this points to a sort of 'plateau effect' or 'regional trend' in the spatial distribution of the manganese content. An anisotropy of this kind, of course, renders the construction of a suitable stochastic model more difficult.

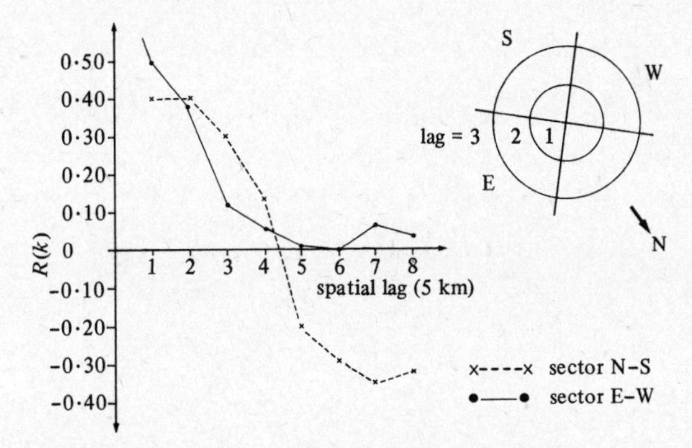

Figure 12.1. Spatial autocorrelation coefficients, $R(k)$ of manganese concentration in the Northern Eifel.

Example 3

The third example is taken from a mapping of the blackberry flora (*Rubus*) performed by Wittig and Weber (1978) in the 'Münsterländer Bucht' on a regular grid basis (grid size 2·8 km). Numerical analysis was based on a portion comprising 225 of these grid areas. This example differs from the two preceeding ones; here, the spatial autocorrelation of a binary variable is subjected to analysis, that is, the existence or nonexistence of the respective species.

To estimate the spatial autocorrelation in binary variables, Cliff and Ord (1973) recommend the so-called 'black–white' statistic,

$$BW(k) = \tfrac{1}{2} \sum_{i=1}^{n} \sum_{\substack{j=1 \\ i \neq j}}^{n} w_{ij}^{(k)} (y_i - y_j)^2 , \qquad (12.4)$$

where

$$y_i \text{ or } y_j = \begin{cases} 0 \text{ if the grid square does not contain } \textit{Rubus}, \\ 1 \text{ if the grid square does contain } \textit{Rubus}. \end{cases}$$

Neighbourhood weights are determined for this regular grid according to the criterion of common edges and, for higher spatial lags, according to the principle of shortest paths. (Definition of weights according to the alternative criterion of 'edges and corners' did not cause significant changes in the results.)

This strictly geometric view of the spatial neighbourhood effect obviously originates from a more conventional spatial scheme of thinking rather than from the basis of well-founded hypotheses relating to the spatial distribution pattern of the different *Rubus* species. Figure 12.2 shows the 'black–white' statistic converted into normalized z-values for two of the most abundant *Rubus* species in the 'Münsterland', *adspersus* and *gratus*; both occur preferentially in areas occupied by the *Quercion roboris-petraeae* (oak forests growing on acidic soil). At the first spatial step, both show a high positive autocorrelation; however, whereas *R gratus* preserves this spatial persistence, even over large spatial lags, it rapidly disappears in the case of *R adspersus*. This strongly indicates that physiological or ecological factors acting differentially may be involved. For example, more careful geobotanical analyses have shown that *R gratus* may grow on a wide variety of soils, even on very poor quartz-sandy soils or on pseudogleys occupied by the *Stellario carpinetum*, while *R adspersus* rarely spreads to different habitats.

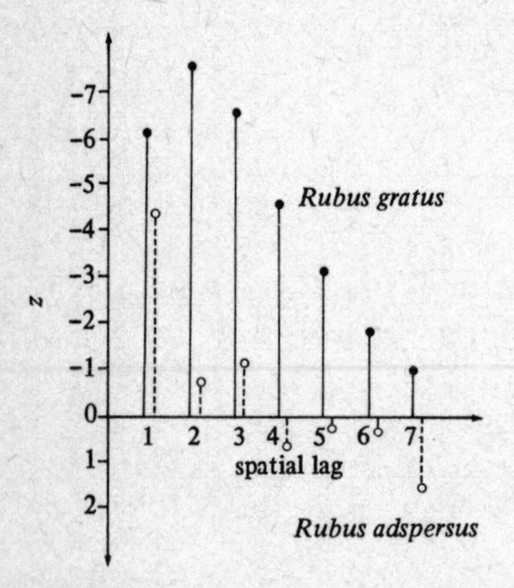

Figure 12.2. Spatial autocorrelation of *Rubus gratus* and *Rubus adspersus* in the 'Münsterländer Bucht' (z is the scores of the black–white statistic).

Spatial variograms

An alternative concept of autocorrelation is the stochastic theory of 'regionalized variables' and their application to 'Kriging' as developed by the French geostatistician Matheron (1963).

Let us consider a euclidean space where n sample points are represented by vectors $\boldsymbol{x}_1, ..., \boldsymbol{x}_n$. Instead of the stringent requirement of stationarity, we make the somewhat weaker assumption that the spatial process $Y(\boldsymbol{x})$ is 'intrinsic'; for each directional distance $\boldsymbol{h}$ the following relationships apply (David, 1977):

$$\mathrm{E}[Y(\boldsymbol{x})-Y(\boldsymbol{x}+\boldsymbol{h})] = 0\,, \tag{12.5}$$

that is, the process is on the same level at all points;

$$\mathrm{var}[Y(\boldsymbol{x})-Y(\boldsymbol{x}+\boldsymbol{h})] = 2\gamma(\boldsymbol{h})\,, \tag{12.6}$$

that is, the variance of the first differences depends on the increment $\boldsymbol{h}$, not on the absolute position. The function $\gamma(\boldsymbol{h})$ is called a (semi) variogram; for stationary processes, it equals the difference between the variance and autocovariance function,

$$\gamma(\boldsymbol{h}) = \mathrm{var}(Y)-\mathrm{cov}(\boldsymbol{h})\,. \tag{12.7}$$

To obtain an estimate of this two-dimensional function from the localised sampling values, it is a prerequisite to obtain discrete spatial values with respect to the vector $\boldsymbol{h}$; this corresponds to the structural pattern given by defining neighbourhood weights in the concept of spatial autocorrelation. Since a euclidean space is assumed, discrete values are obtained by dividing the infinite planar directions into regular sectors, and distances into intervals of equal length. The empirical variogram value, $\gamma(s, k)$, in a direction s and with a spatial lag k, can be calculated from the expression

$$\gamma(s, k) = \frac{1}{2N}\sum_{i=1}^{n}\sum_{j=1}^{p_i}[y(\boldsymbol{x}_i)-y(\boldsymbol{x}_i+\boldsymbol{h}_j)]^2\,, \tag{12.8}$$

where

p_i is the number of sample points $(\boldsymbol{x}_i+\boldsymbol{h}_j)$, located with respect to $\boldsymbol{x}_i$ in sector s and in the distance interval k;

$N = \sum_{i=1}^{n} p_i$, n is the total number of sample points.

Figure 12.3 shows sector-specific empirical variograms of daily precipitations (mm) measured on 7th July 1979 at 150 meteorological stations. Small variogram values point to a positive persistence while, for those above the variance (S^2), negative autocorrelation occurs. In the SSW–NNE direction the similarity of the precipitations decreases more rapidly than in the WNW–ESE sector (ignoring higher spatial lags). Hence, one may conceive the neighbourhood relations for each point to be shaped elliptically. The ellipses are tilted by approximately 157° against the horizontal line and compressed along the SSW–NNE axis by a factor of approximately 1·8.

A characteristic feature of quite a number of variograms is their noticeable discontinuity at zero: for small distances $|\boldsymbol{h}|$ they show a tendency towards values greater than zero. This so-called 'nugget effect' can be regarded as the variance of a purely random component superimposed on the otherwise continuously varying spatial process (see figure 12.5 below).

Estimating the variogram function, as well as Kriging itself, becomes increasingly problematical if the intrinsic hypothesis is violated; for example, if there is a trend or, according to Matheron, a 'drift'. Figure 12.4 shows an example of this kind. Groundwater levels as measured at 80 observation wells in the 'Hessische Ried' (between Frankfurt and Darmstadt), coincide with an area of deep relief gradient in the WSW

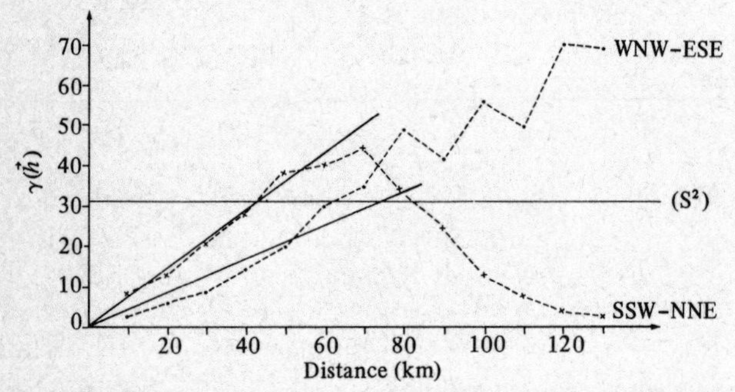

Figure 12.3. Linear anisotropic variogram of daily precipitation on 7 July 1979, measured at 150 stations in the 'Münsterland' and 'Rheinisches Schiefergebirge'. - - - - empirical; —— theoretical.

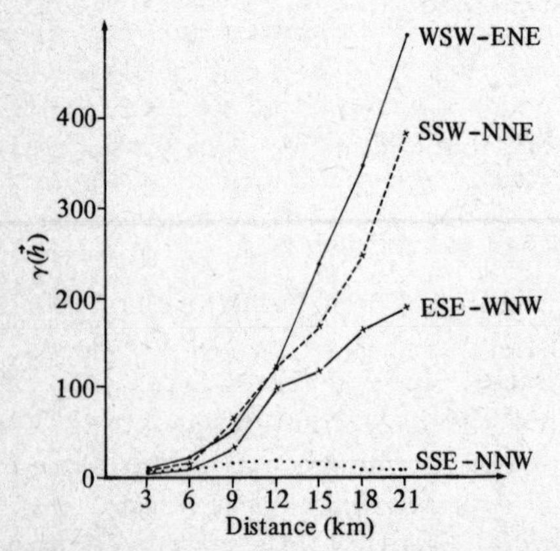

Figure 12.4. Directional empirical variograms of groundwater level (December 1970), measured at 80 wells in the 'Hessisches Ried'.

direction towards the River Rhine. Accordingly, the empirical variogram of this sector shows a parabolic increase. However, the variogram for the direction perpendicular to it (SSE–NNW) is, for all distances, close to zero because of the close similarity of neighbouring values. In this case, for each unit of space, the drift has first to be estimated by taking into account the values of the immediate neighbourhood, and then subsequently it has to be extracted. A (probably biased) estimate of the variogram of the remaining residuals may then be obtained in the usual way. The problem is, however, that to calculate the drift some knowledge must be available not only about its type (usually a polynomial model) but also about this residual variogram. The iterative technique recommended for this purpose has some strongly subjective elements and, as well as the closely related 'universal Kriging', will not be discussed here (but see David, 1977, page 266).

Kriging

The examples discussed above, which employ spatial correlograms and variograms, strongly emphasize the exploratory character of stochastic methods. In the last part of this account, I will discuss briefly some examples demonstrating the applicability of spatial stochastic models. In doing so, I shall restrict myself to dealing with Kriging (or Krigeage), although spatial autoregressive models (Cliff and Ord, 1975; Ord, 1975), especially as they belong to the more comprehensive class of spatiotemporal STARIMA models (Martin and Oeppen, 1975; Bennett, 1979), without any doubt deserve considerable attention. My own comparative investigations indicate that with metric data in euclidean spaces Kriging is at least not inferior to spatial ARIMA models.

A difference which is especially important in pursuing pragmatic objectives has been stressed by Nipper (1981). Using the spatial autoregressive model he has performed a best linear unbiased (BLU) estimate of the model parameters and made a spatial forecast for a specific unit of space thus providing the mean value to be expected. Kriging, in contrast, allows, for each single value of the regionalized variable, a point-specific BLU estimate.

The possibility of interpolating and extrapolating spatial data is of practical importance to a number of problems in the geosciences and regional planning. To give but one example, stimulated by the Canadian Geographical Information System (CGIS) and the American GRID and IMGRID systems, computer-aided 'Landschaftsinformationssysteme' (landscape information systems) are under examination in German-speaking countries, and they are increasingly used in regional planning, above all in ecologically-oriented landscape planning.

Filling up spatial information gaps and smoothing information obtained from irregularly distributed sampling points, as with data on precipitation, evaporation, or groundwater levels, presents a special problem. Since it is frequently impossible, because of the absence of data, to make an estimate of such variables from factors generally accepted to influence them, conventional procedures for obtaining weighted means and polynomial trend surfaces are applied almost exclusively. Bicubic splines are also increasingly useful, although they are restricted to lattice data. These methods have, at least partially, serious disadvantages. Weighted means have a strong heuristic component; also they, as well as trend surfaces, do not interpolate exactly, and they do not, as neither do splines, provide reliable information concerning the errors of estimation. If we consider trend-surface analysis as a statistical procedure, difficulties may arise through the incorrect estimation of the standard error and the coefficient of determination in the case of spatially autocorrelated residuals (Martin, 1974). However, if we consider trend-surface analysis as a purely numerical method, there is still the problem of excessive overshooting between the nodes with high-order polynomials.

Kriging based on variogram analysis offers, from my point of view, an interesting alternative: I restrict myself here to giving just a brief outline of 'point Kriging', and I should like to quote in this context the publications by Olea (1975) and David (1977), which deal comprehensively with this topic.

Formally, Kriging is a refined method of obtaining spatial means over nearest-neighbour points:

$$y(\boldsymbol{x}_0) \sim \hat{y}(\boldsymbol{x}_0) = \sum_{i=1}^{m(\boldsymbol{x}_0)} \lambda_i(\boldsymbol{x}_0) y(\boldsymbol{x}_i) , \qquad \boldsymbol{x}_i \in U(\boldsymbol{x}_0) \tag{12.9}$$

surrounding the point $\boldsymbol{x}_0$.

The λ-weights for the m neighbouring points $\boldsymbol{x}_i$ are to be determined such that the estimate of the unknown value $y(\boldsymbol{x}_0)$ is the BLUE, that is

$$\mathrm{E}[Y(\boldsymbol{x}_0) - \hat{Y}(\boldsymbol{x}_0)] = 0 , \qquad \left(\Rightarrow \sum_{i=1}^{m} \lambda_i(\boldsymbol{x}_0) = 1 \right) \tag{12.10}$$

$$\mathrm{var}[Y(\boldsymbol{x}_0) - \hat{Y}(\boldsymbol{x}_0)] = \text{a minimum} . \tag{12.11}$$

Applying the Lagrange principle gives a system of linear equations,

$$\begin{bmatrix} \gamma(\boldsymbol{x}_1 - \boldsymbol{x}_1) & \ldots & \gamma(\boldsymbol{x}_1 - \boldsymbol{x}_m) & 1 \\ \cdot & & \cdot & \cdot \\ \cdot & & \cdot & \cdot \\ \cdot & & \cdot & \cdot \\ \gamma(\boldsymbol{x}_m - \boldsymbol{x}_1) & \ldots & \gamma(\boldsymbol{x}_m - \boldsymbol{x}_m) & 1 \\ 1 & \ldots & 1 & 0 \end{bmatrix} \begin{bmatrix} \lambda_1 \\ \cdot \\ \cdot \\ \cdot \\ \lambda_m \\ \mu_0 \end{bmatrix} = \begin{bmatrix} \gamma(\boldsymbol{x}_0 - \boldsymbol{x}_1) \\ \cdot \\ \cdot \\ \cdot \\ \gamma(\boldsymbol{x}_0 - \boldsymbol{x}_m) \\ 1 \end{bmatrix} , \tag{12.12}$$

where

$\boldsymbol{x}_i - \boldsymbol{x}_j = \boldsymbol{h}_{ij}$;

μ_0 is the Lagrange multiplier.

This system can be used to calculate the λ-weights provided the theoretical variogram is known; generally, the procedure has to be repeated for each point to be estimated.

To obtain the theoretical variogram, which is a prerequisite, a suitable analytical function is fitted to the empirical values; in practice, one takes into consideration only a few standard possibilities, such as the spherical variogram shown in figure 12.5 or linear variograms. However, for anisotropic processes, more complicated functions are needed. In performing the fitting, one can restrict oneself to the first four to ten spatial lags, since with increasing distances persistence may rapidly decrease, and since, for estimating the unknown value $y(\boldsymbol{x}_0)$, usually only the nearest neighbours are used.

The question as to how many neighbouring points are to be included in the estimation can, in most cases, be answered only in a pragmatic way. With spherical variograms converging to the variance, a range can be determined (in figure 12.5 for instance a range of approximately 55 km is used) which allows all neighbouring points within that range to be taken into consideration. With linear or otherwise infinitely increasing variograms, however, it is recommended that a maximum number of neighbouring points be preset. Because of the screen effect, the maximum number can be set lower the smaller is the nugget effect. This screen effect causes a strong reduction of the λ-weights for all those points which are 'shadowed' by closely located neighbours (which corresponds to the intervening opportunities case in interaction models).

As an example demonstrating the application of Kriging, an attempt to obtain a spatial interpolation of daily amounts of precipitation for 7th July 1979 is presented (see figure 12.6). The area investigated by means of 150 sample points is the 'Münsterland' and the adjacent 'Rheinische Schiefergebirge'. On this particular day, with a typical cyclonic situation prevailing, a cold front spread to Central Europe, leading to rain showers amounting on average to approximately 8 mm. A field of maximum

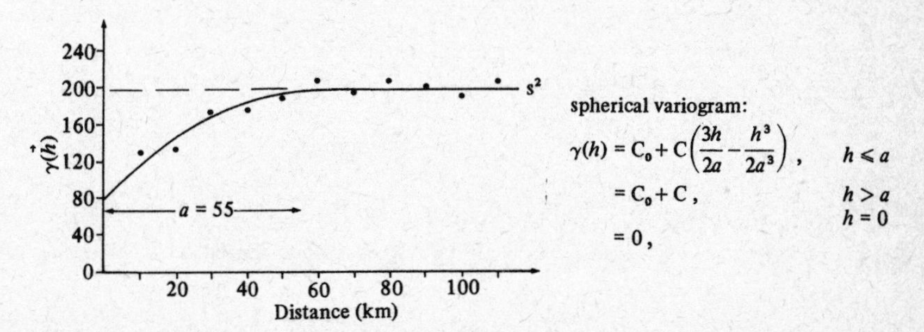

Figure 12.5. Isotropic variogram of daily precipitation on 5 June 1979 (150 stations, 'Münsterland'-'Rheinisches Schiefergebirge'). ••• empirical; —— theoretical ($C_0 = 80$, $C = 119$, $a = 55$).

precipitation reached from WNW to ESE separating the two spatial minima which were approximately 130 kms apart. As a result, the empirical variogram (see figure 12.3) shows a significant anisotropy with a more rapid decrease of persistence in the SSW–NNE direction and a reapproach to smaller variogram values at a distance of approximately 130 km. In this case, a cone function has been chosen as a theoretical variogram. In horizontal sections this is represented by ellipses, and in vertical sections

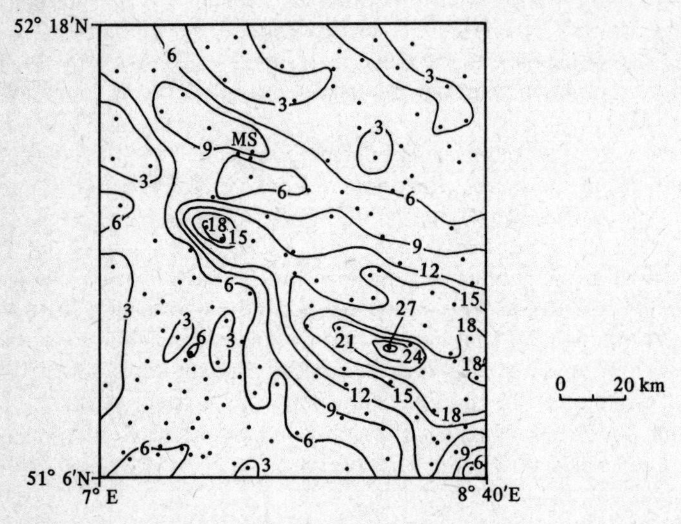

Figure 12.6. Measured value (mm) of daily precipitation on 7 July 1979. • 150 stations in the 'Münsterland' and 'Rheinisches Schiefergebirge'; MS—Münster.

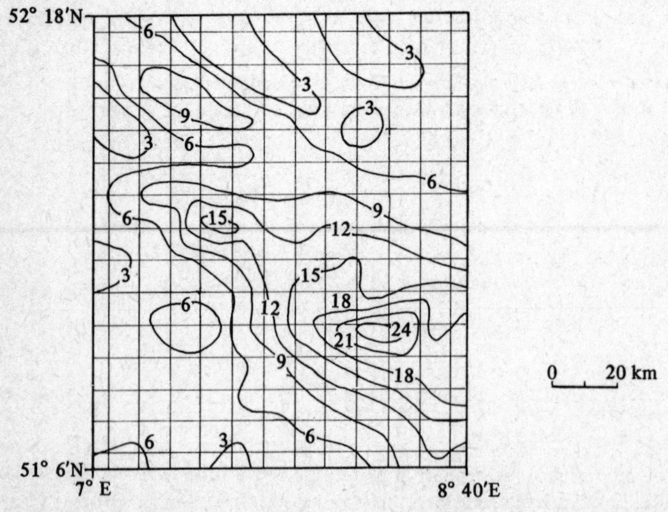

Figure 12.7. Kriging. Estimated values of daily precipitation on 7 July 1979 (mm) for a quadratic grid of ~10 km□. (See figure 12.6.)

by linear variograms of directional specificity (linear variogram with 'geometrical' anisotropy; David, 1977, page 137). Interpolation on a square grid with a grid length of 10 km (figure 12.7) agrees fairly well with the isoline-map of the measured data. Singular extreme values have been slightly smoothed as a result of being estimated as weighted means of neighbouring values $\{$smoothing factor $\mathrm{var}[\hat{Y}(\boldsymbol{x})]/\mathrm{var}[Y(\boldsymbol{x})] = 0{\cdot}84\}$. Attention should also be given to the map showing the errors of estimation (figure 12.8): minimum values of 1 mm and less (mean = 1·8) always occur where the information density is particularly high; the smaller number of sample points in the Western part of the 'Münsterland', however, gives rise to standard errors larger than 2·5 mm.

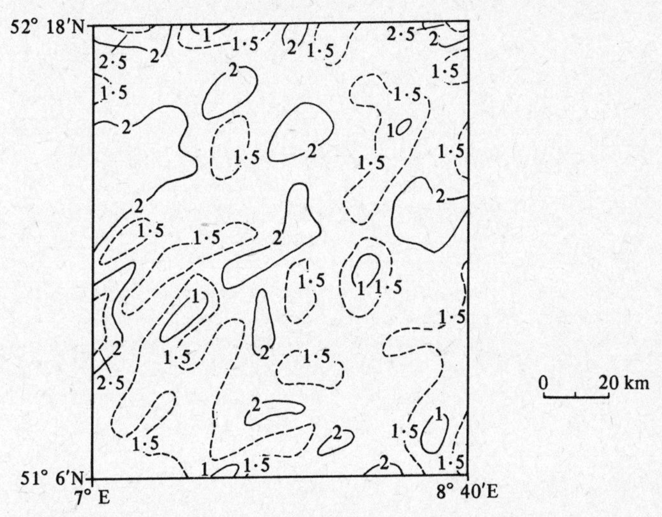

Figure 12.8. Kriging. Standard error of estimation (mm) of the daily precipitation on 7 July 1979, for a quadratic grid of ~10 km□. (See figure 12.7.)

Finally, it should be mentioned that for each point Kriging provides an equation which can be used for estimation. It does not provide, as for instance, does trend-surface analysis, an analytical function for the whole area or surrounding areas of some size. Hence, a combination of Kriging with bicubic splines offers, in my opinion, an interesting possibility for future development: following a lattice interpolation by means of Kriging, these spline functions can be calculated easily.

References

Bennett R J, 1979 *Spatial Time Series: Analysis-Forecasting-Control* (Pion, London)
Cliff A D, Ord J K, 1973 *Spatial Autocorrelation* (Pion, London)
Cliff A D, Ord J K, 1975 "Space-time modelling with an application to regional forecasting" *Transactions, Institute of British Geographers* **64** 119-128
David M, 1977 *Geostatistical Ore Reserve Estimation* (Elsevier, Amsterdam)
Haggett P, Cliff A D, Frey A, 1977 *Locational Analysis in Human Geography. Volume 2 Locational Methods* (Edward Arnold, London)

Klemes V, 1974 "The Hurst phenomenon: a puzzle?" *Water Resources Research* **10** 675-688
Martin R L, 1974 "On spatial dependence, bias and use of first differences in regression analysis" *Area* **6** 185-194
Martin R L, Oeppen J E, 1975 "The identification of regional forecasting models using space-time correlation functions" *Transactions, Institute of British Geographers* **66** 95-118
Matheron G, 1963 "Principles of geostatistics" *Economic Geology* **58** 1246-1266
Nipper J, 1981 "Autoregressiv- und Kriging-Modelle. Zwei Ansätze zur Erfassung raumvarianter Strukturen" in *Theorie und Quantitative Methodik in der Geographie* Eds M Ostheider, D Steiner; Geographisches Institut der ETH Zürich, Zürich
Nipper J, Streit U, 1978 "Modellkonzepte zur Analyse, Simulation und Prognose raum-zeit-varianter stochastischer Prozesse" in *Bremer Beiträge zur Geographie und Raumplanung* Heft 1, Universität Bremen, Schwerpunkt Geographie, Bremen, pp 1-17
Olea R A, 1975 *Optimum Mapping Techniques using Regionalized Variable Theory* Kansas Geological Survey, Series on Spatial Analysis number 2, Lawrence, Kansas
Ord J K, 1975 "Estimation methods for models of spatial interaction" *Journal of the American Statistical Association* **70** 120-126
Streit U, 1979 "Raumvariante Erweiterung von Zeitreihenmodellen. Ein Konzept zur Synthetisierung monatlicher Abflußdaten von Fliessgeweissern unter Berücksichtigung von Erfordernissen der wasserwirtschaftlichen Planung" *Gießener Geographische Schriften* Heft 46, Geographisches Institut der Universität, Giessen
Symader W, 1979 "Trendflächenanalyse zur Beurteilung raumvarianter Prozesse am Beispiel des Schwermetallgehaltes in Oberböden" *Verhandlungen der Gesellschaft für Ökologie* 7, Institut für Geographie, Universität Münster, pp 209-214
Unwin D J, 1977 "Statistical methods in physical geography" *Progress in Physical Geography* **1** 185-221
Wittig R, Weber H E, 1978 "Die Verbreitung der Brombeeren (Gattung *Rubus* L., *Rosaceae*) in der Westfälischen Bucht" *Decheniana* **131** 87-128

A model of the decrease of temperature with altitude in mountains

A Douguédroit

Introduction

In studies of the physical environment in mountains the description of the climate, especially from the thermal energy point of view, is often poor. Usually, the data for the most interesting areas are not available and there is no model of the distribution of temperature to generate them. I shall present here just such a model for mountains of high and middle latitudes.

In this account a model is estimated for a homogeneous-climate region which is divided into four main topoclimatic zones: sunny slopes, shaded slopes, the bottoms of valleys, and the tops of ridges. The values of their temperature gradients are different, and hence so are their temperatures at the same altitude.

Description of the model

The hypothesis of a constant decrease in temperature with altitude has been well known for a long time, but the values of the gradient of decrease have generally been estimated only roughly.

The usual method of calculation of the gradient of temperature

During the second part of the 19th century, geographers tackled the question of how to estimate decrease in temperature with altitude by proposing a method for calculating the lapse rate. Angot has used for France the method established in Switzerland by Hirsch (quoted by Angot, 1903, page 119) which is still favoured by most workers. The value of the gradient is calculated empirically by using pairs of points located close together in space but separated by significant differences in level (Clermont-Ferrand and Puy-de-Dôme, Bagnéres-de-Bigorre and Pic du Midi, for example).

For a general formula, the monthly gradients of each pair are calculated and then averaged. For Puy-de-Dôme, Angot has obtained

$$\Delta/°\mathrm{C}\ (100\ \mathrm{m})^{-1} = 0{\cdot}597 + 0{\cdot}195\sin(m+292) + 0{\cdot}050\sin(2m+250) \qquad (13.1)$$

where all angles are measured in degrees of arc,

Δ is the monthly temperature gradient [°C $(100\ \mathrm{m})^{-1}$],

the mean annual calculated gradient is 0·597 °C $(100\ \mathrm{m})^{-1}$,

m is the time, measured as an angle from 1st January, with the whole year set at 360°.

Sines are needed to obtain gradients which are lower in the cold months than in the warm months.

Angot has then gone further and aggregated the formulae for the whole of France and the neighbouring countries into a single formula. This he then uses for Europe in general and France in particular, namely

$$\Delta/°\mathrm{C}\ (100\ \mathrm{m})^{-1} = 0{\cdot}55 + 0{\cdot}15\sin(m+300) + 0{\cdot}050\sin(2m+260)\ . \qquad (13.2)$$

However, the results obtained by this method are rather unsatisfactory when one tries to calculate the monthly or annual mean temperatures for real weather stations.

The proposed model

The hypothesis of a single regional gradient is too simple a model. In regions considered homogeneous from a thermal energy point of view, the temperatures are distributed as a mosaic of topoclimates strictly determined by the topography.

Four main topoclimates are induced in this way: the slopes, divided into two cases, viz (1) the exposed or sunny slopes ('adrets') and (2) the shaded slopes ('ubacs'); (3) the bottoms of the valleys and the plains; and (4) the tops of the ridges and the table-lands. At any specific moment in time, each topoclimate can be distinguished from the others by a special reading of temperatures. According to the hour of the day and the month of the year, the temperature readings will be close to or far apart from one another.

In mountains, the altitude causes the temperature to decrease everywhere, but the gradients vary with the topoclimates. There is no single lapse rate but several, and these also change with the statistics of the temperature—maxima, minima, means, etc. I have chosen Provence and the French Southern Alps to show how they may be calculated. Two methods are proposed, the choice depending on the number of available data. The first, a regression method, is used when data are numerous; the second, a correlation method, is used when data are sparse.

Calculation by simple regression

The position of every weather station in the study area was defined by using topographic maps (scale 1/20000 or 1/25000), and then classified in one of the four topoclimates. Nearly all of the weather stations are in the 'adrets' or the valley bottoms because they are located near inhabited places. The adrets group together all the slopes facing to the south (from east to west), and the valley bottoms include the lower parts of the slopes, up to 30 m above the river. This limit has been determined *a posteriori*. Each type includes, for the area of Provence–French Southern Alps, twenty-four stations.

The method

The mean monthly values of the maximum, minimum, and mean temperature have been calculated for each weather station; first for 7 years (1959–1965) and then for 20 years (1959–1978) (Douguédroit, 1976;

Douguédroit and de Saintignon, 1970; 1974; 1981). The homogeneity of the series of data has been tested by double-mass curves (Douguédroit and de Saintignon, 1970) established with reference to the main weather station. Corrections were needed when important modifications had occurred to the location or the kind of the weather station.

The simple statistical method of linear regression has been used to calculate the gradients. The decrease of temperature with altitude is supposed to be represented by a first-degree equation of the form

$$T = b - \Delta z \,.$$

This is estimated using the ordinary least-squares method, where

T is the temperature at a stated altitude (in °C),
Δ is the temperature gradient for 100 m,
z is the altitude in units of 100 m, and
b is the temperature at the zero level.

The results

The equations obtained are significantly different. The value of the coefficient of determination fluctuates between 0·78 and 0·99. Student t-tests are good; table 13.1 shows the main results. I have tried to combine all the records for the region, but keeping temperature maxima, minima, and means separated, and have found a coefficient of determination of 0·25.

These results support the two hypotheses of the model, viz that the decrease of temperature with altitude is linear, and that the values of the

Table 13.1. Gradients for the adrets and the bottoms of the valleys in the region Provence-French Southern Alps.

	Adrets, maxima		Adrets, minima	
	equations	IV[a]	equations	IV[a]
January	$T = 12{\cdot}8 - 0{\cdot}57z$	1·6	$T = 3{\cdot}9 - 0{\cdot}60z$	1·9
April	$T = 20{\cdot}8 - 0{\cdot}65z$	1·1	$T = 9{\cdot}3 - 0{\cdot}58z$	1·4
July	$T = 31{\cdot}4 - 0{\cdot}60z$	1·5	$T = 18{\cdot}4 - 0{\cdot}58z$	1·6
October	$T = 21{\cdot}7 - 0{\cdot}56z$	0·9	$T = 11{\cdot}6 - 0{\cdot}57z$	1·6
Year	$T = 21{\cdot}6 - 0{\cdot}61z$	1·0	$T = 10{\cdot}4 - 0{\cdot}55z$	1·5
	valley bottoms, maxima		Valley bottoms, minima	
	equations	IV[a]	equations	IV[a]
January	$T = 10{\cdot}4 - 0{\cdot}51z$	1·4	$T = 1{\cdot}7 - 0{\cdot}76z$	1·7
April	$T = 21{\cdot}0 - 0{\cdot}67z$	1·3	$T = 7{\cdot}9 - 0{\cdot}62z$	0·7
July	$T = 31{\cdot}4 - 0{\cdot}60z$	1·5	$T = 16{\cdot}6 - 0{\cdot}67z$	0·8
October	$T = 21{\cdot}0 - 0{\cdot}53z$	0·6	$T = 9{\cdot}7 - 0{\cdot}63z$	0·9
Year	$T = 21{\cdot}1 - 0{\cdot}59z$	0·8	$T = 8{\cdot}8 - 0{\cdot}64z$	0·9

[a] Interval values to 80%.

gradient change with each topoclimate. As a result we have the four-equation model:

$$T_{max, A} = b_{max, A} - \Delta_{max, A} z \tag{13.3}$$

$$T_{min, A} = b_{min, A} - \Delta_{min, A} z \tag{13.4}$$

$$T_{max, VB} = b_{max, VB} - \Delta_{max, VB} z \tag{13.5}$$

$$T_{min, VB} = b_{min, VB} - \Delta_{min, VB} z \tag{13.6}$$

where

T_{max}, T_{min} are the maximum and minimum temperatures, respectively,
A and VB are adrets and valley bottoms, and
Δ is the gradient (regression slope parameter).

These differences in temperature gradient can be explained in terms of the effects of the directions and gradients of the slopes on the energy balance. But this is not our main purpose here. Our focus of interest is prediction of temperature decrease with altitude. The quasi-Gaussian distribution of the residuals enables us to calculate the required temperatures, to within standard errors of 80%, from one part of the regression line to another. Figure 13.1 shows the estimates of temperature for the month of January.

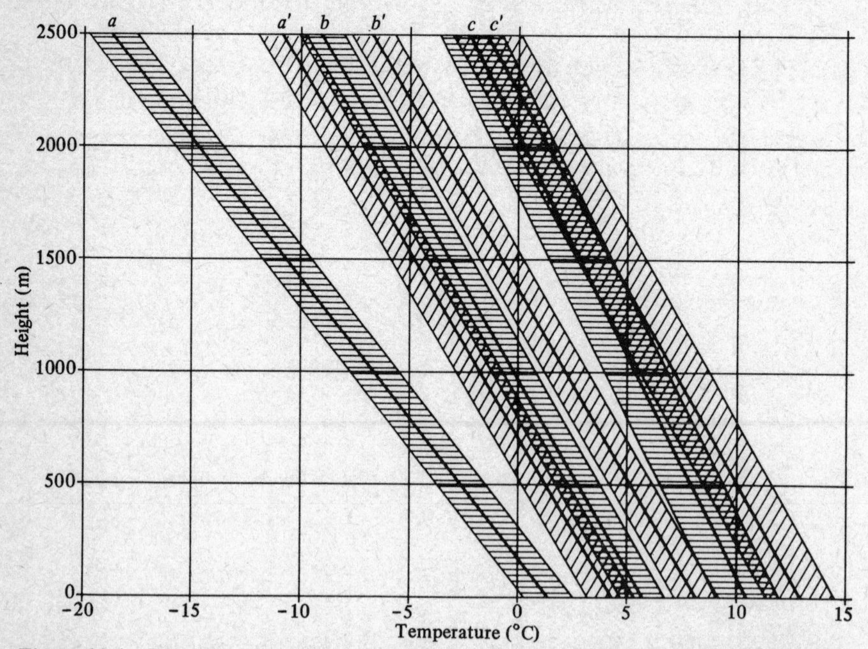

Figure 13.1. Decrease of temperature with altitude of the adrets and the bottoms of the valleys in January (with the standard errors shaded): *a*, *b*, *c* are minima, means, and maxima of the bottoms of the valleys; *a'*, *b'*, *c'* are minima, means, and maxima of the adrets.

Calculation by correlation

As the weather stations are rarely located on shaded slopes (ubacs) or on the tops of ridges, there is insufficient data and the previous method cannot be used. Another method, using correlation, has thus been tried. This employs fewer records and hence will be somewhat less reliable (Douguédroit, 1980).

The method

This method relies on the comparison between the thermal energy patterns known for the whole region (as estimated using the previous method for the adrets and valley bottoms) and the other values recorded at only a few points. The data used in this case are from a group of stations specially set up near St Martin-Vésubie (Alpes-Maritimes) in the valley of the Boreon river between 1500 and 1650 m altitude. Four stations have been in operation from 1966 to 1971: number 1 (adret, 1500 m), number 2 (adret, 1530 m), number 3 (bottom of valley, 1530 m), number 4 (ubac, 1650 m), and two from 1971 to 1978: number 5 (adret, 1650 m), and number 6 (ubac, 1650 m).

First, the local monthly records were compared with the mean regional record of the same topoclimate at the same altitude. The local records are included in the interval values of the regional record, even when different periods are concerned. Then, second, the data were correlated in pairs, using each time the adret as the independent variable, x. The resulting equations are given in table 13.2.

Table 13.2. Correlations between the temperatures of three topoclimates.

Correlation number	Comparison	Maxima		Minima	
		equations	IV[a]	equations	IV[a]
1	regional	$y = 1{\cdot}08x - 1{\cdot}35$	0·95	$y = 1{\cdot}09x - 3{\cdot}3$	0·55
2	station 3 with 2	$y = 1{\cdot}15x - 4{\cdot}3$	1·3	$y = x - 2$	0·9
3	station 3 with 1	$y = 1{\cdot}15x - 1{\cdot}7$	1·5	$y = 0{\cdot}96x - 2{\cdot}1$	1·1
4	station 4 with 1	$y = 1{\cdot}1x - 2{\cdot}5$	1·0	$y = x - 1{\cdot}1$	0·9
5	station 4 with 5	$y = 1{\cdot}09x - 3{\cdot}0$	1·5	$y = x - 0{\cdot}8$	0·75

[a] Interval values to 80%.

The results

The correlation lines obtained for the temperature maxima of the adrets and the valley bottoms are quite close to each other and are included inside the interval values of the regional regression (figure 13.2). These have both a regional and a local meaning. With such correlations, when one knows the temperature record for a stated month, it is possible to obtain the unknown temperature at the bottom of the valley during the same month, and *vice versa*.

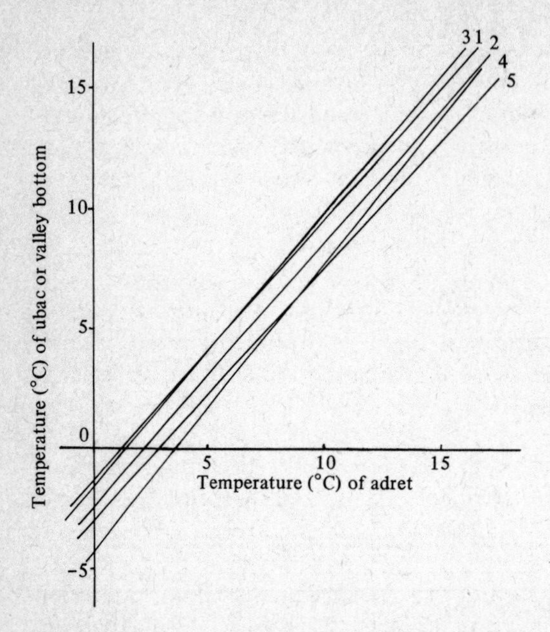

Figure 13.2. Correlations between the temperature maxima of the three topoclimates. A: 'adrets', U: 'ubacs', VB: bottoms of the valleys. For the numbers of the lines, see table 13.2 and the text.

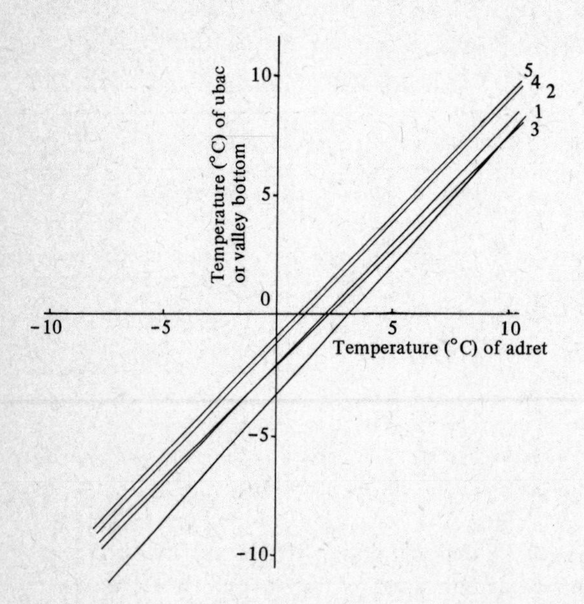

Figure 13.3. Correlations between the temperature minima of three topoclimates. A: 'adrets', U: 'ubacs', VB: bottoms of the valleys. For the numbers of the lines, see table 13.2 and the text.

The correlation between the maxima of the ubacs and the adrets was calculated by reducing the records of the latter to the 1650 m level by using the regional-gradient regression line previously calculated. The lines are parallel to the others, but 1·5 °C lower. So we can conclude that the maximum temperatures at the ubacs can most likely be represented by a formula of the kind:

$$T_{\mathrm{max,\,U}} = T_{\mathrm{max,\,VB}} - 2\,. \tag{13.7}$$

For the altitudes between 1500 and 1650 m, the correlations of the minima are given by temperatures between 10 °C and −10 °C in figure 13.3. For reasons still unknown, the local line thus obtained differs from the regional line below 0 °C, where it goes beyond the standard error limit. The relation of the monthly temperature of the local topoclimates with the regional topoclimates is not as good for the minima as for the maxima. For the ubacs, this can be represented by a line parallel to the regional regression, adrets–valley bottoms, but 2 °C higher. This is given by

$$T_{\mathrm{min,\,U}} = T_{\mathrm{min,\,VB}} - 2\,. \tag{13.8}$$

Conclusion

A region with a homogeneous climate can be considered as a mosaic of four topoclimates. Their temperature gradients are all different and can be calculated for mountain areas if enough information is available. The equations (13.3) to (13.6) given here can be generalised to every region of high and middle latitudes where the record of the same kinds of topoclimates are sufficiently numerous. Equations (13.7) and (13.8) given above are specific to the region studied here and cannot be generalised readily, but they do show a way to complete the model in the case of poor amounts of information.

References

Angot A, 1903 "La variation de la température moyenne avec l'altitude" in *Etudes sur le climat de la France: la température moyenne, Annales du Bureau Central Météorologique de France* pp 119–121

Douguédroit A, 1976 *Les paysages forestiers de Haute-Provence et des Alpes Maritimes* (Edisud, Aix-en-Provence)

Douguédroit A, 1980 "Les topoclimats de la Haute-Vésubie (Alpes-Maritimes, France)" *Méditerranée* **4** 3–11

Douguédroit A, de Saintignon M F, 1970 "Méthode d'étude de la décroissance des températures en montagne de latitudes moyennes: exemple des Alpes du Sud" *Revue de Géographie Alpine* **58** 453–472

Douguédroit A, de Saintignon M F, 1974 "A propos des Alpes Françaises du Sud, un nouveau mode de représentation des températures en montagnes: l'orothermogramme" *Revue de Géographie Alpine* **62** 205–217

Douguédroit A, de Saintignon M F, 1981 "Les régimes thermiques de la Région Provence-Alpes su Sud" *Mélanges Péguy* ER numero 30 du CNRS (Université Scientifique et Médicale de Grenoble)

Part 3

Applications in human geography

The evolution of urban spatial structure: the evolution of theory

A G Wilson

Introduction
The theory of the structure of urban areas, and the evolution of that structure, poses one of the most important problems in human geography. In this chapter, a very broad review is undertaken of the evolution of this theory of evolution. The argument for doing this is threefold: first, by examining past contributions to theory within a broad framework, it is possible to see linkages which otherwise might be missed; second, we can seek to identify the most important contributions for the present from past work—and we find that some issues raised and partially dealt with in the past are not effectively handled today (and conversely); and third, if a pattern of evolution of theory emerges, this can offer useful pointers for future research.

The argument is organised as follows: in the first section, some broad concepts are introduced which provide an overall framework; the next three sections contain reviews of theory in three roughly chronological phases—the classical approaches, the products of the second 'mathematical' phase, and the rapidly-developing concern with dynamics and system evolution. Some speculations about future phases are presented in the conclusion.

A basic framework
Introduction
It is argued that to approach theory building in relation to urban systems, six 'design' issues are involved. These are concerned with: entitation, scale, partialness/comprehensiveness, form of spatial representation, hypotheses, and methods.

Entitation—Chapman's (1977) notion—is concerned with the definition and articulation of the elements and systems which make up the overall object of study. Scale relates to the level of resolution at which the system of interest is observed. The approach to theory can then be on a spectrum from very partial, relating to particular elements, to totally comprehensive. It is argued below that the form of spatial representation is particularly crucial and is not usually considered explicitly. These four steps then provide the subject matter and its description, about which it is possible to build hypotheses and to select from a variety of methods to build theories and models. Each of these six design issues is outlined briefly in turn below.

Entitation

The main types of elements with which we are concerned are: people, organisations, commodities, goods, and services, land, and physical structures and facilities.

Some of the characteristics which are important for describing people are listed in table 14.1 and some types of organisation are listed in table 14.2. The latter includes, by implication, a broad description of commodities, goods, and services.

Our theoretical questions are concerned with the activities and processes associated with these elements, and in particular the *location* and *spatial interaction* patterns associated with them. A theory of dynamics is concerned with change in such patterns.

Table 14.1. Some characteristics of people.

Age	House type and residential location
Sex	Shopping frequencies
Education	Baskets of goods purchased and location
Position in a household	Recreational activities and locations
Job description and location	Services used and locations
Income	Allocation of time amongst different activities
Wealth	

Table 14.2. Types of organisation (by activity).

Primary	Housing
Manufacturing	Distributive trades:
Industrial and domestic services:	Wholesaling
Utilities	Retailing
Construction (except housing)	Governmental personal services:
Transport	Education
Communications	Health
Professional and scientific	Social
Financial, etc	Fire
Legal	Police
General governmental services to industry	Miscellaneous
and the population:	Cultural and recreational—indoor
Administration	Cultural and recreational—outdoor
Defence	
Justice	
Other	

Scale

A useful broad division of possible levels of resolution is micro, meso, and macro. At the microscale, at least some individual elements can be distinguished. The mesoscale, in a geographical context, usually implies a sufficiently fine spatial level of resolution that spatial patterns can be identified, but it probably deals with elements in groups rather than as individuals. The macroscale also relates to groups and will have at most a very coarse scale of spatial resolution.

The important point to note at this stage is that many hypotheses, about the behaviour of individual people or firms, for example, involve a microscale representation. However, there are some properties and patterns in urban systems—such as transport flows and densities—which are essentially meso in scale, and yet others—such as nationally determined commodity prices—which are essentially macro.

It is also important to emphasise that the notions of micro, meso, and macro are approximate: more subdivisions may be involved. However, there *are* different scales, and one of the most important and difficult tasks of the theoretician is to relate ideas and concepts at different scales—the so-called aggregation problem.

The partial-comprehensive spectrum

The simplest (at least conceptually) theoretical problem would involve the investigation of the locational behaviour of an individual household or firm with the rest of the overall system being taken as a 'given' environment. This is an important problem in its own right and can generate valuable insights for bigger problems, as we shall see. However, it is impossible to aggregate theories built in this way to add up to a theory of the structure of the system as a whole.

The next step along a spectrum towards total comprehensiveness would be to involve, say, two or more individuals (or firms) of the same type (in the same sector), perhaps because they are in competition. A further step would be to model competing classes or sectors, and so on. We shall explore a wide range of examples below.

Form of spatial representation

It is helpful to begin this discussion with a useful, if approximate, distinction made by Paelinck and Nijkamp (1975). They define consumers (roughly our 'people', but also including some activities of organisations) and producers (our organisations) and make a distinction between dispersed activities and concentrated ones. For relationships between consumers and producers, this generates the four possible combinations presented in table 14.3.

Examples of each case are:

I a firm and its suppliers (both concentrated);

II people (dispersed in residences) and producers of major services, say (concentrated);

Table 14.3. A spatial classification of activities.

		Producers	
		Concentrated	Dispersed
Consumers	Concentrated	I	III
	Dispersed	II	IV

III markets (concentrated consumers) and farmers (dispersed producers, using a lot of land);
IV people (dispersed) and small service facilities (dispersed—like corner shops).

The important argument here is that many activities can be assumed to be concentrated at a *point* in space, whereas others (housing and agriculture in particular) are large land users. (Note that the 'dispersed' category covers the second type of example just mentioned, but also the case where there are a lot of points.)

There are three kinds of element which can make up spatial representation in relation to land use: points, zones, and continuous space. To these, for completeness, we might add network links in relation to spatial interaction.

Points carry the implications of particular spatial addresses and are obviously used in relation to concentrated activities. Continuous-space representations have been much used—for example in theories about densities, which are obviously 'continuous' properties. Zoning systems are interesting in that they can be used in both ways: quantities associated with a zone (such as population) can be notionally considered to be located at the zone centroid, and is thus the basis of a point representation, at least approximately; or continuous-space properties (like densities) can be calculated as an average for each zone, and the pattern for all zones then provides an approximate representation of quantities which would more usually be dealt with in a continuous-space representation.

Network links connect vertices which are, of course, points. These may be a subset of real points at which there may also be concentrated activities; or the network may be a notional one, connecting zone centroids (a 'spider network', offering a coarser scale for the network representation).

One further distinction has to be introduced: when point representations are used, the points can be considered fixed—for example on a lattice, or as the vertices of a network—or they may be allowed to vary in continuous space. A location question would then be stated in terms of 'at which point?' in the former case, or 'where?' in the latter. If zone centroids are being treated as points, then this representation is of the first type, and 'at which point?' or 'where?' questions then become 'in which zone?'.

Classical point representations, using a lattice as in central place theory for example, often involve a large number of points together with some undesirable regularities which are imposed by the form of the lattice. Continuous-space representations are usually very difficult to handle in relation to the mathematical tools which are available for them.

It is argued here that zone systems usually offer the best of all worlds, even though approximations are sometimes involved, for the geographical theorist. They include all the benefits of point representations (because they can be interpreted as such if appropriate); they can be regular or not (which may be useful in relation to data availability); and they provide the basis of a mathematical representation which is very powerful—simply by

Table 14.4. A basis for the formulation of theoretical problems.

Scale	Element/subsystem	Partial/comprehensive		Relevant spatial representations	Structure (static)	Process (dispersed)
Micro	Households	Single:	Multiple:	Points relating to other	Location	Birth
	Farms	given	(a) fixed	points; possibly varying		Death
		environment	environment in	continuously		Migration/
			other sectors	Fields	Flows	relocation
	Firms—private		(b) changing	Points relating to other		Change in level
	public		environment in	points		of activity
	Services—private		other sectors	Points relating to 'markets'		
	public					
	Land plots			Land units		Objectives/
						constraints
Meso	Residential	Single sector,	Interdependent	Continuous zones		
	Agricultural	given others	sectors	Continuous zones		
	Industrial			Continuous zones		Level changes
	Services—private			Continuous zones	Location	
	public				patterns	
	Settlement patterns			Continuous zones		
	(boundary problems)					Densities
	Land-use mix			Continuous zones		
	Transport/communications			Networks	Flows	Accessibilities, etc
Macro	Population	Demography			Population	Prices, etc
			Interdependent	Aggregated	totals:	Level changes
	Economy	Economy,			urban,	Growth/decline
		given			demand for	processes
		population			goods and	
					services	
					Economy:	
					supply of goods	
					and services	

labelling zones consecutively and using the label as a zone subscript. Thus P_i may be the population of zone i, and T_{ij} the number of trips from zone i to zone j.

For present purposes, the main point of this argument is that choice of spatial representation should be explicit; and also that it is useful to scrutinise the corresponding choice by other theorists to see if alternatives would be more fruitful.

Hypotheses
At the microscale, the assumptions of neoclassical economics usually dominate theory building. It is assumed in the theory of consumers' behaviour that preferences can be recorded in some way, usually in some form of utility function. The firm is correspondingly assumed to maximise profits in relation to a production function which expresses the technical possibilities. The broad frameworks of these theories contain relatively weak assumptions; what is usually missing is the detail. This is a point to which we shall return in the final section.

At the mesoscale and at the macroscale, theories can be constructed either by applying microtype assumptions to groups or sectors, or by using statistical averaging procedures, or through some more ad hoc phenomenological approach.

Further discussion about hypothesis formulation is reserved for the presentation of the various examples below.

Methods
A more detailed account of methods for theory building is given elsewhere (Wilson, 1981a). Here, I shall simply present a list and some brief comments and again take the discussion further in relation to the examples below. The main headings include: algebraic modelling, entropy maximising, account-based statistical averaging, optimisation, network analysis, and dynamical systems theory (including bifurcation theory).

Essentially, this is a list of some of the more important mathematical techniques which are available to turn hypotheses into theories in the form of operational models. Many of these have become available in relatively recent years and it is important to bear this in mind in assessing the contributions of the classical theorists.

Synthesis
A presentation of the main range of theoretical problems based on the argument of the preceding subsections is offered as table 14.4.

Phase 1: the classical approaches
Introduction
In this section, I shall review some of the major classical contributions to the theory of spatial structure, taking their content as well-known and therefore focusing on the implicit theory–design decisions in relation to the six main headings of the previous section.

Agricultural land use—von Thünen

Von Thünen (1826) was concerned with the intensity and type of agricultural land use around a single point which constituted the market. His analysis generated the well-known concentric rings. The scale for the analysis is basically meso—he does not distinguish individual farms for instance (though in another context, he does apply his method to a single farm). The approach is partial in being concerned with only one sector, and the more so because he does not deal explicitly with competition within that sector. The system being analysed is of the (single) point (consumer)-dispersed (producer) type, and space is treated continuously for producers. This illustrates a feature of geographical theory in such a treatment: that the task is to delimit the *boundaries* of different uses.

One of von Thünen's most important contributions was the basis of his main hypothesis: that land use is determined by the maximisation of economic (sometimes called 'location') rent. This notion of the concept of rent has had implications far beyond von Thünen's use of it. His results can be derived using elementary algebra—and indeed the original presentation is mainly in terms of the arithmetic of a large number of cases. However, the most valuable methodological treatment from a modern point of view is that developed by Stevens (1968), who showed how the analysis could be cast as a mathematical programming problem. It is also possible to use an alternative spatial representation.

Industrial location—Weber

Weber's (1909) main problem was the optimum location of a firm, given the locations of its inputs and of its market (assuming all consumption is at a single point). So this is a microscale but very partial approach with a spatial representation based on points. There is the element of continuous space about the treatment as noted earlier, however, in that the position of the firm can vary continuously. The main hypothesis is based on transport-cost minimisation (since it is assumed that all other costs and prices are location-independent). Weber used a mechanical device due to Varignon to 'solve' his problem, or, at least in the special case of two input points, a geometrical construction.

Weber was well aware of the limitations created by his assumptions and showed in a number of ways how his basic problem could be generalised. He used his 'materials index' to make observations about the likely behaviour of a whole industry, thereby making the approach less partial (and at a more mesoscale). He showed how to introduce variation in labour costs (though with some confusion in relation to spatial representation, since it is not clear whether he restricts labour to points or not); and he made a valuable contribution about the nature of agglomeration economies. He was always aware of the possibilities of further extension and in his last two chapters writes like a modern systems analyst.

The most direct application of Weberian ideas in the present day is probably to the location of public facilities in a cost-minimising way. And the development of mathematical tools has produced a more effective formulation and solution of the problem, in particular, the use of mathematical programming and iterative computer algorithms. A simpler programming version of the model can also be obtained through a slight shift in spatial representation, namely by restricting the possible location of the firm (or other facility) to one of a set of fixed points.

Competition and market areas—Palander, Hoover, and Hotelling

The next step in the argument, which was first taken by Palander (1935), is to make the Weber problem less partial by considering two firms competing for consumers. The problem then becomes one of delimiting their respective market areas. This is now a dispersed (consumers)–concentrated (producers) problem, with consumers being handled in a continuous-space representation and producers again as points. A further and important extension introduced at this time was to use economic theory to model the demand for goods as a function of basic price plus transport costs. Palander considered a variety of cases. Hoover (1937) extended the argument to more than one firm. Note that what is now a market-area problem now has a common feature with von Thünen's approach—the task of delimiting regions in continuous space.

Hotelling (1929) extended Weber's argument in a different way by allowing the two competing firms to change their locations in response to each other and he considered the problem of 'stability under competition'. His famous example is the two ice-cream men on a linear beach. He showed that in the case of inelastic demand, the two would each locate at the centre, sharing the market, rather than at the quartiles, where they would also share the market in locations which would, in addition, be optimum for consumers (by minimising their transport costs).

Central place theory—Christaller and Lösch

Central place theory is concerned with two kinds of consumers—a continuously-distributed rural population, and people in settlements. The settlements are considered as points, and incorporate the producers of goods and services. This is, therefore, a coarse mesoscale picture, but it is much more comprehensive in approach than the examples considered previously. The theoretical basis of the model is built on the different kinds of market areas for different kinds of goods. Low-order goods have small market areas, and conversely.

In Christaller's (1933) system, a level is chosen to start the analysis, and an investigation of market areas determines the centre-spacing at that level. Because of the nested nature of the spatial system, this determines the spacing of all other levels, up and down; and, hence, since the range of goods can vary widely, so is the mix of goods sold at different types of centres.

Lösch's (1940) system is built from the lowest levels of the hierarchy. Increasingly large market areas are considered with increasing orders of goods. Many systems are then superposed around a given metropolitan centre to determine a hierarchical system which has, in terms of mixes of goods, more variety than Christaller's system, but which has its own rigidities in other respects.

The theoretical basis of each model is built on the notion of the demand function for different types of good and the transport cost to the consumer of collecting them. As we have seen, the hypotheses determine the location of the settlement points. Relatively simple assumptions are made about the nature of the competition which generates the market-area patterns. It can be argued that the form of spatial representation, linked with the hypotheses, creates spatial systems in each case which are too rigid to represent reality.

The urban ecologists—Burgess, Hoyt, Harris and Ullman

The approaches represented under this heading are very different in their disciplinary backgrounds. The overall title comes from the work of Burgess and others such as Park and McKenzie from the 1920s school of Chicago sociologists. The ecological label stems from their use of analogy with plant ecology as the basis for determining the patterns of urban land use. The main emphasis is on residential land use, and a continuous space representation is used. The task then becomes one of delimiting boundaries. Burgess's (1927) model generates concentric rings based on notions of succession and dominance of the outward movement of upwardly mobile people. Hoyt (1939) qualifies this by noting major sectoral differentiation; and Harris and Ullman (1945) by the influence of multiple nuclei. The three sets of ideas can in principle be combined (Mann, 1965).

The methods involve elementary geometry and the hypotheses are not really sharp enough to make further progress without new tools.

The gravity model—Carey and Ravenstein onwards

The use of the 'gravity model' analogy in the social sciences has a history which goes back to the nineteenth century (for example, Carey, 1858; Ravenstein, 1885). The first applications were to the flow of migration between cities. Interestingly, the applications up to the 1950s were based on point-to-point flows, and in the first applications in retailing there was a greater concern with the demarcation of market areas than with the flows themselves (Reilly, 1931). Simple terms were used for the mass terms and impedance functions (usually population and inverse distance or inverse distance square, respectively), so that, typically, the model took the form

$$T_{ij} = \frac{KP_iP_j}{d_{ij}^2} \ . \tag{14.1}$$

Phase 2: computers, operational research, mathematical programming, and statistical averaging

The origins of the second phase

It is far beyond the scope of this account to attempt a detailed analysis of the origins of the current phase of theory in human geography. What will be attempted is a number of broad observations, conjectures really, about the generating impulses.

The beginnings of the transition can be found from about the middle 1950s, and the decade of rapid development was the 1960s. At least four different impulses can be identified, although they are closely related. They mostly stem from outside geography. First, one should note the development of large electronic computers. It became possible to tackle much bigger problems and to relate theory to observation. This effectively dates from the late 1950s. Preceding this chronologically, but perhaps a second 'cause' in relation to the 'new geography', was the development of operational research during the Second World War. This generated skills of mathematical modelling and provided a tradition within which much theory could be treated more formally and explicitly by using mathematical tools. Third, although operational research produced a range of methods, it is worthwhile singling out one which turned out to be particularly important for geographical theory: that is, mathematical programming, both linear and nonlinear. The importance of this lies in the *behavioural* optimising processes which underly much geographical theory—utility or 'location rent' maximisation, profit maximisation or consumers' surplus maximisation providing examples. This also connects to an important subsidiary role of geographical theory in providing the analytical basis of much planning, and this involves a different kind of optimisation. Fourth, it can be argued that geographers learned the techniques for modelling certain types of large system—Weaver's (1958) systems of disorganised complexity—either by account-based statistical averaging methods (applied to populations or economics) or by entropy-maximising methods (applied to spatial interaction and location problems).

There are, of course, many other starts to theoretical development; another, for example, is the push from urban and regional economics from the early 1960s onwards. It is argued here, however, that it is useful to focus on the broad background in order to see the context within which geographical theory made substantial strides forward.

In the rest of this section, I shall review examples of some of the products of this phase. It is impossible to be comprehensive and I shall concentrate on showing particularly the benefits of new representations and new methods. It is convenient to deal with spatial interaction first, and then with aspects of location theory.

Spatial interaction

The main influence on the development of spatial interaction models was the work of civil engineers employed on large-scale transportation studies from the mid-1950s onwards. They conceptualised the structure of transport patterns in terms of generation (production and attraction of trips, from origins to destinations), distribution, modal split, and network assignment. Possibly because of the nature of the large-scale surveys associated with these studies, the data analysis and modelling were based on discrete zoning systems.

A number of key advances can be identified (see, for example, Wilson, 1974, chapter 9):

(1) Because it was thought that total trip productions and attractions were 'sounder' quantities than flows, trip matrices were 'adjusted' so that origin and destination flows summed to give (or modelled) totals. In effect, this replaced the old constant of proportionality in the gravity model by multiplicative balancing factors.

(2) A corollary of this step was to seek more sophisticated models of 'mass' terms.

(3) A wide range of distance-decay functions were explored.

(4) Trips were categorised by purpose, mode, person type, and so on.

(5) New measures of 'distance' were introduced, culminating in the notion of 'generalised cost'.

(6) Network congestion was treated explicitly by coupling certain elements in the generalised cost terms with link costs estimated from network assignment submodels.

A typical spatial interaction model might then be

$$T_{ij}^n = A_i^n B_j^n O_i^n D_j^n \mathrm{f}^n(c_{ij}) \,, \qquad (14.2)$$

for trips of category n between zones i and j. O_i^n and D_j^n are known origin and destination totals, f^n is a suitable function, c_{ij} is the generalised cost. A_i^n and B_j^n are balancing factors which ensure

$$\sum_j T_{ij}^n = O_i^n \,, \qquad (14.3)$$

$$\sum_i T_{ij}^n = D_j^n \,, \qquad (14.4)$$

and hence are given by

$$A_i^n = \left[\sum_j B_j^n D_j^n \mathrm{f}^n(c_{ij})\right]^{-1} , \qquad (14.5)$$

$$B_j^n = \left[\sum_i A_i^n O_i^n \mathrm{f}^n(c_{ij})\right]^{-1} . \qquad (14.6)$$

These ideas were eventually absorbed into a range of other disciplines, including geography. A wealth of empirical experience became available, mainly through the transportation studies and theoretical understanding developed in a number of directions. By the late 1970s, for example, there were a large number of alternative theoretical derivations of spatial interaction models, ranging from entropy maximising (Wilson, 1970) as a form of spatial interaction model to various forms of utility maximising models (Williams, 1977), derived from microeconomic principles. These represent alternative ways of solving the aggregation problem. For a general review, see Wilson (1974, chapter 9).

The basic model was also related to the transportation problem of linear programming, and the balancing factors could then be seen as transformations of the dual variables of a nonlinear mathematical program and interpreted in terms of comparative advantage (Evans, 1973; Wilson and Senior, 1974).

The outcome of much research can be seen as the development of an extended *family* of spatial interaction models, some of which have implications for location theory as we will see in the next section (Wilson, 1971).

Location theory

Two broad approaches to location theory will be sketched in this section. The first is, at least initially, essentially phenomenological—likely empirical relationships form the basis of the models; the second is economic. The two approaches were at first based on different spatial representations—the use of discrete zones and continuous space respectively. But when discrete zoning systems are used in the economic approach, it turns out that the two approaches come very close together in an interesting way.

The phenomenological approach can be seen in the first instance as based on the *singly*-constrained spatial interaction model. This takes the form

$$T_{ij} = A_i O_i W_j \mathrm{f}(c_{ij}) , \qquad (14.7)$$

where

$$A_i = \left[\sum_j W_j \mathrm{f}(c_{ij})\right]^{-1} \qquad (14.8)$$

to ensure that

$$\sum_j T_{ij} = O_i . \qquad (14.9)$$

T_{ij} is the flow, O_i the known origin total, W_j a measure of the attractiveness of zone j, c_{ij} the interzonal cost, and f the impedance function.

The total flow attracted to each zone j, say D_j, can then be predicted by the model as

$$D_j = \sum_i T_{ij} . \qquad (14.10)$$

Lowry (1964) and Lakshmanan and Hansen (1965) were among early authors who observed that such an interaction model was also a *location* model—the predicted variable D_j is a locational property of the system.

This model is appropriate if it can be argued that such locational distributions are mainly determined by an interaction. Examples are the use of shopping centres, where T_{ij} is the flow from residences to centres, with O_i as the given expenditure by residential zone, and W_j the attractiveness of the shopping centre at j; or residential location, where O_i would be a given job distribution, and W_j the attractiveness of j for housing. The 'shopping' example can easily be extended to other services; and since these services are a substantial source of employment, service models and residential models can be linked as was achieved in Lowry's (1964) *Model of Metropolis*.

The economic approach hinges on defining utility functions for people, both in relation to residence, purchase of goods, and use of services; and to profit functions for firms. A turning point in the development of the theory was Alonso's (1960; 1964) use of the concept of 'bid rent' in urban analysis—in some ways an extension of von Thünen's notion of rent. Bid rent is, in effect, a representation of the utility function. It is the maximum amount a consumer would be prepared to bid—not necessarily what he actually pays—for a particular combination of housing, goods, and services *at particular locations* (where this is relevant). Most of the development of the theory has used continuous-space representations and is well reviewed by Richardson (1977).

Here we concentrate on the theory as expressed in zoning-system representations. This was achieved early in the Herbert and Stevens (1960) model which was a mathematical programming representation of Alonso's theory. Using the same kinds of variables as above, we can write this as

$$\text{maximise } Z = \sum T_{ij}(b_{ij} - c_{ij}) , \tag{14.11}$$

subject to

$$\sum_j T_{ij} = O_i . \tag{14.12}$$

This is a linear programming problem.

It turns out that there is the same relationship between the spatial interaction model of residential location, as between the doubly-constrained spatial interaction model and the transportation problem of linear programming (Senior and Wilson, 1974). If the spatial interaction model is written in the form,

$$T_{ij} = A_i O_i \exp[\beta(b_{ij} - c_{ij})] , \tag{14.13}$$

with the constraint (14.12), then the linear programme is the limit of (14.13) as $\beta \to \infty$. In practice, of course, there are more constraints and the model has to be disaggregated to introduce person-type and house-type categories.

There are two major implications of this convergence of modelling styles. First, the attractiveness terms in singly-constrained interaction models can, with suitable transformation, be interpreted in terms of benefits, and these are measures of bid rents, utility, or preferences. Second, the interaction model can be seen as incorporating the same behavioural *optimising* principles as the economic model, but where the optimum optimorum is not achieved. This *dispersion* of behaviour from the optimum is realistic in terms of market imperfections, variations in preferences which are not recorded in the simple functions used, and so on. It can be considered as generated from an entropy-maximising philosophy or from, say, random utility theory. Thus, the phenomenological model can now be reinterpreted as an economic model, and indeed in such a way that the traditional economic model is a special case (achieved when one or more parameters tend to infinity).

An important implication for model development is that great attention should be paid to the design of attractiveness functions. The W_j in equation (14.7) can be taken in a function of a number of variables—those which are the components of the appropriate utility function. Formally, we can write

$$W_j = X_{1j}^{\alpha_1} X_{2j}^{\alpha_2} X_{3j}^{\alpha_3} \dots , \tag{14.14}$$

and an important research task is to identify the variables X_{kj} and the parameters α_{kj}. To illustrate the potential complexities, we can note that these variables themselves may be composite. For example, part of the residential attractiveness function may be related to 'accessibility to shops'. This can be written as

$$X_j = \sum_i W_i^{\text{sh}} \mathrm{f}(c_{ij}) , \tag{14.15}$$

where W_i^{sh} is the shopping-centre attractiveness function. The term X_j can now be used in an equation like (14.14).

The argument so far essentially is that there is a rich set of models available for modelling the distribution of population and population activities. However, many crucial terms in attractiveness (or utility) functions have to be taken as given in the system. The obvious examples are housing supply and shopping-centre supply. These represent the physical structures which contain the various activities. The next step in the argument, therefore, is to see how to model these.

In residential location models, it has been common to assume that housing supply 'follows' the population allocated by residential location models. This may be adequate for new development but does not describe the process of change adequately. The turnover of different kinds of people in housing is obviously more rapid than the change in housing stock itself, for example.

In service models, there have been essentially two approaches. The first and simplest involves inputting a *trial* set of exogenous structural variables, like shopping-centre locations and size, and using the activity models to calculate the revenues which would be attracted. Normal adjustments can then be made within a planning process. The alternative is to try to model the supply side. This in turn subdivides into two approaches. If the private sector is being modelled, as with shopping centres, this involves modelling entrepreneurial behaviour. In the case of public-sector facilities, a mathematical programming formulation—say minimising costs or maximising consumers' surplus—can be used. This can be done using location–allocation models [as reviewed by Leonardi (1980) for example] or by *embedding* spatial-interaction type models within an overall programming framework (Coelho and Wilson, 1977; Coelho et al, 1978). An extensive discussion of many of these models can be found in Wilson et al, 1981. These approaches all involve, in different ways, the study of stability and system dynamics, and this takes us into the third phase of modelling the evolution of urban spatial structure; in the next section we explore some possible approaches.

Phase 3: towards effective dynamic modelling

Origins

Phase-2 models could be, and were, used for forecasting; but in a comparative-static mode. It was assumed that the systems of interest were always in equilibrium, though in practice, of course, this is unlikely to be the case. The first major break away from this position came with the publication of Forrester's (1969) *Urban Dynamics*. His model was a set of simultaneous difference equations which seemed to show quite complicated behaviour. The model was constructed on an ad hoc basis with inadequate theory and even more inadequate data. Nonetheless, it served to remind modellers that it was complacent to assume that strong equilibrium conditions and comparative-static methods were always appropriate. However, possibly because of the difficulties of assembling data even for comparative-static modelling let alone anything more complicated, there have been relatively few alternative developments.

The phase-2 models—particularly the Lowry model—have been developed into difference-equations dynamic models by Batty (1971). Cordey Hayes (1972) has shown how to write 'kinetic' equations—a form of accounting equations—for urban and regional systems. Essentially these take the form,

$$\frac{\mathrm{d}x_i}{\mathrm{d}t} = \sum_j (a_{ji}x_j - a_{ij}x_i) , \qquad (14.16)$$

for a set of system-state variables $\{x_i\}$ and transition coefficients $\{a_{ij}\}$. Such equation systems can be large and complicated, first because indices like i and j can themselves be lists, $(i_1, i_2, i_3, ..., j_1, j_2, j_3, ...)$ say, and second because the transition coefficients can be complicated functions of many other variables, including state variables. The last condition makes the equations nonlinear, and this is important in another context as we will see shortly.

Another approach to dynamic modelling has its origins in the work of Orcutt et al (1961). This involves the microsimulation of the states of individuals in the system. These states are changed and updated by Monte Carlo methods. The main theoretical content of the approach is in the specification of the probability functions which determine the possible transitions. This method can be seen as solving numerically large simultaneous equation systems of the type (14.16). It is likely to be increasingly important, especially as more relevant data becomes available. Current large-scale projects employing this method as a basis are described by Clarke et al (1979), and Kain et al (1976).

Another current approach is to focus on some basic growth equations and to elaborate the parameters within them to form the basis of dynamic urban models. A family of such equations is

$$\dot{X}_i = \epsilon_i[D_i(X_1, X_2, ..., X_N) - X_i]X_i^n \tag{14.17}$$

for parameters ϵ_i and n, and a set of functions D_i which represent some form of capacity or demand. Allen et al (1978) have developed large models on this basis, linked to what they call the 'order from fluctuations' approach. A deeper understanding of the possibilities of such an approach, and indeed any approach to dynamic modelling which involves nonlinear components to the equations, involves a knowledge of bifurcation theory, and it is to this that we now turn, first in a comparative-static context, then in a more fully dynamic one.

The existence and stability of equilibrium states

In this section, the argument will be carried through using the shopping model as an example. A phase-2 production-constrained spatial-interaction shopping model takes the form

$$S_{ij} = A_i e_i P_i W_j^{\alpha} \exp(-\beta c_{ij}), \tag{14.18}$$

where

$$A_i = \left[\sum_j W_j^{\alpha} \exp(-\beta c_{ij})\right]^{-1}. \tag{14.19}$$

S_{ij} is the flow of cash from zone i to zone j, e_i is the per capita expenditure in zone i, P_i is the population of zone i, W_j is the attractiveness of shops in j, c_{ij} is the interzonal travel cost, and α and β are parameters. A_i in condition (14.19) ensures that

$$\sum_j S_{ij} = e_i P_i ,$$

and the model can be used as a location model to predict the revenue attracted to each centre, D_j:

$$D_j = \sum_i S_{ij} = \sum_i \frac{e_i P_i W_j^{\alpha} \exp(-\beta c_{ij})}{\sum_j W_j^{\alpha} \exp(-\beta c_{ij})}, \tag{14.20}$$

where we have used equations (14.18) and (14.19) to obtain the version of equation (14.20) on the right-hand side. This shows how D_j is a nonlinear function of all the W_j's.

I explained under the discussion of phase 2 that it was customary to use this model with trial sets of given W_j's. Suppose we now add a hypothesis on suppliers' behaviour, viz that a W_j grows if D_j exceeds the cost of supply, say kW_j (where k is a unit cost and we are assuming that W_j is measured in size units, say floorspace) and vice versa. This can be expressed as

$$\dot{W}_j = \epsilon(D_j - kW_j) \tag{14.21}$$

where D_j, recall, is given by equation (14.20). The equations (14.21) can therefore be written in full as

$$\dot{W}_j = \epsilon\left[\sum_i \frac{e_i P_i W_j^{\alpha} \exp(-\beta c_{ij})}{\sum\limits_j W_j^{\alpha} \exp(-\beta c_{ij})} - kW_j\right] . \tag{14.22}$$

We now have a set of equations which model the dynamics of *structural* variables such as $\{W_j\}$. In the next section, we shall focus on the properties of these equations directly; here, we shall concentrate on the equilibrium properties. The argument below is in accordance with Harris and Wilson (1978).

The equilibrium position is achieved when

$$\dot{W}_j = 0 , \tag{14.23}$$

that is, from expression (14.22), when

$$D_j = kW_j , \tag{14.24}$$

or, in full, by using equations (14.22), when

$$\sum_i \frac{e_i P_i W_j^{\alpha} \exp(-\beta c_{ij})}{\sum\limits_j W_j^{\alpha} \exp(-\beta c_{ij})} = kW_j . \tag{14.25}$$

The left- and right-hand sides of condition (14.25) are essentially different ways of calculating revenue at equilibrium. It is convenient to write

$$D_j^{(1)} = \sum_i \frac{e_i P_i W_j^{\alpha} \exp(-\beta c_{ij})}{\sum\limits_j W_j^{\alpha} \exp(-\beta c_{ij})} , \tag{14.26}$$

and

$$D_j^{(2)} = kW_j , \tag{14.27}$$

and then the equilibrium condition is

$$D_j^{(1)} = D_j^{(2)} . \tag{14.28}$$

We can now investigate the nature of the equilibrium using a trick.

Consider $D_j^{(1)}$ and $D_j^{(2)}$ as different functions of W_j. They can each be plotted ($D_j^{(2)}$ easily because it is a line of slope k). Various combinations for different parameter values are shown in figure 14.1.

A *bifurcation point* is a point in *parameter space* at which the nature of the solution to the differential equations changes; and here we are focusing on equilibrium solutions only. The parameter values at bifurcation points are *critical* values. Thus, we show for $\alpha > 1$ that there are critical values of k, labelled k_j^{crit}, at which the nonzero stable equilibrium disappears. Since this *only* happens for $\alpha > 1$, $\alpha = 1$ is also a critical value. It turns out that the $D_j^{(1)}$ curves move as functions of α and β, and so if the system is in a critical state, it is critical with respect to α and β as well as to k—and indeed all other exogenous variables like $\{e_i\}$, $\{P_i\}$, and $\{c_{ij}\}$. There is thus a surface in parameter space of critical parameter values. On one 'side' of this surface, development in zone j is possible (a DP-state); on the other, it is not (an NDP-state). This notion is very important for locational analysis. As the whole system develops, zones will 'cross-into' the DP-state, or vice versa. The complexity arises from the mutual interdependence of zones. In particular, at a critical state for zone j, exogenous 'parameters' include all the W_k, $k \neq j$, the structural variables for other zones. A numerical example of this argument is given by Wilson and Clarke (1979).

It is highly likely that, at any time, there are multiple stable-equilibrium states (by taking different combinations of W_j's to be zero). Thus, although this analysis gives a lot of insight into the mechanisms of development, it does not necessarily determine a particular path. This could be influenced in a major way by 'historical accidents' for example. We have the beginnings of a general theory which also demands knowledge of particular circumstances in particular places.

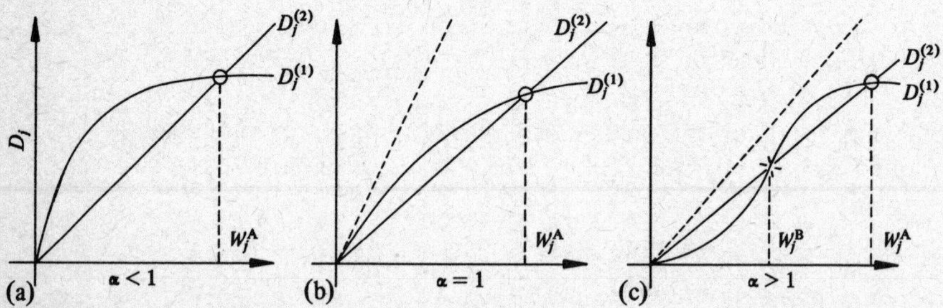

Figure 14.1. Revenue-centre-size relationship (w_j^A are nonzero stable equilibrium values; w_j^B is an unstable state).

Extensions to nonequilibrium states: difference equations and bifurcation

We must now return to equation (14.21) and explore bifurcation properties of a system which is not in equilibrium as a function of the parameter ϵ. Here, we follow the argument of Wilson (1979) as illustrated by Beaumont et al (1980a; 1980b). The simplest illustration arises when equation

(14.21) is transferred into difference-equation form. This is done by writing

$$W_{jt+1} - W_{jt} = \epsilon(D_{jt} - kW_{jt})W_{jt} \,, \tag{14.29}$$

and using an obvious modification of notation. A factor W_{jt} has been added so that it represents logistic growth. Thus a unit step length is assumed, or that the step length has been absorbed into the parameter ϵ. Equation (14.29) can be written

$$W_{jt+1} = [(1+\epsilon D_{jt})W_{jt} - \epsilon k W_{jt}^2] \,. \tag{14.30}$$

May (1976) has shown that a condition for a stable equilibrium point to exist is

$$0 < \epsilon D_{jt} < 2 \,, \tag{14.31}$$

and that for,

$$2 < \epsilon D_{jt} < 3 \,, \tag{14.32}$$

the equation has various kinds of oscillatory solutions ranging from 2-cycles for ϵD_{jt} near to 2, to 'chaotic' behaviour as it nears 3. Thus ϵD_{jt} equal to 2 is a critical point, and there are other bifurcation points up to ϵD_{jt} equal to 3. This illustrates a new kind of bifurcation: a transition from a stable equilibrium point to a periodic solution. Examples of this behaviour for a hypothetical case are shown in figure 14.2.

The bifurcation point arises when ϵ becomes so large that the system 'overresponds' to $(D_j - kW_j)$ differences. There are the additional complications that ϵD_{jt} itself varies with time and, for a particular value

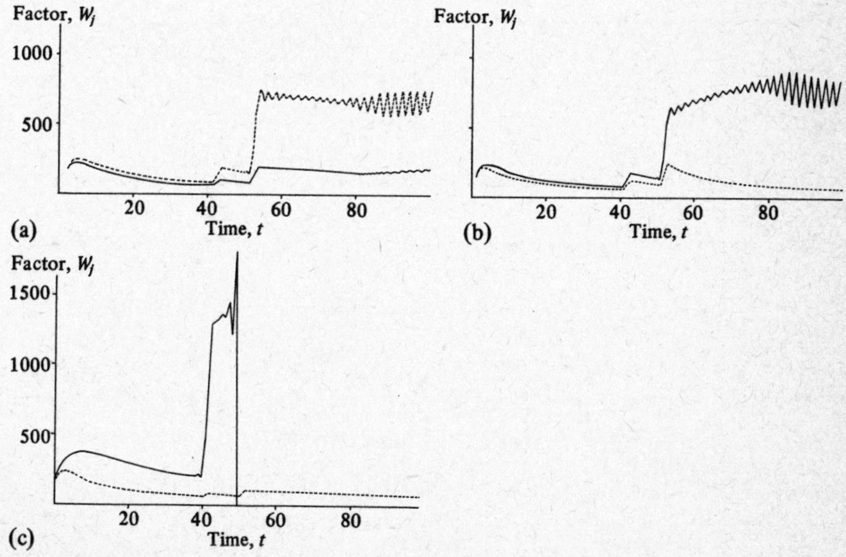

Figure 14.2. (a) and (b) Equilibrium to periodic transition; (c) divergence to zero.

of j, is a function of what is happening in other zones, with both of these effects arising through D_{jt}.

This kind of bifurcation has been illustrated in relation to one example. It has wider implications as we shall see in the next subsection.

Towards a comprehensive dynamic model

The example used in the preceding two subsections is obviously partial. Models could be built on similar principles for most sectors and it is clear from the example considered that they would be coupled. For instance, the shopping model would be coupled to a residential model (through the P_i's) and vice versa. This obviously increases the complexity of the overall model and the range of bifurcation possibilities. Some of the coupling also arises through disaggregation: if retailing is divided into sectors, then attractiveness for lower order goods may be enhanced by the presence of higher order goods. These complexities and couplings often enter through attractiveness terms, and hence into D_j terms in the differential or difference equation.

Examples of much more complex models are presented elsewhere (Wilson, 1981b). The main point to make here is that the kind of bifurcation behaviour which has been identified, both in relation to equilibrium points and to dynamics, can be expected even when the underlying submodels differ from those used to illustrate the argument here.

Future phases: towards evolutionary models

The need for more operational research

Phase 2 has generated an immense variety of models for particular subsystems and for interdependent subsystems. Although most of these models have been 'tested' in some form or other, the quantity of empirical work per model is relatively small (see Wilson et al, 1977, for examples). The only submodel which has been extensively used is the transport model. This work has been generated by the large investment in conurbation-transport studies all over the world. These have led to the development of the model in a number of directions and a much better understanding of its potential usefulness (as, for example, in Williams and Ortuzar, 1980), though *this* understanding is not yet widely put into practice. Work in other sectors, and with comprehensive models, has suffered from an ill-informed backlash against the use of models in planning. This tide will, in time, turn. Perhaps the most effective operational work which could be carried out to assist this would be on relatively partial problems, for example the location of public facilities such as schools and health services, where the scope seems substantial but is relatively untapped. This may provide the confidence for larger-scale operational research in urban and regional modelling.

There remains the question of empirical work for phase 3 and future-phase models, and these questions are taken up separately below.

The rewriting of phase 1

The essential argument given in relation to phase-2 models is that great progress has been possible through a combination of (mainly) two factors: first, a switch to discrete zoning as a form of spatial representation; second, the availability for the first time of certain powerful mathematical tools, particularly those of mathematical programming. The developments of phase 3, particularly the understanding of possible stability and bifurcation properties of systems, has similarly come about through the availability of new mathematical tools, and these are built solidly onto phase-2 foundations.

It is interesting in the light of this to look back to the classical theories which were identified in relation to phase 1. In most instances their essential contributions can be written, often very powerfully, in the new frameworks. Let us discuss each briefly in turn.

It is a straightforward matter to rewrite von Thünen's model as a mathematical programming model which employs discrete zoning. This process has not been taken very far, the seminal contribution being that of Stevens (1968).

Weber's main problem, similarly, can be rewritten as a mathematical programming problem by using point locations (which are broadly equivalent to discrete zoning systems, taking the points as zone centroids). It is interesting that its greatest potential usefulness in a modern context is possibly for the optimal location of public facilities. Indeed the one area which has not been effectively taken up in the phase-2 analysis above is that of industrial location. Much more sophisticated analyses of the relevant factors are now available (as, for example, in Smith, 1971). Some linear programming models have been built, for example by Stevens (1961), but as Smith notes, agglomeration economies and demand factors cannot easily be incorporated into such models. There seems to be no reason in principle why industrial location theory should not leap into phase 3 with models of the form of equation (14.29), given suitable definitions of the variables.

Palander's and Hoover's work, and to some extent that of Christaller and Lösch was based on theories which involved the demarcation of market areas. These are totally explicit in models based on spatial interaction, and the models are no longer tied to the restrictive assumption that market areas do not overlap. However, because the phase-2 models are mesoscale models, individual firms are not represented and so results like Hotelling's are not incorporated. There may be scope for further research here, particularly in sectors where a relatively small number of organisations are involved.

Central place theory, as we saw, is very much at a coarse mesoscale. With the phase-2 models, it is possible to offer more precise measures of settlement structure: like E_j^r, the number of employees by zone by sector; W_j^g, the service facilities in sector g by zone; and P_i^w, the number of type-w people in zone i. And as we have noted, the market areas are all explicit. Further, the phase-3 methods provide the potential for building a dynamic central place theory and of avoiding the rigidities of the original formulations in this respect.

The urban ecologists' theories have been overtaken by factorial ecology —see Rees (1979) for example—again using discrete zoning systems, and by the various approaches to residential location described under phase 2. There is one interesting sidelight here: the original work was based on analogies with plant ecology. It turns out that the dynamic models sketched under phase 3 above have similar structures to corresponding models in present-day ecology. This is not surprising in that both systems are made up of elements competing with each other for resources.

Finally, we noted under phase 2 that the gravity model, based on point-to-point or discrete zoning systems, has been overtaken by a much broader family of models with wider-ranging theoretical properties which can be applied in a variety of circumstances.

Alternative aggregations and submodels

In this review, I have perhaps overconcentrated on particular models. It is worth emphasising explicitly therefore that there is a great variety of possible models for different subsystems, especially at a detailed level. The alternatives often turn on different ways of solving the aggregation problem, relating microscale hypotheses to the final models at the mesoscale. It can be argued, however, that the basic methods and design choices which form the basis of phase-2 and phase-3 approaches remain the same across most of these alternatives. We have to recall the beginning of our discussion on future phases above and argue that the differences can be sorted out only in the context of detailed empirical work.

How comprehensive?

Most models are developed for particular subsystems. These can be linked through common elements. Variables which are exogenous in one may be endogenous in another. How important are the links? This raises a research question which has not yet been satisfactorily answered. To what extent are there systemic effects at the whole-system level which can be picked up only by a comprehensive model in which all the links are represented? This question, as with many others, can only be tackled by further empirical work. Intuition and the arguments under phase 3, however, suggest that there may be more systemic effects than were hitherto thought arising from the nonlinearities introduced by the variety of variables in attractiveness functions (or utility functions) and the possible bifurcation points which these can generate.

Developmental dynamics: phase 3 into practice

The methods and models described under phase 3 provide us with the potential to reproduce change from known mechanisms. They predict the size of housing developments and organisations, given submodels of preference structures, demand functions, and technical production possibilities. As in some other areas noted above, what is now urgently needed is some relevant empirical work. This involves measurement of quantities with which we are unfamiliar, such as response times (implicit in the ϵ-parameters of the illustrative models above) and relaxation times. There is also the hard problem of identifying bifurcation phenomena empirically. It may be possible to do this in a broad way fairly easily—for example in the transition from corner-shop retailing to supermarkets—but much more difficult at the zonal level.

Towards evolutionary models

The headings in this subsection and the preceding one reflect the distinction used in biology between development and evolution. In developmental biology, the concern is with the growth of a known organism through understood (or sought for) mechanisms. Evolution is concerned with the emergence of new species, usually of a 'higher order' in some sense. Is there an analogue in urban studies? It would have to relate to new forms of organisation or new modes of behaviour. The problem in modelling such phenomena is that the 'technical possibilities' are not easily (if at all) known before the form has evolved. So this may pose both an exciting research question and a limit on what can be achieved.

References

Allen P M, Deneubourg J L, Sanglier M, Boon F, de Palma A, 1978 *The Dynamics of Urban Evolution, Volume 1: Interurban Evolution; Volume 2: Intraurban Evolution* Final Report to the US Department of Transportation, Washington, DC

Alonso W, 1960 "A theory of the urban land market" *Papers, Regional Science Association* 6 149-157

Alonso W, 1964 *Location and Land Use* (Harvard University Press, Cambridge, Mass)

Batty M, 1971 "Modelling cities as dynamic systems" *Nature* **321** 425-428

Beaumont J R, Clarke M, Wilson A G, 1980a "Changing energy parameters and the evolution of urban spatial structure" *Regional Science and Urban Economics* (forthcoming) (WP-279, School of Geography, University of Leeds, Leeds)

Beaumont J R, Clarke M, Wilson A G, 1980b "The dynamics of urban spatial structure: some exploratory results using difference equations and bifurcation theory" Paper presented at the IBG Quantitative Methods Study Group Meeting, Cambridge, September 1980, *Environment and Planning A* **13**(12) forthcoming

Burgess E W, 1927 "The determination of gradients in the growth of the city" *Publications, American Sociological Society* **21** 178-184

Carey H, 1858 *Principles of Social Science* (Lippincott, Philadelphia)

Chapman G P, 1977 *Human and Environmental Systems: A Geographer's Appraisal* (Academic Press, London)

Christaller W, 1933 *Die centralen Orte in Suddeutschland* (Gustav Fischer, Jena; English translation by C W Baskin, *Central Places in Southern Germany* Prentice-Hall, Englewood Cliffs, NJ)

Clarke M, Keys P, Williams H C W L, 1979 "Household dynamics and economic forecasting: a micro-simulation approach" Paper presented at the International Meeting of the Regional Science Association, London, August 1979

Coelho J D, Wilson A G, 1977 "Some equivalence theorems to integrate entropy maximising submodels within overall mathematical programming frameworks" *Geographical Analysis* **9** 160-173
Coelho J D, Williams H C W L, Wilson A G, 1978 "Entropy maximising submodels within overall mathematical programming frameworks: a correction" *Geographical Analysis* **9** 195-201
Cordey Hayes M, 1972 "Dynamic framework for spatial models" *Socio-Economic Planning Sciences* **6** 365-385
Evans S P, 1973 "A relationship between the gravity model for trip distribution and the transportation problem in linear programming" *Transportation Research* **7** 39-61
Forrester J W, 1969 *Urban Dynamics* (MIT Press, Cambridge, Mass)
Harris B, Wilson A G, 1978 "Equilibrium values and dynamics of attractiveness terms in production-constrained spatial-interaction models" *Environment and Planning A* **10** 371-388
Harris C D, Ullman E L, 1945 "The nature of cities" *Annals of the American Academy of Political and Social Science* **242** 7-17
Herbert D J, Stevens B H, 1960 "A model for the distribution of residential activity in urban areas" *Journal of Regional Science* **2** 21-36
Hoover E M, 1937 *Location Theory and the Shoe and Leather Industries* (Harvard University Press, Cambridge, Mass)
Hotelling H, 1929 "Stability in competition" *Economic Journal* **39** 41-57
Hoyt H, 1939 *The Structure and Growth of Residential Neighborhoods in American Cities* (Federal Housing Administration, Washington, DC)
Kain J, Apjar W C, Jr, Ginn J R, 1976 "Simulation of the market effects of housing allowances, Volume 1. Description of the NBER urban simulation model" RR R77-2, Department of City and Regional Planning, Harvard University, Cambridge, Mass
Lakshmanan T R, Hansen W G, 1965 "A retail market potential model" *Journal of the American Institute of Planners* **31** 134-143
Leonardi G, 1980 "A unifying framework for public facility location problems" WP-80.79, International Institute for Applied Systems Analysis, Laxenburg, Austria
Lösch A, 1940 *Die räumliche Ordnung der Wirtschaft* (Gustav Fischer, Jena; English translation by W H Woglam, W F Stolper *The Economics of Location* Yale University Press, New Haven, Conn.)
Lowry I S, 1964 *A Model of Metropolis* RM-4035-RC (Rand Corporation, Santa Monica)
Mann P H, 1965 *An Approach to Urban Sociology* (Routledge and Kegan Paul, Henley-on-Thames, Oxon)
May R M, 1976 "Simple mathematical models with very complicated dynamics" *Nature* **261** 459-467
Orcutt G H, Greenberger M, Korbel J, Rivelen A M, 1961 *Microanalysis of Socio-economic Systems: A Simulation Study* (Harper and Row, New York)
Paelinck J H, Nijkamp P, 1978 *Operational Theory and Method in Regional Economics* (Saxon House, Teakfield, Farnborough, Hants)
Palander T, 1935 *Beitrage zur Standortstheorie* (Almqvist and Wiksells Boktryckeri AG, Uppsala)
Ravenstein E G, 1885 "The laws of immigration" *Journal of the Royal Statistical Society* **4** 165-235, 241-305
Rees P H, 1979 *Residential Patterns in American Cities* RP-189, Department of Geography, University of Chicago
Reilly W J, 1931 *The Law of Retail Gravitation* (Pilsbury, New York)
Richardson H W, 1977 *The New Urban Economics: And Alternatives* (Pion, London)

Senior M L, Wilson A G, 1974 "Some explorations and syntheses of linear programming and spatial interaction models of residential location" *Geographical Analysis* **6** 209-237

Smith D M, 1971 *Industrial Location* (John Wiley, Chichester, Sussex)

Stevens B H, 1961 "Linear programming and location rent" *Journal of Regional Science* **3** 15-26

Stevens B H, 1968 "Location theory and programming models: the von Thünen case" *Papers, Regional Science Association* **21** 19-34

Thünen J H von, 1826 *Der isolierte Staat in Beziehung auf Landwirtschaft und Nationalökonomie* (Gustav Fischer, Stuttgart; English translation *The Isolated State* C M Wartenburg with a translation by P Hall, Oxford, 1966)

Weaver W, 1958 "A quarter century in the natural sciences" *Annual Report* The Rockefeller Foundation, New York, pp 7-122

Weber A, 1909 *Über den standort der industrien* (Tubingen; English translation by C J Friedrich, University of Chicago Press, Chicago)

Williams H C W L, 1977 "On the formation of travel demand models and economic evaluation measures of user benefit" *Environment and Planning A* **9** 285-344

Williams H C W L, Ortuzar J D, 1980 "Travel demand and response analysis—some integrating themes" presented at the International Conference on Research and Applications of the Disaggregate Travel Demand Model, University of Leeds, July 1980

Wilson A G, 1970 *Entropy in Urban and Regional Modelling* (Pion, London)

Wilson A G, 1971 "A family of spatial interaction models, and associated developments" *Environment and Planning* **3** 1-32

Wilson A G, 1974 *Urban and Regional Models in Geography and Planning* (John Wiley, Chichester, Sussex)

Wilson A G, 1979 "Some new sources of instability and oscillation in dynamic models of shopping centres and other urban structures" WP-267, School of Geography, University of Leeds, Leeds

Wilson A G, 1981a *Geography and the Environment: Systems Analytical Approaches* (John Wiley, Chichester, Sussex)

Wilson A G, 1981b *Catastrophe Theory and Bifurcation: Applications to Urban and Regional Systems* (Croom Helm, London)

Wilson A G, Clarke M, 1979 "Some illustrations of catastrophe theory applied to urban retailing structures" in *Developments in Urban and Regional Analysis* Ed. M J Breheny (Pion, London) pp 5-27

Wilson A G, Coelho J D, Macgill S M, Williams H C W L, 1981 *Optimisation in Locational and Transport Analysis* (John Wiley, Chichester, Sussex) forthcoming

Wilson A G, Rees P H, Leigh C M (Eds), 1977 *Models of Cities and Regions* (John Wiley, Chichester, Sussex)

Wilson A G, Senior M L, 1974 "Some relationships between entropy maximising models, mathematical programming models and their duals" *Journal of Regional Science* **14** 207-215

New developments in multidimensional geographical data and policy analysis

P Nijkamp

Introduction

Phenomena and problems in modern societies are characterised by complexity, variation, and interwoven relationships. This also holds true for spatial patterns and processes. Quantitative geography aims at providing theories and methods which describe such spatial structures and developments in a mathematical and/or statistical way in order to analyse in an operational way the dispersion and coherence of phenomena in regional and urban systems.

The picture of spatial systems is, in general, rather complicated, and hardly any phenomenon in such systems can be described or represented adequately by means of a simple attribute such as a single scalar variable. Normally, such phenomena have a whole set of attributes (aspects, criteria, features) which give a representative mapping of these phenomena. Such a multidimensional representation of phenomena in spatial systems requires adjusted operational methods for an appropriate regional and urban data analysis as well as for a satisfactory policy analysis.

During the seventies a wide variety of multidimensional methods have been developed which are extremely useful for data and decision analysis. Many of them are able to provide an operational framework for the analysis of spatial behaviour and for planning and decision problems. In the chapter the value of multidimensional methods will be set out by providing a selected survey of these methods and of their potential or actual applications. First, various multidimensional techniques for data analysis will be discussed and much emphasis placed on the treatment of soft (ordinal or qualitative) data. Next, several multidimensional methods for policy analysis (multiobjective programming, multicriteria analysis) will be examined. Here again much attention is paid to the treatment of soft information; several parts of the chapter have been based on the discussion in Nijkamp (1979).

Multidimensional analysis

The plurality of spatial patterns and of changes therein require very often a multidimensional analytical framework. This is a prerequisite for arriving at an operational and comprehensive insight into complex phenomena such as residential location decisions, evaluation of intangibles, the existence of interregional inequalities, decline in environmental quality, spatial interaction and attractiveness, and so forth.

In formal terms, a multidimensional approach implies that a certain variable x is characterised by a vector profile $\boldsymbol{v}$ with elements v_i $(i = 1, ..., I)$. In other words,

$$x \rightarrow \boldsymbol{v} = \begin{bmatrix} v_1 \\ . \\ . \\ . \\ v_I \end{bmatrix} . \tag{15.1}$$

Usually, the elements of $\boldsymbol{v}$ are measured in different dimensions. Sometimes, it is common to standardise the elements of $\boldsymbol{v}$. Clearly an ordinal measurement of the elements of $\boldsymbol{v}$ implies already a certain dimensionless standardisation.

Multidimensional data analysis aims at detecting a structure in data presented in vector profiles. To some extent, this modern analysis can be regarded as a straightforward extension of traditional unidimensional methods, although several specific problems may emerge in treating multidimensional profiles. Sometimes rather cumbersome statistical and mathematical problems may arise, and therefore it is important to employ a set of advanced techniques which can tackle these problems. Examples of such techniques are interdependence analysis, canonical correlation, etc.

A multidimensional approach may also lead to considerable complications in decision and planning problems [cf the well-known multiattribute utility developed by Lancaster (1971)]. In this respect, it is extremely important to develop operational methods for policy analysis which are able to take into account conflicts between groups, issues, goals, decision levels, etc. Furthermore, uncertainties (lack of reliable information, for example) have to be taken into account. The last part of this chapter will focus on multidimensional methods for policy analysis.

Pattern and impact analysis

Phenomena such as residential quality, environmental pollution, and spatial congestion have to be represented by means of a multidimensional pattern. In this way, various units of a spatial system (districts, regions, etc) can also be depicted in a comprehensive profile. Suppose, for example, that a residential attractiveness profile $\boldsymbol{a}$ is composed of the following elements:

$$\boldsymbol{a} = \begin{bmatrix} a_1 \\ . \\ . \\ . \\ a_I \end{bmatrix} = \begin{bmatrix} \text{quantity of dwellings} \\ \text{quality of dwellings} \\ \text{size of recreation areas} \\ \text{availability of shops} \\ \text{cultural facilities} \\ . \\ . \\ . \end{bmatrix} . \tag{15.2}$$

Then, for a set of regions 1, ..., R (cities, districts, etc), the following multidimensional matrix representation can be constructed:

$$\mathbf{A} = \begin{bmatrix} a_{11} & \cdots & a_{1R} \\ \cdot & & \cdot \\ \cdot & & \cdot \\ \cdot & & \cdot \\ a_{I1} & \cdots & a_{IR} \end{bmatrix} . \tag{15.3}$$

A standardisation of the elements of **A** may be carried out in several ways. A rather easy standardisation is

$$a_{ir}^{*} = \frac{a_{ir} - a_i^{\min}}{a_i^{\max} - a_i^{\min}} , \tag{15.4}$$

where $a_i^{\min}$ and $a_i^{\max}$ are the minimum and maximum values of a_i over all R regions. In expression (15.4), the ith indicator is supposed to be a benefit indicator (the higher, the better); otherwise, a reverse standardisation has to be used.

The (unweighted) distance $d_{rr'}$ between the attractiveness profiles of regions r and r' can be calculated *inter alia* via a generalised Minkowski p-metric:

$$d_{rr'} = \left[\sum_{i=1}^{I} (a_{ir}^{*} - a_{ir'}^{*})^{p} \right]^{1/p} , \qquad p \geqslant 1 . \tag{15.5}$$

A similarity index, $s_{rr'}$, between any two profiles can then be defined as

$$s_{rr'} = \frac{1}{1 + d_{rr'}} . \tag{15.6}$$

The above-mentioned pattern analysis of multidimensional phenomena can be extended by making a distinction between main profiles and subprofiles. In this way, one may divide a into main categories (for instance, residential quality, recréation, medical care, etc); next, each main category may be divided into subcategories,

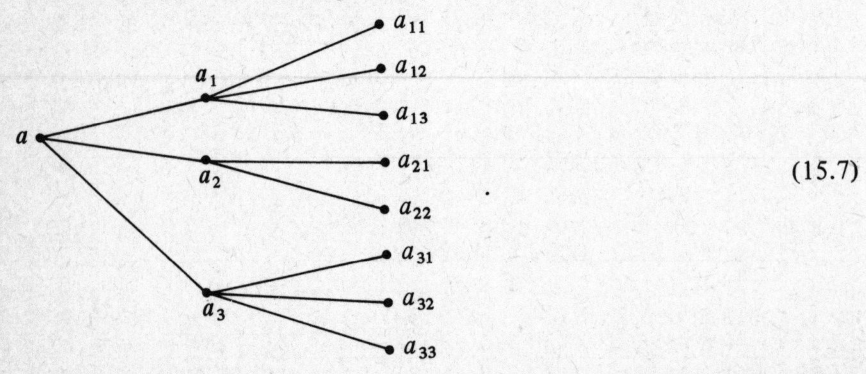

Such an hierarchical pattern representation is also very relevant in detailed multidimensional impact analyses (among others, in environmental impact statements). In this way, the coherence and variety of the key variables of a spatial system can also be taken into account. The same holds true for shifts in the main categories of a system as a consequence of a change in a main determinant (for example, the decision to build a new road will have a variety of economic, physical, environmental, and infrastructural repercussions).

A more comprehensive impact analysis can be based on a stimulus-response approach. The stimuli $\boldsymbol{\sigma}$ can be defined on the basis of the elements of a main determinant of a system which exerts a substantial influence on a set of responses $\boldsymbol{\rho}$ (for instance, the elements of the above-mentioned attractiveness profiles),

$$\boldsymbol{\sigma} \rightarrow \boldsymbol{\rho} \tag{15.8}$$

or

$$\boldsymbol{\rho} = f(\boldsymbol{\sigma}) \tag{15.9}$$

where f is a so-called impact function. There are several ways to operationalise f; for instance, multiple regression analysis, partial least squares, canonical correlation, etc (see below).

Interdependence analysis

A drawback of the multidimensional profile approach is its extensive information pattern, so that a straightforward interpretation is sometimes less easy. Furthermore, many attributes in a profile may contain redundant information. Since a lower number of attributes may facilitate the interpretation of the results, it is worthwhile undertaking an attempt at reducing the data contained in multidimensional profiles.

A traditional data-reduction technique is principal component analysis. This is a transformation from a set of originally mutually correlated variables to a new set of independent variables (based on an orthogonal data transformation in which the original variables are substituted for independent factors). A drawback of these techniques is the fact that new artificial variables are created which can be interpreted on the basis of factor loadings, but which have no clear direct meaning *per se*.

In respect to this, a more recently developed technique, called *interdependence analysis*, is more appropriate. Interdependence analysis is an optimal-subset selection technique, by means of which a subset of variables that best represents an entire variable set can be chosen (see Boyce et al, 1974). The advantage of interdependence analysis is that an optimal subset of original variables is selected, so that a data transformation is not necessary.

Suppose we have matrix $\mathbf{A}$ from expression (15.4) with R observations (profiles) on I variables. Next, P variables are to be selected from the I

variables such that this subset of P variables demonstrates an optimal correspondence with respect to the original data set. Hence $(I\text{-}P)$ variables are to be eliminated.

Now interdependence analysis is based on a series of successive regression analyses between the individual 'dependent' $(I\text{-}P)$ variables to be eliminated and the 'independent' variables to be retained. Given $(I\text{-}P)$ regressions, the minimum correlation coefficient can be calculated. Next, for all permutations of P in $(I\text{-}P)$ variables, a similar operation can be carried out. Then the optimal subset is defined as that subset which maximises over all permutations the values of the above-mentioned minimum correlation coefficient. This max–min solution bears a correspondence to the equilibrium solution of a game procedure, in which the information contained in a data matrix is reduced so that the selected variables constitute a best representation of the information pattern. For details of alternative subset-selection criteria see Nijkamp (1979).

Applications of interdependence analysis can be found, among others, in the field of optimal network algorithms, multicriteria evaluation methods, attractiveness analyses of human settlements, spatial complex analyses, and spatial inequality analyses.

Multidimensional scaling analysis

In addition to principal component techniques and interdependence analysis, several other data-reduction techniques have been developed during the last decade, such as correspondence analysis [see Benzécri (1973)] and multidimensional scaling (MDS) analysis. MDS techniques especially have found many applications.

The original rationale behind the use of MDS techniques was to transform ordinal data into cardinal units. Suppose that matrix **A** from expression (15.4) is measured in ordinal units. Then a transformation toward a metric system can be made by assuming that each region r can be represented as a point in a P-dimensional Euclidean space. Since there are R such points, one might interpret the distances between each pair of these R points as a measure for the discrepancy between each pair of profiles. Clearly, the Euclidean coordinates are unknown, but they can be gauged by a similarity rule stating that the R points have to be located in the Euclidean space in such a way that their positions correspond to a maximum extent with the ordinal information on the original R profiles. It is clear that the only way to derive metric profiles for each region is to reduce the dimensionality of these profiles. In fact, the degrees of freedom resulting from this reduction in dimensionality are used to transform nonmetric data into cardinal units. In other words, the following transition takes place ($P < I$):

$$\overset{\mathbf{A}}{\begin{bmatrix} a_{11} & \cdots & a_{1R} \\ \cdot & & \cdot \\ \cdot & & \cdot \\ \cdot & & \cdot \\ a_{I1} & \cdots & a_{IR} \end{bmatrix}} \rightarrow \overset{\hat{\mathbf{A}}}{\begin{bmatrix} \hat{a}_{11} & \cdots & \hat{a}_{1R} \\ \cdot & & \cdot \\ \cdot & & \cdot \\ \cdot & & \cdot \\ \hat{a}_{P1} & \cdots & \hat{a}_{PR} \end{bmatrix}}, \qquad (15.10)$$

where $\mathbf{A}$ and $\hat{\mathbf{A}}$ are measured in nonmetric and metric units respectively.

MDS methods can be regarded as extremely powerful tools in spatial data analysis. First of all, they can be used as a data-reduction technique as such, but they are especially important in the case of unreliable or soft data (such as ordinal information). MDS techniques allow researchers to draw metric inferences from nonmetric input.

There are many geographical applications in the field of MDS techniques: individual perception and preference analyses, mental maps, recreation behaviour, environmental quality analysis, urban renewal projects, and multicriteria analysis.

Canonical correlation

Canonical correlation is especially developed to identify correlations between sets of variables. In contrast to regression techniques, which aim at explaining one single variable from an underlying set of variables, canonical correlation attempts to link a profile of variables to another profile of variables.

Canonical correlation attempts to identify the degree of connection between sets of attributes of the same population via a generalised linear correlation analysis. Suppose we have a set of R regions which are characterised by two different profiles; for example, a socioeconomic profile composed of I indicators, and a spatial-infrastructural profile composed of J indicators. Then one may try to find a correlation between these two multidimensional profiles.

Next, a canonical correlation analysis attempts to identify a relationship between a linear combination of the elements of the first profile and a linear combination of the elements of the second profile, such that the underlying linear model demonstrates a maximum correlation between both linear expressions. Canonical correlation analysis can be used to test whether or not different profiles characterising the same phenomenon show a high degree of similarity.

The number of applications of canonical correlation in geographical research is fairly limited, but some applications have been made in the field of regional income analysis, unemployment analysis, and spatial pattern analysis. Sometimes canonical correlation can also be combined with spectral analysis, especially for time-series–cross-section problems.

In the field of canonical correlation analysis there are two related techniques, viz partial least squares analysis (Wold, 1977) and discriminant analysis (Anderson, 1958).

Partial least squares analysis is a special kind of path model technique which attempts to identify a block structure for latent variables and their indicators, as well as between the latent variables themselves (the 'inner' structure), on the basis of iterative regression analysis. To a certain extent, partial least squares can be regarded as an extension of canonical correlation towards more than two profiles.

Discriminant analysis is essentially an assignment method which aims at assigning a certain unit (person, district) to a certain class on the basis of a multidimensional profile of attributes of this unit. Stability tests on the results of a discriminant analysis can be carried out via canonical correlation.

Spatial correlation and econometrics

Spatial (auto)correlation is another phenomenon which frequently occurs in spatial systems. Several test statistics have been developed in order to identify spatial autocorrelation or cross-section correlation (among others by Moran, Geary, Cliff and Ord, and Hordijk). Given a multidimensional profile for a set of regions, and given the connectivity structure of the spatial system concerned, several measures for autocorrelation can be defined. These measures can easily be extended for spatiotemporal profiles and for different spatial and temporal lag structures.

A similar approach can be used to detect spatial correlation among the disturbances of a linear spatiotemporal model, so that adjusted econometric techniques can be used to produce consistent parameter estimates [see Nijkamp (1979)]. Some appropriate techniques may be a Zellner generalised least squares method or a Markov scheme method.

In many cases, regional modelling is characterised by the existence of latent (indirectly observed) variables. Such latent variables have usually only a soft or qualitative meaning, but they can be approximated by means of a vector profile of indirect indicators. An appropriate technique for dealing with latent variables is Lisrel; this is based on a maximum likelihood approach and it needs precise information concerning the distribution of the observed variables and the specification of the theoretical model [for some applications see, among others, Jöreskog (1977) and Folmer (1979)].

A more difficult problem arises, if (parts of) the explanatory variables are measured only in ordinal terms. In that case an MDS approach can be used. Assume the following model:

$$y = \mathrm{f}(\boldsymbol{x}, \boldsymbol{z}) \,, \tag{15.11}$$

where $\boldsymbol{x}$ and $\boldsymbol{z}$ are profiles with metric and nonmetric attributes, respectively.

Thus z contains ordinal information. Suppose the number of elements of z is K, and the number of observations is R. Then an MDS technique can be applied in order to transform the $R \times I$ matrix of ordinal observations into an $R \times P$ matrix of metric data ($I > P$). Next, a normal regression procedure can be applied to the transformed data set [see Nijkamp (1980a)]. Tests on autocorrelation can again be performed via the above-mentioned statistics.

Ordinal multidimensional data

The major part of mathematical and statistical data techniques is based on metric data, although it is surprising that, in practical research, soft information is very often a rule rather than an exception.

In the past, only a few techniques for ordinal data treatment have been developed. The best known examples are the Spearman and Kendall rank-correlation coefficients for ordinal data. In regression analysis, dummy variable techniques have become rather popular in order to deal with nominal or qualitative information. In addition, path models (and more recently partial least squares and Lisrel techniques) have been developed for latent variables.

It has already been explained under multidimensional scaling analysis that MDS techniques are powerful techniques in geographical research. Some caveats, however, relate to the number of dimensions to be taken into account (just as in factor analysis) and the value of a satisfactory goodness-of-fit. Although MDS methods are in general extremely useful tools for a wide variety of ordinal multidimensional-data problems, it may be worthwhile examining whether for certain specific problems alternative techniques may also be appropriate.

Beside MDS techniques, it is in some cases useful to make use of order statistics (either in an analytical way or via random generators).

A special problem arises in the case of model (15.11) when the endogenous variable y is also measured in nonmetric units. In that case, it is difficult to apply MDS techniques, since there is only one vector of ordinal data which cannot be reduced to a lower dimension. There are then two possibilities. First, one may—in a manner analogous to a metric-regression analysis—write the estimator entirely in terms of (Kendall rank) correlation coefficients. The justification for this analogy is, however, hard to prove.

A second approach may be to make a pairwise comparison of the nonmetric data and to assign a zero–one dummy depending on whether or not a certain ordinal number is higher than the other one. In this case, one may apply a probit analysis in order to estimate the probability that a certain outcome of the endogenous variable is higher than another one, given certain zero–one values for the explanatory part of the regression equation.

The latter result also means that ordinal interdependence analyses can be carried out in various ways: (1) via MDS methods, (2) via Kendall rank-correlation coefficients, and (3) via soft regression techniques.

The same holds true for canonical correlation, partial least squares, and spatial (auto)correlation statistics. Adjusted techniques can also be developed for discriminant analysis and clustering techniques.

Multidimensional preference and perception analysis

The multidimensionality principle can also be used to assess individual preference and perception patterns. In regard to this, one may ask individuals to rank their priorities or perceptions (or both) concerning a multidimensional set of items (for example, different shops or recreation areas) by means of ordinal numbers. Next, an MDS approach can be used to draw metric inferences concerning the relative preferences (or perceptions) of the individuals, the discrepancies among individuals, and the differences among the items. In this case, a joint MDS procedure may be useful, because such a joint configuration of individuals and items gives a comprehensive representation of the entire preference (or perception) pattern.

Such a preference analysis can also be applied to the supply side of commodities, so that an MDS technique may also provide insight into the relative disequilibrium between the demand and supply side [see, for an application to shopping behaviour, Blommestein et al (1980)].

A next step of such a multidimensional analysis may be to construct a behavioural model which tries to explain individual behaviour by means of a multidimensional set of explanatory variables. If the information concerned is nominal or ordinal, a wide variety of disaggregated choice models can be used to assess individual behavioural parameters [see, for a survey, van Lierop and Nijkamp (1980a; 1980b)]. Some well-known analytical tools for disaggregated choice models are logit and probit analysis. Both techniques have found many applications in spatial interaction models.

Multidimensional policy analysis

In the seventies, economists and operations researchers have paid much attention to multidimensional optimisation methods as a tool in modern decisionmaking. The background to this interest in depth in new decision analyses is the lack of operationality of traditional decision techniques. A frequently felt shortcoming of almost all these techniques is the fact that all dimensions of a decision problem have to be translated into a common denominator (like income, profit, efficiency, etc) or at least have to be made commensurate with the primary objective of a decision problem.

The awareness of a multiplicitiy of different objectives in decisionmaking and management has evoked the need for more adequate techniques, ones which take into account the multidimensionality and heterogeneity of individual, social, or entrepreneurial behaviour. The need for such adjusted methods is even more apparent owing to the mutually conflicting or noncommensurable nature of many objectives. The presence of (partially) incompatible priorities can be considered as an essential characteristic of a wide variety of modern planning and decision problems.

Therefore, several attempts have been made recently to develop more adequate theories and methods which take into account explicitly the existence of multiple criteria in decisionmaking. The basic feature of these techniques is that a wide variety of relevant decision aspects is included without translating them into monetary units or any other common denominator. These multidimensional optimisation methods are also able to integrate intangibles normally falling outside the realm of the traditional price and market system. Expositions of multidimensional optimisation theory can be found among others in the following books: Bell et al (1977), Blair (1979), Boyce et al (1970), Cochrane and Zeleny (1973), Cohon (1978), van Delft and Nijkamp (1977), Fayette and Nijkamp (1980), Haimes et al (1975), Hill (1973), Hwang and Masud (1979), Keeney and Raiffa (1976), Nijkamp (1977; 1979; 1980b), Nijkamp and Spronk (1980), Rietveld (1980), Thiriez and Zionts (1976), and Zeleny (1976).

These new approaches are extremely relevant for private and public decisionmaking in the sphere of production, resources, investment, location, marketing, etc. In all these cases, pecuniary elements (like profitability) play an important role, but in addition several other elements are equally important (like social aspects, environmental impacts of production, use of scarce natural resources, risk characteristics, labour conditions, etc). The multidimensional optimisation methods also have a great relevance for regional and urban policy analysis owing to the conflicting nature of many goals (either within a region or between regions).

The general format of a multidimensional optimisation model is

$$\text{maximise } \boldsymbol{w}(\boldsymbol{x}) , \tag{15.12}$$

subject to

$$\boldsymbol{x} \in K ,$$

where $\boldsymbol{w}$ is an $I \times 1$ vector of objective functions, $\boldsymbol{x}$ is a $J \times 1$ vector of decision arguments, and K is a feasible area. An example of such a multidimensional programming problem may be: maximise production *and* employment *and* environmental quality *and* energy savings *and* systems accessibility, subject to the side conditions set by the economy and technology.

It should be noted that decisionmaking in a multigroup or multiregional context is fairly complicated, because a part of the one system is under control of another system. Without a master control for the entire system at hand, a compromise choice between conflicting options has to be based on a negotiation or bargaining process between all participating decisionmakers.

In this respect the notion of interactive decision strategies is very important (see later).

The concept of an efficiency curve plays a central role in multidimensional optimisation problems, because an efficiency curve precisely reflects the degree of conflict or complementarity between diverging options. The problem of a multidimensional optimisation model is to find efficient points, $\boldsymbol{x}^*$, such that no other feasible point $\boldsymbol{x}$ will exist where

$$w_i(\boldsymbol{x}) > w_i(\boldsymbol{x}^*) , \qquad \forall i , \tag{15.13}$$

and

$$w_{i'}(\boldsymbol{x}) \neq w_{i'}(\boldsymbol{x}^*) , \qquad \text{for at least one } i' \ (i' = 1, ..., I) .$$

The computationally equivalent problem is

$$\text{maximise } \varphi = \sum_{i=1}^{I} \lambda_i w_i(\boldsymbol{x}) , \tag{15.14}$$

subject to

$$\boldsymbol{x} \in K$$

$$\sum_{i=1}^{I} \lambda_i = 1 , \qquad \lambda_i \geqslant 0 .$$

By means of a parametrisation of the λs, the entire set of Pareto solutions can, in principle, be identified (at least for a convex programming problem), although this may be a rather time-consuming procedure for large problems. Since the λs are a set of weights (trade-offs) associated with the efficient solutions, any ultimate compromise solution between the diverging objectives can be related *ex post* to these λs (note that any optimal solution is efficient).

Another important concept in multidimensional optimisation theory is the *ideal point*. The ideal point, $\boldsymbol{w}^0$, is an $I \times 1$ vector of maximum values of the successive individual objective functions; in other words, the elements w_i^0 of $\boldsymbol{w}^0$ are defined as:

$$w_i^0 = \max w_i(\boldsymbol{x}) , \qquad \text{subject to } \boldsymbol{x} \in K . \tag{15.15}$$

It is clear that an ideal point is not a feasible point, but it may serve as an important frame of reference for evaluating points on the efficiency frontier (see later).

Since the aim of this chapter is to provide a survey of multiobjective modeling, it may be meaningful to make some classifications. A first typology may be based on a subdivision into *continuous* and *discrete* (integer) decision models. Discrete models are characterised by a finite number of feasible alternative choices or strategies (for example, in the case of plan-evaluation or project-evaluation problems); discrete models are often called *multicriteria models*. Continuous models are based on an infinite number of possible values for the decision arguments and hence for the objective functions; they are usually called *multiobjective optimisation models*.

Another distinction between multidimensional choice models may be according to the degree of accuracy of information. In this respect one may subdivide such choice models into *soft information models* (based, for example, on qualitative, fuzzy, or ordinal information) and *hard information models* (based, for example, on deterministic cardinal data input). Thus, figure 15.1 may be constructed, and in the next sections, the categories I–IV of the figure will be further discussed.

	hard information	soft information
multicriteria models	I	III
multiobjective optimisation models	II	IV

Figure 15.1. A typology of multidimensional optimisation models.

Hard multicriteria models

Hard multicriteria models are based on reliable metric information on discrete alternatives (plans, projects, or strategies). A first step in all these methods is the construction of an *impact* matrix which reflects the outcomes of all alternatives with respect to all I relevant decision criteria: (table 15.1).

Table 15.1. An impact matrix.

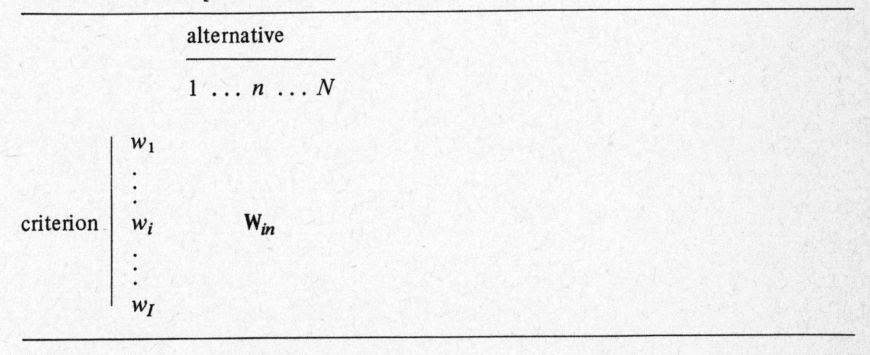

		alternative $1 \ldots n \ldots N$
	w_1	
	$\vdots$	
criterion	w_i	$\mathbf{W}_{in}$
	$\vdots$	
	w_I	

The element $\mathbf{W}_{in}$ reflects the value of the ith criterion with regard to the nth plan; it is assumed that $\mathbf{W}_{in}$ is measured in a normal metric system.

In the past, cost–benefit analysis has been a favourite method to evaluate discrete alternatives. Because of the unpriced nature of several commodities, this method is inappropriate for most urban and regional planning problems [for an extensive criticism see Nijkamp (1977)]. Some adjusted methods, such as the planning-balance-sheet method, the cost-effectiveness analysis, and the shadow-project approach, can be regarded as a significant improvement of traditional cost–benefit analysis, but they do not provide a solution for the problem of judging unpriced and intangible goods.

Instead of using (artificial) prices for unpriced and intangible goods, multicriteria models assign political priorities to certain decision criteria. These weights reflect the relative importance attached by the decision-maker(s) to the outcome of each criterion. These weights reflect the priority scheme of the decisionmaker and may be linear or nonlinear.

It is often a hard task to infer political weighting schemes by means of revealed preferences or questionnaires. When such weights cannot be assessed *a priori*, two ways are open to proceed with a multicriteria-evaluation model:
First, to use general alternative scenarios as the basis for deriving alternative sets of weights for future policy choices [see Nijkamp and Voogd (1980)];
Second, to use an interactive learning procedure during which relative priorities are specified in a stepwise manner [see van Delft and Nijkamp (1977) and Rietveld (1980); see also the section on interactive decision models below].

The following multicriteria methods for discrete decision and evaluation problems may be distinguished:

- trade-off analysis
- expected value method
- correspondence analysis
- entropy analysis
- discrepancy analysis
- concordance analysis
- goals-achievement method

The general feature of these multicriteria methods is that they include a multiplicity of decision criteria, so that they are rather appropriate for modern planning and management problems in which unpriced goods play an important role.

Hard multiobjective optimisation models

Hard multiobjective optimisation models are based on metric information regarding continuous objective functions and constraints. There is also a wide variety of different multiobjective optimisation models: for example, utility models, penalty models, constraint models, goal-programming models, hierarchical models, min–max models, and ideal point models.

The general conclusion concerning these hard multiobjective optimisation models is that the last four categories especially may be appropriate for economic–environmental models, because they use the best available information without too many arbitrary assumptions or too much additional information.

Soft multicriteria models

Soft multicriteria models are based on ordinal information or even qualitative information ('good, better, best'). The following soft multicriteria models may be distinguished:

expected value method	metagame analysis
lexicographic method	eigenvalue method
ordinal concordance method	frequency method
permutation method	multidimensional scaling method

It must be concluded that there is a whole series of ordinal evaluation techniques, starting from simple but dubious solution methods to complex but satisfactory solution methods. These evaluation techniques are especially useful for regional, urban, and environmental management problems because usually many data on the impacts concerned are uncertain, fuzzy, or biased.

Soft multiobjective optimisation models

Soft multiobjective models are a less developed category of multidimensional choice models. They are characterised by qualitative objective functions (for example, a systems performance measured in ordinal units) and/or qualitative constraints (for example, qualitative impact statements such as 'good, better, best').

The number of ways to deal with such choice models is rather limited so far. The following three alternative approaches can be distinguished: fuzzy set models, stochastic models, and soft econometric models.

The conclusion is that the field of soft multiobjective optimisation models is still underdeveloped. For the moment its use for regional and urban decisionmaking is very limited, although especially the soft econometric models seem to be fairly promising.

Interactive decision models

The multidimensional choice models discussed so far were based on a certain technique or algorithm to identify a compromise between conflicting objectives. In this respect, these models can be regarded as a fruitful contribution to environmental, regional, and urban policy models.

However, in many planning and decision problems the first (compromise) solution obtained by one of the above-mentioned methods is not considered to be entirely satisfactory. Therefore, instead of regarding the compromise solutions as a final equilibrium point, one may develop a certain interactive learning procedure in order to reach in a series of steps such a satisfactory final compromise solution. This implies that the first compromise solution is only a trial solution which has to be presented to the decisionmaker(s) as a frame of reference for judging alternative efficient solutions. The easiest way to carry out such an interactive procedure is to ask the decisionmaker(s) which values of objective functions are satisfactory and which ones are unsatisfactory (and hence have to be improved).

This can easily be done by using a checklist encompassing all first compromise values of the I objective functions—table 15.2.

Let S represent the set of objective functions which are to be increased in value. Then the decisionmaker's judgement concerning the trial compromise solution can be taken into account by specifying the following constraint,

$$w_i \geqslant \hat{w}_i\,, \qquad \forall i \in S\,. \tag{15.16}$$

In consequence, the following model has to be solved for the next stage of analysis [cf model (15.12)]:

$$\begin{aligned} &\text{maximise } w_i(\boldsymbol{x})\,, \\ &\text{subject to} \\ &\boldsymbol{x} \in K\,, \\ &w_i(\boldsymbol{x}) \geqslant \hat{w}_i\,, \qquad \forall i \in S\,. \end{aligned} \tag{15.17}$$

By specifying model (15.17), a new multidimensional choice problem emerges. This can again be dealt with by means of one of the methods set out in the previous sections. After the calculation of the outcomes of this model, a new (trial) compromise solution emerges which can again be checked with the decisionmaker(s) by means of the table 15.2 checklist, and so forth. The procedure has to be repeated, until a final satisfactory compromise solution has been identified by the decisionmaker(s). [See, for applications, van Delft and Nijkamp (1977), Fayette and Nijkamp (1980), and Rietveld (1980) among others.]

The steps of an interactive procedure are briefly summarised in figure 15.2.

The advantages of such interactive procedures are evident: they provide information to the decisionmaker(s) in a stepwise manner, they lead to an active role for the decisionmaker(s), and they avoid the prior specification of trade-offs (although they can be inferred *ex post*).

In my opinion, the use of interactive multidimensional choice models is extremely important for environmental, urban, and regional decisionmaking and management, because it enables decisionmaker(s) to assign a clear role to intangibles and incommensurables in evaluation and decision problems.

Table 15.2. A checklist for the interactive learning process.

Values of first compromise solutions	Satisfactory
$\hat{w}_1$	
$\vdots$	+ or −
$\hat{w}_I$	

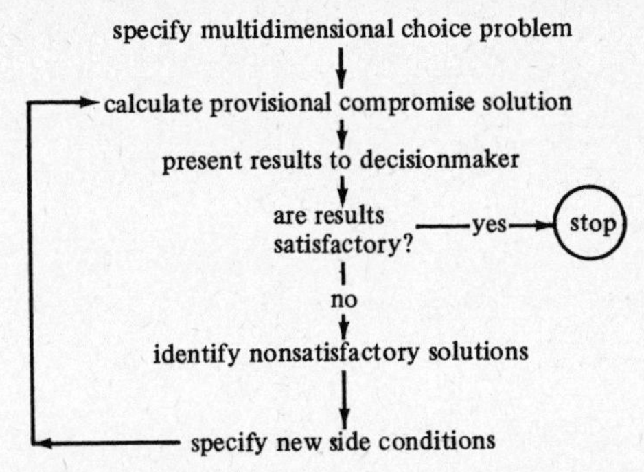

Figure 15.2. Steps of an interactive multidimensional choice problem.

References

Anderson T W, 1958 *An Introduction to Multivariate Statistical Analysis* (John Wiley, New York)

Bell D E, Keeney R L, Raiffa H, 1977 *Conflicting Objectives in Decisions* (John Wiley, New York)

Benzécri J P, 1973 *L'Analyse des Données* (Dunod, Paris)

Blair P D, 1979 *Multiobjective Regional Energy Planning* (Martinus Nijhoff, Boston)

Blommestein J H, Nijkamp P, Veenendaal W M, 1980 "Shopping perceptions and preferences" *Economic Geography* **56**(2) 155-174

Boyce D E, Day M D, McDonald C, 1970 *Metropolitan Plan Making* Monograph Series number 4, Regional Science Research Institute, Philadelphia

Boyce D E, Fahri A, Weischedel R, 1974 *Optimal Subset Selection* (Springer, Berlin)

Cochrane J L, Zeleny M (Eds), 1973 *Multiple Criteria Decision Making* (University of South Carolina Press, Columbia)

Cohon J L, 1978 *Multiobjective Programming and Planning* (Academic Press, New York)

Delft A van, Nijkamp P, 1977 *Multicriteria Analysis and Regional Decision-Making* (Martinus Nijhoff, The Hague)

Fayette J, Nijkamp P, 1980 *Project Evaluation for Decision-Makers* (OECD, Paris)

Folmer H, 1979 "Measurement of the effects of regional economic policy instruments" Paper presented at the International Conference of the Regional Science Association, London, *Papers and Proceedings, Regional Science Association* (forthcoming)

Haimes Y Y, Hall W A, Freedman H T, 1975 *Multi-Objective Optimization in Water Resource Systems* (Elsevier, Amsterdam)

Hill M, 1973 *Planning for Multiple Objectives* Monograph Series number 5, Regional Science Research Institute, Philadelphia

Hwang C L, Masud A S M, 1979 *Multiple Objective Decision Making* (Springer, Berlin)

Jöreskog K G, 1977 "Structural equation models in the social sciences" in *Application of Statistics* Ed. P R Krishnaiah (North-Holland, Amsterdam) pp 265-286

Keeney R L, Raiffa H, 1976 *Decision Analysis with Multiple Conflicting Objectives* (John Wiley, New York)

Lancaster K, 1971 *Consumer Demand: A New Approach* (Columbia University Press, New York)

Lierop W J F van, Nijkamp P, 1980a "Spatial choice and interaction models" *Urban Studies* (forthcoming)

Lierop W J F van, Nijkamp P, 1980b "A survey of disaggregated choice models" Research Memorandum 1980-8, Department of Economics, Free University, Amsterdam

Nijkamp P, 1977 *Theory and Application of Environmental Economics* (North-Holland, Amsterdam)
Nijkamp P, 1979 *Multidimensional Spatial Data and Decision Analysis* (John Wiley, New York)
Nijkamp P, 1980a "Soft econometric models" Research Memorandum 1980-5, Department of Economics, Free University, Amsterdam
Nijkamp P, 1980b *Environmental Policy Analysis* (John Wiley, Chichester, Sussex)
Nijkamp P, Spronk J, 1980 *Multicriteria Analysis: Practical Methods* (Gower Press, Farnborough, Hants)
Nijkamp P, Voogd J H, 1980 "The use of psychometric techniques in evaluation procedures" *Papers of the Regional Science Association* **42** 119-138
Rietveld P, 1980 *Multiple Objective Decision Methods and Regional Planning* (North-Holland, Amsterdam)
Thiriez H, Zionts S (Eds), 1976 *Multiple Criteria Decision Making* (Springer, Berlin)
Wold H, 1977 "On the transition from pattern recognition to model building" in *Mathematical Economics and Game Theory* Eds R Henn, O Moeschlin (Springer, Berlin) pp 536-549
Zeleny M (Ed.), 1976 *Multiple Criteria Decision Making* (Springer, Berlin)

A decade of developments in spatial interaction and travel-demand modelling: a brief overview

M L Senior

General themes and a classification of model developments

The key issues in travel-demand modelling in the 1970s have been those of aggregation and behavioural content, and these twin themes help to identify three main streams of model development:

(1) *Aggregate models*, developed using nonbehavioural entropy maximising (EM) procedures (Wilson, 1970a), have been rejigged and extended to accord with an underlying theory of rational economic choice (for example, Champernowne et al, 1976; Williams and Senior, 1978).

(2) *Disaggregate behavioural (individual choice) models* (Hensher and Stopher, 1979) have proved most seductive because of their potential for recording and calibrating behaviour at a more appropriate level of travel decisionmaking; for using data efficiently; for their increased policy relevance and transferability over space and time; and for approaching the aggregation–disaggregation problem from the 'micro end' of the spectrum.

(3) For those finding the charms of the latter models more superficial than real (Heggie, 1978), *alternative behavioural frameworks*, especially those based on *human activity principles* (Burnett and Thrift, 1979), hold more promise for capturing the subtleties of tripmaking within a broader household-activity context.

Development of aggregate models

Because "entropy is associated not with the system itself but with information about the system known to the analyst" (Wilson, 1970b, page 249), and because the analyst usually maximises his uncertainty about individual behaviour, EM models lack any explicit microbehavioural content. Thus the variety of flows from each origin that an EM model of spatial interaction generates is a logical expression of the analyst's uncertainty about individual interaction behaviour; no behavioural motivation is attributed to individual tripmakers to explain such flows. The EM method should, therefore, be seen as a useful *exploratory* procedure for model generation in situations where behavioural theories are either absent or deemed unacceptable. However, EM models cannot be ends in themselves if the analyst desires more than descriptive, opaque models; rather they should, if empirically valid, stimulate a search for suitable behavioural theories with which to underpin the modelling process. In other words, EM methods will tend to create the conditions for their own demise in each situation, to be replaced by behaviourally-based modelling procedures.

This behavioural issue pervades the developments of aggregate models within a mathematical programming framework. EM formulations have been exploited as nonlinear programming (NLP) problems, especially in terms of their limiting relationships with equivalent linear programming (LP) problems (Evans, 1973) and in terms of their dual properties (Wilson and Senior, 1974). However, comparative interpretations of equivalent EM and LP formulations are complicated because the former has a nonbehavioural entropy objective function, while the LP objective usually has a behavioural interpretation, such as minimising transport costs. In turn this implies that dual entropy variables have no intrinsic behavioural content, whereas the LP dual variables have; for example, a location-rent interpretation. It is tempting to interpret EM formulations as introducing a realistic dose of suboptimality into the optimality solutions derived from equivalent LP problems (Senior and Wilson, 1974), but to be convincing such notions need enveloping in a behavioural theory, which might take the form of a substantially modified utility maximising theory.

An alternative approach is to retain the assumption of optimising behaviour from the LP models and restructure the EM formulation by bringing constraints containing benefit and/or cost (utility) elements (for example, the total cost constraint) into the objective function, and expressing the 'entropy' term in the same units of measurement by dividing it by the dispersion or elasticity parameter. Although this restructured NLP problem will generate virtually identical model structures as the EM formulation, a behavioural watershed has been crossed, as the revised objective function is now capable of an interpretation in terms of groups of tripmakers maximising group net utilities or surpluses (Champernowne et al, 1976). This interpretation originated from a recognition (Neuberger, 1971; Williams, 1976) of the inconsistent use, in many transport studies, of nonbehavioural travel-demand forecasting models, based on EM principles, being used to provide inputs to economic evaluation procedures assuming utility maximising travel-demand behaviour and based on Hotelling's generalised surplus measure of benefit. This surplus maximising approach can generate all those spatial interaction, activity-location, and related models which EM methods produce, and it has been extended to develop location–allocation design models, with continuous or discrete supply variables (Coelho and Williams, 1978; Beaumont, 1980), and which incorporate more realistic spatial interaction behaviour and evaluation criteria than conventional cost-minimising location–allocation representations.

The implication of these developments is that a range of spatially aggregate models can be underpinned by theories of rational economic choice, although the aggregation process by which such models "are derived from the utility maximising behaviour of individual decision-makers within a spatial context remains unspecified" (Champernowne et al, 1976, page 276).

Disaggregate behavioural models based on choice theoretic approaches
Unlike the preceding models, disaggregate models focus on individual tripmakers as the basic decisionmaking units rather than on groups of tripmakers identified by zonal labels. In principle, therefore, they seek to embrace the greatest observable variability in tripmaking (that is, between individuals rather than between zones). During the 1970s, considerable efforts have been devoted to deriving such models from economic and psychological theories of individual choice behaviour. Aggregation issues arise, however, if such models are to be used in policy forecasting/ evaluation contexts.

By treating trips to various destinations as goods in an individual's utility function, Niedercorn and Bechdolt (1969), Golob et al (1973), and others have applied *deterministic utility theory* to spatial interaction behaviour. Such a theory imposes restrictive aggregation conditions as it is permissible to group together only those tripmakers with identical utility functions. Intuitively, therefore, such a theory appears to be consistent with LP models of spatial interaction, where homogeneous tripmakers from the same origin zone choose the same highest utility destination(s) subject to any 'capacity' constraints. Conversely, the claims to have generated gravity-like models from this deterministic theory seem implausible, as the theory would appear to be incapable of explaining the variety of trip-destination choices made by persons from the same origin zone.

Probabilistic choice theories relax the restrictive condition that tripmakers with the same *observed* characteristics should make identical travel choices. In the *constant utility approach* (Smith, 1975) the net utilities associated with, say, possible destination alternatives are considered to be invariant, but instead of each tripmaker from the same tripmaker group choosing the same alternative (namely the one affording the highest net utility), a probabilistic choice process is assumed. This idea comes from Luce's (1959) work in mathematical psychology, where it was used to 'explain' apparently inconsistent behaviour (Sheppard, 1978). Thus, this approach modifies deterministic utility-maximising behaviour by adding chance selection processes, but thereby offers a descriptive rather than explanatory model of the variability of travel behaviour within tripmaking populations, however defined. Central to the approach is Luce's choice axiom, a probabilistic version of the *independence from irrelevant alternatives* (IIA) principle, which states that the relative probability of choice among any two alternatives is unaffected by the addition or deletion of any other choice options. If this axiom holds, it is possible to derive the multinomial logit (MNL) model. The realism of the IIA condition, and thus of the MNL model which incorporates it, depends on the strength of the correlation between the attributes of the choice alternatives (Williams, 1977). If all the alternatives are distinctively different (that is, with little or no attribute correlation) the MNL model is applicable; with increasingly strong correlation between the characteristics of some alternatives, the model is likely to become less and less appropriate (examples are given in Williams, 1977; Sheppard, 1978).

The MNL model may be derived from the immensely popular branch of probabilistic choice theory known as *random utility theory*. Unlike the constant utility approach, each individual in the tripmaking population is deemed to choose his maximum net utility alternative. However, the perceived utility of alternatives is assumed to vary over the tripmaking population, on the grounds that the *analyst* cannot hope to observe all the characteristics of tripmakers and all interpersonal variations in tastes which affect travel behaviour. Thus variation in travel behaviour between tripmakers with the same *observable* characteristics is attributed to 'factors' which are either difficult or impossible for the analyst to detect. Clearly, this does *not* deny that each individual tripmaker pursues utility-maximising behaviour. To handle this situation formally, the net utility of each alternative is specified as a 'representative' utility component, common to all members of the tripmaking group (identified on the basis of the same observable characteristics), plus a random 'individual' utility term. The representative utility is treated as a function (usually linear) of the measured attributes of the choice alternatives, and a distribution is assumed by the analyst for the random utility component which will reflect the unobserved influences mentioned above. If these latter random variables are assumed to be identically and independently distributed according to a Weibull distribution, the MNL model is generated.

This particular model is not a unique structural form derivable from random utility theory, as the analyst is not logically compelled to assume a Weibull distribution. The MNL model is, however, the most tractable and convenient form, but it has been increasingly realised that these desirable properties may be outweighed by the model's adherence to the restrictive IIA axiom. Consequently a number of model structures have been developed, largely within a random utility framework, to allow for attribute correlation between alternatives. These include the hierarchical and cross-correlated logit models of Williams (1977), McFadden's (1979) general extreme value model, the dogit model (Gaudry and Dagenais, 1979), and the multinomial probit (MNP) model, which is generated by assuming normal distributions for the random utilities. Indeed, because of its generality the MNP model is conceptually appealing, and this has stimulated research concerning the computational problems of calculating choice probabilities for more than four choice alternatives (Daganzo et al, 1977). This model is also attracting attention because of its potential for aggregation; for example, Kitamura et al (1979) have sought an interzonal spatial interaction model by aggregating over individual tripmakers and destination alternatives in a disaggregate MNP model of destination choice.

Of course, the price for increasing generality of models within a probabilistic choice framework is increased complexity from a practical standpoint (Williams and Ortuzar, 1979). This is one reason for seeking alternative approaches to travel-demand modelling.

Alternative behavioural viewpoints

Although the individual choice models of the last section are regarded as improvements on the aggregate models described above, there have been criticisms of their behavioural adequacy and policy relevance (Heggie, 1978). Critics have emphasised the derived nature of travel demand, with the implication that travel activities cannot be isolated from the whole range of household activities. Travel provides the linkages between some of these houshold activities and is in part dictated by their timing and location (Hägerstrand, 1970). Moreover, in a household situation, there is often interdependence between each member's tripmaking decisions, so that household structure, stage in the life cycle, and the ownership of driving licences are likely to be important determinants of travel behaviour. Further complexities are added because activity and travel alternatives are not equally well known or perceived by all tripmakers, and because activity–travel patterns tend to display some inertia, with the result that 'incentives' of sufficient magnitude are required to induce a change in habit. Given such considerations, it is argued that a wider range of behavioural adaptations to policy stimuli need to be catered for in travel-demand models; for example, the rescheduling of linked activities and the consolidation of trips.

To capture such intricacies and subtleties of household activity–travel behaviour, interactive simulation has been put forward as being a particularly apposite method. By contrast with the aggregate and disaggregate models already considered, such a method relies more heavily on stated rather than revealed preferences as a guide to tripmaker response to policy stimuli. The HATS technique (Jones, 1979) is a particularly important innovation for it involves all the members of each household interviewed, and makes explicit to them the spatial and temporal constraints on household behaviour. Participants can be confronted with the consequences of their intended activity decisions, and various conflicts can be mutually resolved. It therefore avoids the naive market-research approach to determining people's preferences and future courses of action. Heggie (1978) identifies its main disadvantage as its inability to allow for the longer term adaptations and experience of the household.

This style of travel modelling tends to be useful for increasing our understanding of travel behaviour and for guiding policy design. Perhaps the weakest feature of the whole household-activity approach is the lack of broadly-based and operational forecasting models, although micro-simulation methods are attracting increasing interest.

Some concluding issues

The issues in the previous section concerning a more adequate representation of travel decisionmaking processes provide an immediate stimulus to further model development. Indeed improvements to disaggregate choice models in the areas of trip chaining behaviour (for example, Lerman, 1979) and notions of habit and threshold change (for example, Krishnan, 1977) are already relaxing those 'secondary assumptions' of the random utility approach (Williams, 1979).

However, both from scientific and from planning standpoints there is certainly far too little basic spatial and temporal validation of existing models to provide more precise and specific stimuli to model development and improvement. Clearly the existence and adequacy of comparable data sets, especially over time, is a severe constraint. Yet such data shortages do not rule out more extensive sensitivity analysis of models. For example, Bonsall et al (1977) trace the sensitivity of aggregate transport-model forecasts to changes in exogenous and policy-control variables; Senior and Williams (1977) and Williams and Senior (1977) have examined the forecasting implications of alternative aggregate trip-distribution and mode-choice models calibrated from the same data; and Williams and Ortuzar (1979) have simulated the effects of behavioural misrepresentation (in terms of varying knowledge of travel alternatives and varying degrees of habit in travel behaviour) on tripmaker response to policy stimuli.

Yet sensitivity analysis of cross-sectionally calibrated models, although informative, is not a substitute for genuine validation of such models by using data revealing tripmakers' responses over time. As Williams and Ortuzar (1979) emphasise, reliance on cross-sectional data for a single time horizon does not allow us to discriminate effectively between alternative explanations of tripmaker behaviour.

References

Beaumont J R, 1980 "Spatial interaction models and the location-allocation problem" *Journal of Regional Science* **20** 37-50

Bonsall P W, Champernowne A F, Mason A C, Wilson A G, 1977 "Transport modelling: sensitivity analysis and policy testing" *Progress in Planning* 7(3) 153-237 (Pergamon Press, Oxford)

Burnett K P, Thrift N J, 1979 "New approaches to understanding traveller behaviour" in *Behavioural Travel Modelling* Eds D A Hensher, P R Stopher (Croom Helm, London) chapter 5

Champernowne A F, Williams H C W L, Coelho J D, 1976 "Some comments on urban travel demand analysis, model calibration and the economic evaluation of transport plans" *Journal of Transport Economics and Policy* **10** 267-285

Coelho J D, Williams, H C W L, 1978 "On the design of land use plans through locational surplus maximisation" *Papers of the Regional Science Association* **40** 71-85

Daganzo C F, Bouthelier F, Sheffi Y, 1977 "Multinomial probit and qualitative choice: a computationally efficient algorithm" *Transportation Science* **11** 338-358

Evans S P, 1973 "A relationship between the gravity model for trip distribution and the transportation problem in linear programming" *Transportation Research* **7** 39-61

Gaudry M J I, Dagenais M G, 1979 "The dogit model" *Transportation Research* **13B** 105-112

Golob T F, Gustafson R L, Beckmann M J, 1973 "An economic utility theory approach to spatial interaction" *Papers of the Regional Science Association* **30** 159-182

Hägerstrand T, 1970 "What about people in regional science?" *Papers of the Regional Science Association* **24** 7-21

Heggie I G, 1978 "Putting behaviour into behavioural models of travel choice" *Journal of the Operational Research Society* **29** 541-550

Hensher D A, Stopher P R, (Eds), 1979 *Behavioural Travel Modelling* (Croom Helm, London)

Jones P M, 1979 "HATS: a technique for investigating household decisions" *Environment and Planning A* **11** 59-70

Kitamura R, Kostyniuk L P, Ting K L, 1979 "Aggregation in spatial choice modelling" *Transportation Science* **13** 325-342

Krishnan K A, 1977 "Incorporating thresholds of indifference in probabilistic choice models" *Management Science* **23** 1224-1233

Lerman S R, 1979 "The use of disaggregate choice models in semi-Markov process models of trip chaining behaviour" *Transportation Science* **13** 273-291

Luce R D, 1959 *Individual Choice Behaviour* (John Wiley, New York)

McFadden D, 1979 "Quantitative methods for analysing travel behaviour of individuals: some recent developments" in *Behavioural Travel Modelling* Eds D A Hensher, P R Stopher (Croom Helm, London) chapter 13

Neuberger H L I, 1971 "User benefit in the evaluation of transport and land use plans" *Journal of Transport Economics and Policy* **5** 52-75

Niedercorn J R, Bechdolt B V, 1969 "An economic derivation of the 'gravity law' of spatial interaction" *Journal of Regional Science* **9** 273-282

Senior M L, Williams H C W L, 1977 "Model based transport policy assessment, 1: the use of alternative forecasting models" *Traffic Engineering and Control* **18** 402-406

Senior M L, Wilson A G, 1974 "Explorations and syntheses of linear programming and spatial interaction models of residential location" *Geographical Analysis* **6** 209-238

Sheppard E S, 1978 "Theoretical underpinnings of the gravity hypothesis" *Geographical Analysis* **10** 386-402

Smith T E, 1975 "A choice theory of spatial interaction" *Regional Science and Urban Economics* **5** 137-176

Williams H C W L, 1976 "Travel demand models, duality relations and user benefit measures" *Journal of Regional Science* **16** 147-166

Williams H C W L, 1977 "On the formation of travel demand models and economic evaluation measures of user benefit" *Environment and Planning A* **9** 285-344

Williams H C W L, 1979 "Travel demand forecasting—an overview of theoretical developments" WP-243, School of Geography, University of Leeds, Leeds

Williams H C W L, Ortuzar J D, 1979 "Behavioural travel theories, model specification and the response error problem" WP-263, School of Geography, University of Leeds, Leeds

Williams H C W L, Senior M L, 1977 "Model based transport policy assessment, 2: removing fundamental inconsistencies from the models" *Traffic Engineering and Control* **18** 464-469

Williams H C W L, Senior M L, 1978 "Accessibility, spatial interaction and the spatial benefit analysis of land use-transportation plans" in *Spatial Interaction Theory and Planning Models* Eds A Karlqvist, L Lundqvist, F Snickars, J W Weibull (North-Holland, Amsterdam)

Wilson A G, 1970a *Entropy in Urban and Regional Modelling* (Pion, London)

Wilson A G, 1970b "The use of the concept of entropy in system modelling" *Operational Research Quarterly* **21** 247-265

Wilson A G, Senior M L, 1974 "Some relationships between entropy maximising models, mathematical programming models and their duals" *Journal of Regional Science* **14** 207-215

17

Factor analysis in geographical research

P J Taylor

Factor analysis has had a very chequered history in geographical research. A decade or so ago its application in urban geography seemed to epitomise the research successes of the new geography. At that time the cleavage within the geographical community was not so much between proponents and critics but rather between those who understood and those who did not (Taylor, 1976). All this had changed by the late 1970s when factor analysis had become unfashionable and even old-fashioned as the new geography—or the spatial school as it was now termed—was new no more. This short account is about this change in fortunes, and it will be suggested that some of the recent criticisms are misplaced. It is not proposed that factor analysis be rehabilitated to its former glories but merely that since it is only a technique it is hardly fair to criticise *it* for how *we* have used it. Like all techniques it does some jobs reasonably well and other jobs poorly. In this chapter I hope to identify those jobs that factor analysis does well, so as to guide our applications in the future.

The development of this argument is organised around three main sections. We begin by considering the changing popularity of factorial ecology by briefly describing two particularly interesting areas of research: general field theory and urban ecology. The criticisms of these studies are then considered both from statistical and geographical perspectives. In the final section, applications of factor analysis in electoral geography are briefly reviewed and some current substantive research on American elections is described. The gist of the argument is that factor analysis is best considered as simply a measurement technique, and any research employing this tool is ultimately to be assessed not on technical grounds but in terms of the overall social model underlying the research.

The rise and fall of factorial ecology

The application of factor analysis in geographical research is closely related to the early pioneering work of Brian Berry. In the early 1960s he published two papers which were to foreshadow the more ambitious uses of the technique in the later 1960s: Berry (1961) proposed an approach for quantitative regionalisation which was to lead on to his 'general field theory', and Berry (1964) formulated 'cities as systems within systems of cities', which introduced urban factorial ecology into geography. Berry was both pioneer and a major developer of these new ideas as he became the dominant personality of early quantitative geography.

General field theory of spatial behaviour

Berry's general field theory represents the most complete use of factor analysis in any geographical research. By this I mean that factor analysis was being employed as a medium for writing theory. In 1963 Burton had emphasised the necessary links between the quantitative revolution and geographical theory, and Berry's work in this area provides the classic example of such integration. Berry was employing factor analysis to define formal and functional regions, and he then synthesised these results in a single canonical model. Hence geography's traditional concern for complex regional patterns was being rewritten into a few matrix-algebra equations (Berry, 1968). The new methodology was illustrated in a large multivariate analysis of the Indian economy (Berry, 1966), which received highly laudatory reviews (for example, Gould, 1968). Strangely this very impressive application came to a stop almost as soon as it began. Although the use of factor analysis in regionalisation was to continue (for example, Spence, 1968), there was no further major applications to write theory as general field theory. This interesting episode in quantitative geography remains a monument to those highly ambitious research goals of the 1960s which developed the habit of becoming *cul-de-sacs*.

Urban social ecology

Berry's second pioneering effort in urban social geography was much more successful in initiating a research school. The term 'factorial ecology' was originally employed to describe the application of factor analysis to small-area data in urban areas, and was first coined in 1965 by the sociologist Sweetzer. Johnston (1971) traces the approach back to an earlier sociological study by Anderson and Bean (1961), although Bell applied a form of factor analysis as an empirical test of social area analysis constructs as early as 1955. From these sociological beginnings there was to occur an almost complete takeover of factorial ecology by geography via Chicago's Center for Urban Studies under Berry's leadership. Factorial ecology became a centrepiece of the new quantitative urban geography, so that by about 1970 it had become one of the most productive research areas in all of geography. The International Geographical Union's Commission on Quantitative Methods commissioned a review and demonstration volume (Berry, 1971), and textbooks began to appear incorporating factorial ecology (Johnston, 1971; Timms, 1971; Herbert, 1972). The establishment of factorial ecology even reached the stage where one department of geography felt it necessary to advertise specifically for a 'factorial ecologist'! Furthermore, this popularity was firmly based on widespread grass-roots support. Rees (1972) was able to compile a list of sixty-five cities in ten different countries which had been factor analysed, and cities from more countries (for example, Timms, 1970; Johnston, 1973) and even in past periods (Goheen, 1970; Warnes, 1973) were continually being added to the list. It is not surprising, therefore, that in 1973 Robson could refer to factorial ecology as the 'meteoric growth node' of urban geography, although he did observe that the three such studies in *Social Patterns in Cities* (Clark and Gleve, 1973) might be the 'tail' of the meteor. This observation was very prophetic. From the mid-1970s factorial ecology studies became less popular, so that by the time of Johnston's (1977) first 'Progress Report' on 'city structures' they warrant a brief passing mention only. Nowadays papers on factorial ecology seem to require an apology to explain their existence, and recently one pioneer factorial ecologist has admitted to feeling a little bit like an academic dinosaur. The rise and fall of factorial ecology in geography has been dramatic indeed!

The demise of factorial ecology cannot be explained as merely part of a wider disenchantment with quantitative approaches in geography. While quantitative geography as a whole has declined relatively in the last decade, it has survived as a lively and productive section of the geographical community. The survival of factorial ecology has been much more problematic as it has experienced a catastrophic absolute decline. It has suffered two very separate but equally debilitating sets of criticisms. First, it has been associated through social area analysis with social theory, which has come to be increasingly found wanting as a basis for explanation in urban geography. As such, factorial ecology has had to bear the brunt of the reaction against posivitism. Second, it has come into serious criticism within the ranks of the quantitative geographers as to whether factor analysis has a respectable mathematical pedigree. These two criticisms have been aired in the pages of *Area* by Gray's (1975) description of "Non-Explanation in Urban Geography" and Williams's (1971) asking "Do you sincerely want to be a factor analyst?" Hence factorial ecology has become unfashionable in urban geography while factor analysis was becoming a little suspect in quantitative geography. No wonder studies of this type declined rapidly in the mid-1970s.

Evaluating the criticism

This two-fold criticism requires separate discussion, first of factor analysis as a technique within the field of statistics and second of the use of factor analysis in geography in the wake of changes in research priorities and purposes.

Three types of statistics

Statistics entered geography in the 1960s as a set of established techniques to be applied to geographical data. In 1963 the first textbook appeared in geography presenting inferential statistics, correlation, and regression (Gregory, 1963), and to these were later added multivariate techniques and stochastic modelling (King, 1969; Johnston, 1978). In these textbooks we are presented with a collection of techniques conveniently labelled 'statistics'. Although these techniques can be formally shown to be interrelated, in this discussion I will emphasise their differences especially in terms of their origins.

The single discipline of statistics which we have found to be available to geographers in the mid-twentieth century in fact has its origins in three distinct developments. Certain research areas of psychology, biology, and economics have responded to their inductive needs by development of psychometrics, biometrics, and econometrics. All three developments involved the establishment of specialist societies with journals specifically devoted to topics in these new 'subdisciplines'. All three have made distinctive contributions to modern statistics because each originally evolved to solve particular problems concerning their own particular subject matter. Let us briefly consider each in turn.

Psychometrics

In the late nineteenth century, attempts to replace speculative philosophy in psychology by a more empirical inductive approach produced, among other developments, a research school of psychometrics (Boring, 1961). The major problems facing these researchers were measurement ones, so that psychometrics became the pioneering area for scaling techniques. Hence Cureton (1968, page 95) defines psychometrics as simply "all aspects of the science of measurement of psychological variables and all research methodologies related to them". One of the products of this study was the theory of scales of measurement (Stevens, 1946), which has become a standard part of all quantitative geography textbooks. The most used technique to be developed in psychometrics was factor analysis (Burt, 1966). Quite simply "the growth and refinement of factor methods owes a great deal to the early explorations of psychologists searching for neat and tidy descriptions of man's intellectual abilities" (Child, 1970, pages 5-6). The common factor model can be traced back directly to Spearman (1904) and his need to separate out the concept of general intelligence from student test scores. The principal components model is usually interpreted as originating with a paper by Pearson (1901), and the earliest suggested application can be found in the work of his associate Macdonnell (1902) as a way of defining physical indices to categorise criminals. In both cases the techniques were developed as solutions to measurement problems.

Biometrics

In contrast, biology has not been overtly concerned with measurement despite its early links with psychometrics through genetics. Biometrics is from the 'hardest' science in our trio and has carried into its work explicit concern for experimentation. This has been reflected in designing agricultural field trials as experimental designs, and developing inferential statistics as necessary adjuncts to such 'non-laboratory experiments'. This is, of course, associated with Fisher in England and Snedecor in the USA. Parallel to this development we can find 'abstract' mathematical experiments in the development of probability distributions as stochastic models. The product of this work has been the numerous probability distributions of quantitative ecology. In both cases, the emphasis is upon the use of probability theory to describe processes either for inferential purposes or as modelling exercises.

Econometrics

Finally the most recent of our trio only acquired "its identity as a distinct approach to the study of economics during the 1920's" (Strotz, 1968, page 351). This culminated in the formation of the Econometric Society in 1930, whose purpose was "to promote studies that aim at a unification of the theoretical-quantitative and empirical-quantitative approach" (Strotz, 1968, page 351). Econometrics therefore attempts to devise techniques for relating existing mathematical economic theory to empirical evidence. The statistics developed are concerned with calibrating functional relationships between variables so that measurement problems and probability processes take a back seat. Here the emphasis is upon the general linear model and its application in regression analysis, with particular concern for time series.

The above discussion is summarised in the first five rows of table 17.1. The emphasis has been upon origins, and clearly each of the current disciplines have delved into statistics beyond these relatively narrow beginnings. Nevertheless three distinct traditions are represented, and in the sixth row of table 17.1 their separate major applications to geography are shown. We can now see where our factorial ecology fits into the picture—out of psychometrics via sociology. The problem with this pedigree is that it is not as mathematically sound as either biometrics or econometrics. Point-pattern analysis has been able to draw upon formal

Table 17.1. Origins and applications of the three statistics.

1 Original subect area	Biology	Psychology	Economics
2 Statistical subdiscipline	Biometrics	Psychometrics	Econometrics
3 Basic concept	Inference	Measurement	Calibration
4 Essential purpose	Testing: experimental design and stochastic models	Quantification: scaling data	Curve fitting: functional relationships
5 Core techniques	Probability distributions and analysis of variance	Correlation and factor analysis	Regression analysis and time-series analysis
6 Basic geographical applications	Point-pattern analysis	Factorial ecology	Areal association
7 Applications in geography of elections	Understanding electoral bias	Measuring the normal vote	Calibrating social bases of voting

mathematical arguments in developing spatial process models (Getis and Boots, 1978), and areal association studies have grown into a geographical econometrics as emphasis upon problems of area data has led, notably, to concern for spatial autocorrelation as a derivative of time-series analysis (Cliff and Ord, 1973).

Psychometrics has never been as formally mathematical as the other two subdisciplines. Although Pearson's Principal Components Analysis is a direct application of mathematics to a correlation matrix, the more popular common factor model is mathematically indeterminate. Hence, for much of its history, factor analysis has failed to satisfy fully the most elementary test of statistics—consistency of results between researchers. Although these problems were overcome in practical terms with the standardisation of approaches emanating from availability of computer packages, the technique remains open to severe criticism (Elffers, 1980). Hence despite Berry's use of matrix algebra in his introduction of factor analysis into geography as a general field theory, factorial ecology has never had the formal deductive properties underlying point-pattern analysis, nor the rigour and precision of recent developments in areal association. Factor analysis may have attracted the first generation of quantitative geographers, but it has failed to satisfy the needs of the second generation of better-trained quantifiers.

Factor analysis and the geographical reaction against positivism

One way of summarising the argument of the last section would be to say that factorial ecology was 'bad positivism'. If that is the case then we might go on to argue that factorial ecology should be less affected by the reaction against positivism in geography. In fact, this unlikely argument is not correct since, as we noted previously, factorial ecology has borne the brunt of critical evaluation of quantitative geography. When Harvey (1973, page 128) talks of "a clear disparity between the sophisticated theoretical and methodological framework we are using and our ability to say anything really meaningful about events as they unfold around us", the most obvious example of this disparity to come to the minds of many geographers was between factorial ecology and urban (and particularly inner-city) problems.

Gregory (1978, page 19) has argued "how truly Victorian were the men who pioneered the New Geography of the 1950's and 1960's" on the basis of their commitment to a 'positivist' view of science which he traces back to August Comte. Despite its declared neutrality, this positivism has distinct and rather predictable political biases both past and present as Gregory documents. In the case of factorial ecology the technical apparatus was intimately linked to a conservative theory of the city which emphasised individual choice at the expense of structural constraints (Gray, 1975). In fact the 'first' factorial ecologist, Wendell Bell, was also the author of the most explicit individual choice interpretation of urban social patterns (Bell, 1955; 1958). This social area analysis formulation came to be the basic source for interpreting factors, and hence factor analysis became the technical area that was to bear the brunt of the reaction against naive positivism.

This conservative use of factor analysis in geography, however, is nothing compared with the original association of the technique with psychological theories justifying the provision of separate and poorer education for the working class in Britain and for limiting southern and eastern European immigration into the USA (Kamin, 1977). This should not be used to imply that, since factor analysis was developed by Victorian minds, it will inevitably always produce conservative results. It is merely a technique, a tool, which has proved useful to conservatives in the past, but that is no reason for nonconservatives to discard it from their battery of techniques to criticise the present world.

In some ways factor analysis is a most strange type of positivist tool. Although the principal components model is a strict mathematical transformation technique, the common factor model involves a theory derived from psychology which hypothesises two sources of variation, common and unique. It is this feature that makes the technique mathematically indeterminate, as discussed previously. This is not so much bad positivism but rather antipositivism, since it posits a conception of causality outside the direct experience of immediate reality (Comte's *le réel*, Gregory, 1978, page 26). In Cattell's (1966, page 191) words, factor analysis is "a generalised method for making invisible influences visible, at least to a first approximation to their form". In the original psychometrics, it was the hidden structure of the mind which was being measured. When factor analysis is applied to groups of individuals it does not seem unreasonable to interpret the aggregate 'invisible influences' as part of a structural explanation "which locates explanatory structures outside the domain of immediate experience" (Gregory, 1978, page 76). The fact that traditional factorial ecology did not involve such interpretation does not preclude this possibility (Taylor and Hadfield, 1981).

The basic lesson to be drawn from the rapid decline of urban factorial ecology is that any technique is only as good as researchers will allow it to be. The most important element of any research design is the social theory in which it is embedded. Unfortunately it is just this feature that is typically ignored as it appears in the undiscussed assumptions of the research. The technique of factor analysis was in some senses unlucky in becoming associated with a very conservative social theory of urban patterns. If, instead of dwelling on past applications, we ask the simple question 'what is it that factor analysis does?' we find that what we are dealing with is a relatively sophisticated measurement tool. Although it is often claimed that the technique has many uses, including formal description of theory—Rummel (1970, pages 29–32) enumerates ten alternative uses of factor analysis for example—I would suggest that we be much more modest in our claims and reduce these uses to just two basic kinds. First, the transformation of data to produce orthogonal indices, which takes us back to Macdonnell (1902) and principal components analysis, and second, the attempt to find hidden structures, which derives from Spearman's (1904) common factor analysis. Ultimately the most important question is what use is made of the measurements produced.

Factor analysis in the geography of elections

In the final row of table 17.1 I have shown the use made of the three statistical traditions in the geography of elections. Although this area of research represents only a minor part of human geography, the general availability of large quantities of regularly published data did generate something of a research boom in the 1970s (Taylor, 1978) in which all three statistical traditions were employed. The use of probability distributions to understand electoral biases is described in Gudgin and Taylor (1979; see also Taylor and Gudgin, 1981), and the linear regression model has been employed to calibrate relations between voting and socioeconomic characteristics of areas, Miller (1977) and Crewe and Payne (1976) providing contributions which tackle problems in spatial analysis of interest beyond the confines of their subject matter. The use of factor analysis has been somewhat less developed in this field, although I will argue below that it may have a major role in measuring what has been termed the normal vote.

A selected review of applications

Our discussion of factor analysis in geographical research so far has concentrated upon what may be termed the more spectacular applications. What of the many applications of the technique outside mainstream factorial ecology which were more modest in their aims? In order to provide a more balanced view of the application of the technique in geography, I will briefly review some uses in the 'less-developed' field of electoral geography. This will provide a broader and probably more typical range of applications than the factorial ecology school discussed previously. The papers are organised in terms of which variables have been factor analysed—voting variables are considered to be dependent, and other socioeconomic variables to be independent.

Factor analysis of independent variables

The simplest use of factor analysis in research is to create orthogonal independent variables. In his study of Conservative voting in London constituencies in 1951, for instance, Cox (1968) carried out a factor analysis on 1951 census data and derived four orthogonal dimensions—social rank, suburbanism, commuting, and age structure—which he then related to Conservative vote and turnout in a causal modelling methodology. A more orthodox application can be found in Kirby and Taylor (1976), where two orthogonal patterns of British economic spatial structure, termed core–periphery and north–south, were derived and then used to predict the 1975 EEC referendum results in a linear regression equation. In both examples factor analysis is employed in a measurement exercise to provide the input for other statistical analysis.

Factor analyses of dependent and independent variables
In another study Cox (1970) defined the 'socio-political milieux' of Wales by factor analysing a mixture of census and electoral data for 1950–1951. Although five factors were identified, the analysis is ultimately used to define two regions: agricultural, nonconformist, Welsh-speaking, Liberal Wales and mining, urban, English-speaking, Labour Wales. In this case the distinction between dependent and independent variables is lost, as the analysis is used to provide regions for studying earlier electoral behaviour. The best example of employing factor analysis to illustrate regional political differences can be found in Allardt and Pesonen's (1967) study of Finland. Social and political variables were included in two separate analyses for the North and East, and for the South and West. The different ways in which election variables loaded on factors for the two regions was used to illustrate how parties can have different appeals in different regions. In Finland, for instance, the modern Social Democrats are part-heirs to the socialist traditions of the South and West but have no such basis of support in the North and East. Finally, we can briefly mention McPhail's (1971) study of the 1969 mayorial race in Los Angeles: a simple factor analysis of both voting and social data produced a description of three basic lines of social and political cleavage—black/white, class, and city/suburb. In all three studies basic concepts are measured which incorporate voting responses and hence illustrate the integration of the political into the socioeconomic environment.

Factor analyses of dependent variables
In certain research designs only the electoral variables are analysed. In such research the basic dimensions of voting are being measured. In one such study (Downing et al, 1980), voting variables for various offices and referenda in the 1978 Massachusetts election are factor-analysed to investigate whether the erosion of party voting in the United States could be found at state level. Four factors were generated which illustrated the separate existence both of strong party voting and of a specific dimension of nonparty voting relating to racial issues. Instead of analysing several voting variables at one election, this approach can be used to analyse one voting variable over a series of elections; in factor-analysis terminology a *T*-mode analysis. MacRae and Meldrum (1969) use this strategy on votes for several offices in Illinois to see whether they could identify any reorientations at the 1896 and 1928 elections. In a much more detailed analysis using sampled 'precincts' as variables (*S*-mode analysis), Dykstra and Reynolds (1978) have investigated the different dimensions of progressivism in Wisconsin in the first half of the twentieth century.

Separate factor analyses were carried out for different time-periods and for different ethnic subsamples, using numerous election results to illustrate how the spatial pattern of progressive support was not a simple ethnic response but fluctuated widely over time. In each of these studies the researchers are searching for basic dimensions in voting patterns and have chosen factor analysis as their measurement tool. I will illustrate this approach in more detail in the final substantive discussion.

Measuring the normal vote

In this final example of a more complete presentation of some factor analysis results, the relationship between their interpretation and the underlying social theory will be discussed. This is particularly appropriate in the realm of elections, since the standard theories emphasise voter choice and party responses to popular will, which are in many ways equivalent to the social area analysis explanations of urban ecology. The alternative theories of elections emphasise parties as constraints upon voter choice and as manipulators of popular will. The interpretations below will be set within these second types of election theories.

Quantitative research on American elections is dominated by an extreme liberal interpretation whereby emphasis has been placed, via survey analysis, upon the individual act of voting (Burnham, 1974). This approach explicitly entered the study of electoral geography in the work of Cox (1969). The paradox is, of course, that elections are essentially aggregate events, with the pay-offs specifically allocated at the group or constituency level rather than at the individual level. Furthermore, a basic finding in electoral geography has been the tendency for voting maps to remain stable over relatively long periods of time. The anarchy which is sometimes portrayed by individual-level studies, with voters seeming to change their minds after every other speech, is a very false picture at the important scale at which political pay-offs occur. Our route into a consideration of US elections will be in terms of defining this stability of voting structure which suggests a need for a factor analytic methodology.

Studies of individual voter responses have developed the concept of the 'normal vote' (Converse, 1966). When the responses of voters at different elections were being researched, it was deemed necessary to define a 'normal response' so that changes between elections could be adequately compared. The normal vote is the response from a group 'all things being equal' as it were, and variations from this level of vote would indicate the specific influences on a particular election (such as Kennedy's religion in 1960). Research in this school then concentrated on these divergences from the normal vote, the particular properties of single elections. We will turn the methodology upside down and deal solely with the normal vote. This will be measured by using a common factor model so that the aggregate vote for a party is divided into a common element, one or more normal votes, and a unique element, the specific character of an election.

Normal votes will be represented by common factors which we will describe by their loadings on the original elections.

The first analysis considers the twenty-five states which were popularly electing presidential college electors by 1836. The first popular election of an American president was that of Andrew Jackson in 1828, and we analyse his Democratic vote percentage over the twenty-five states and all later Democratic presidential vote percentages to 1920. A *T*-mode common

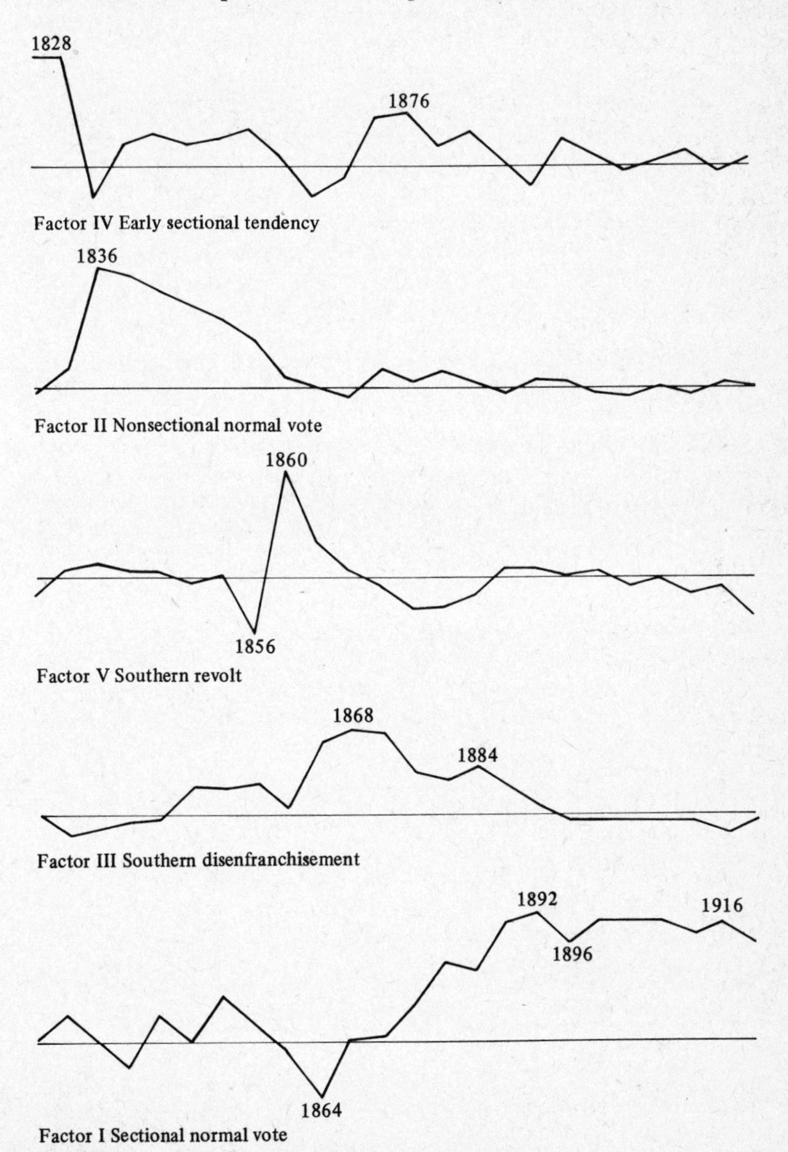

Figure 17.1. Factor loading profiles: 1828–1920.

factor model produces five distinct factors which explain over 90% of the original interstate variation. The results presented here are from an oblique rotation of these factors. For ease of illustration, all results are presented diagrammatically.

The factor loading profiles are shown in figure 17.1, and their main intercorrelations in figure 17.2. The picture to emerge is of two major 'normal votes' either side of the civil war, each surrounded by short sequences of confusing readjustment periods. The two 'normal vote' patterns are referred to as the 'non sectional' and 'sectional normal votes' respectively, since they represent two very different spatial organisations. Two electoral responses as distinctive as this are a surprising result, given the common sectional conflicts in federal politics which were endemic from the initial ratification of the Constitution. Why did this sectionalism not break though, in terms of stable electoral responses, until the 1880s? The answer relates to our interpretation of the role of political parties. If we treat them as democratic institutions reflecting the will of the people then American politics in the nineteenth century becomes very difficult to interpret. On the other hand, we *can* explain our results in terms of political parties as agents of political integration, managing public opinion for the needs of the state and dominant classes. The clearest example is the 'nonsectional normal vote' which defines the second party system of Democrats versus Whigs. This party system was devised specifically to bridge sectional differences and prevent electors expressing their sectional prejudices and interests. Schattsneider (1959) has argued that 'all organisation is bias' and this party system is a classic case whereby public participation was cleverly steered into economically-meaningless parties to prevent sectional conflict. The wonder was not why it collapsed in the 1850s but how it managed to survive for so long. After the Civil War, the North–South sectional pattern was not immediately set up, as war victory initially allowed the North to manipulate Southern elections to their own

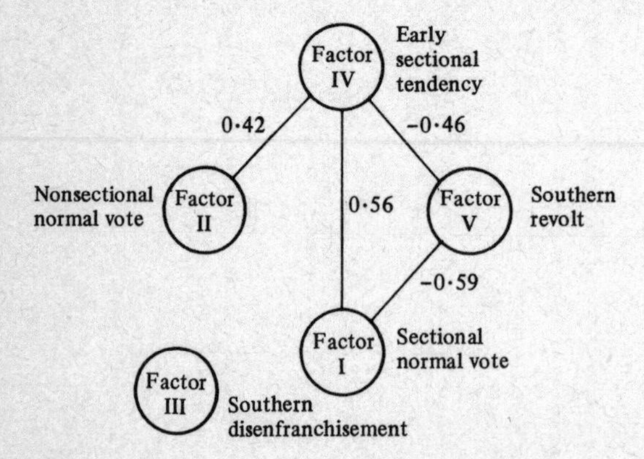

Figure 17.2. Factor correlations: 1828–1920.

ends through Reconstruction. After 1876, this policy was reversed, and the 'normal vote' with which we have become familiar in the twentieth century was established. In order to understand this system, however, we need to consider the trans-Mississippi western states which were entering the Union by this time. This brings us to our second analysis.

The second analysis is very similar to the first but involves all forty-eight contiguous American states covering Democratic vote percentages

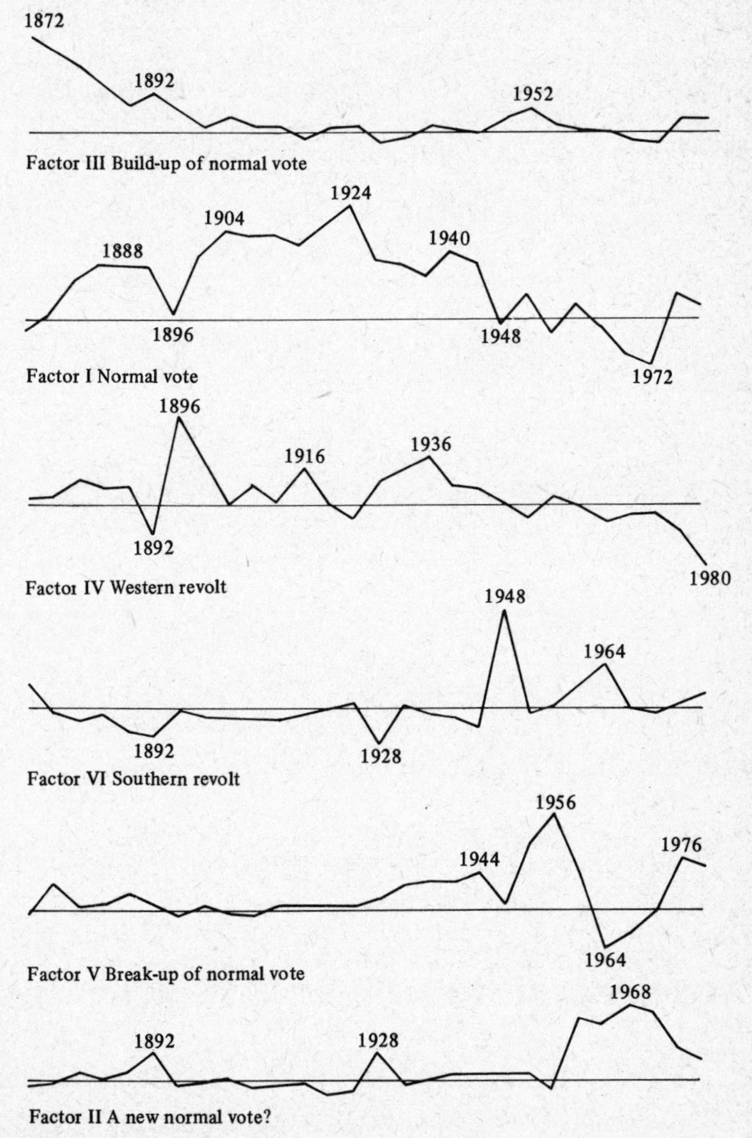

Figure 17.3. Factor-loading profiles: 1872–1980.

from 1872 to 1980. In this case, six factors were extracted which also account for over 90% of interstate variation. The loadings from the oblique rotation and the interfactor correlations are shown in figures 17.3 and 17.4. All of these factors show highly sectional responses, so that management of participation is not so much about avoiding economic sectionalism but ensuring sectional alliances favourable to the state and northern business interests. The method was a threefold division of the country into North, South, and West, with conservative business control of a Republican North, and conservative Democratic control of the South, to isolate progressive and populist challenges from the West. The system was initiated in the 1880s, consolidated in 1896, and was to last, although in modified form, through Roosevelt's New Deal to 1948. The key election is generally taken to be that of 1896, which can be seen to correspond to our 'western revolt' factor in figure 17.3. After this the Democrats are relegated to the southern periphery, and progressive western politics is controlled by the northern Republicans and their business backers. After 1948 this system has broken down. 1948 represents the beginnings of a 'southern revolt' which was to end Democratic dominance in the South for presidential elections. Since this time there has been no single, normal vote pattern. The old sectional arrangements briefly reappeared in the 1950s, and again in 1976 as 'diluted' versions of the previous sectionalism. From 1960 a new normal vote seemed to have emerged as the Democrats forged a liberal alliance, but this has disappeared after 1972. This new volatility in American politics is epitomised by the 1980 result, where Carter's performance is a reversal of the old western revolt pattern of 1896. This is perhaps not as strange as it may seem at first, since the Democrats nominated a western populist (William Jennings Bryan) in 1896, and the Republicans found their own western populist (Ronald Reagan) to

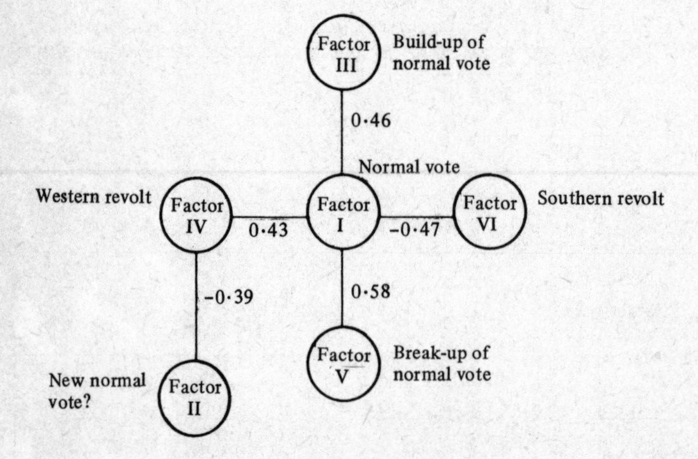

Figure 17.4. Factor correlations: 1872–1980.

forge a similar West-South alliance in 1980. This example illustrates clearly the sensitivity of our measurement tool and how it points to comparisons largely lost in the current political interpretations and analysis. This discussion and the analyses are more fully developed in Archer and Taylor (1981).

My conclusions for this short substantive section will also suffice for the chapter as a whole. What has been illustrated substantively, and what is recommended generally, is the use of factor analysis in a much less ambitious role than in earlier factorial ecology. The technique is merely a measurement tool to define a concept or concepts within a clearly specified social model. In essence, I have presented a relatively simple measurement exercise in which the major structural constraints on American elections have been delineated. This is a job that factor analysis does well and we can ask little more of any tool than that it does its job well.

References

Allardt E, Pesonen P, 1967 "Cleavages in Finnish politics" in *Party Systems and Voter Alignments* Eds S M Lipset, S Rokkan (Free Press, New York)

Anderson T R, Bean L, 1961 "The Shevky-Bell social areas: confirmation of results and a reinterpretation" *Social Forces* **40** 119-124

Archer J C, Taylor P J, 1981 *Section and Party: A Political Geography of American Presidential Elections from Andrew Jackson to Ronald Reagan* (John Wiley, Chichester, Sussex)

Bell W, 1955 "Economic, family and ethnic status: an empirical test" *American Sociological Review* **20** 45-52

Bell W, 1958 "Social choice, life styles and suburban residence" in *The Suburban Community* Ed. W A Dobriner (Free Press, New York)

Berry B J L, 1961 "A method for deriving multifactor uniform regions" *Przeglad Geograficzny* **33** 263-282

Berry B J L, 1964 "Cities as systems within systems of cities" *Papers and Proceedings, Regional Science Association* **13** 147-163

Berry B J L, 1966 *Essays on Commodity Flows and the Spatial Structure of the Indian Economy* RP-111, Department of Geography, University of Chicago, Ill.

Berry B J L, 1968 "Interdependency of spatial structure and spatial behavior: a general field theory formulation" *Papers of the Regional Science Association* **31** 207-227

Berry B J L, (Ed.), 1971 "Comparative factorial ecology" *Economic Geography* **42**(2) supplement, 209-367

Boring E G, 1961 "The beginning and growth of measurement in psychology" *Isis* **52** 238-257

Burnham W D, 1974 "The United States: the politics of heterogeneity" in *Electoral Behavior* Ed. R Rose (Oxford University Press, New York)

Burt C L, 1966 "The early history of multivariate techniques in psychological research" *Multivariate Behavioral Research* **1** 24-42

Burton I, 1963 "The quantitative revolution and theoretical geography" *Canadian Geographer* **7** 151-162

Cattell R B, 1966 "Factor analysis: an introduction to essentials" *Biometrics* **21** 190-215

Child D, 1970 *The Essentials of Factor Analysis* (Holt, Rinehart and Winston, London)

Clark B D, Gleave M B, (Eds), 1973 *Social Patterns in Cities* Special publication number 5 (Institute of British Geographers, London)

Cliff A D, Ord J K, 1973 *Spatial Autocorrelation* (Pion, London)

Converse P E, 1966 "The concept of a normal vote" in *Elections and the Political Order* Eds A Campbell, P E Converse, W E Miller, D E Stokes (John Wiley, New York) pp 9-39

Cox K R, 1968 "Suburbia and voting behavior in the London metropolitan area" *Annals, Association of American Geographers* **58** 111-127

Cox K R, 1969 "The voting decision in a spatial context" *Progress in Geography* **1** 81-117

Cox K R, 1970 "Geography, social contexts, and voting behavior in Wales, 1861-1951" in *Mass Politics* Eds E Allardt, S Rokkan (Free Press, New York)

Crewe I, Payne C, 1976 "Another game with nature: an ecological regression model of the British two-party vote ratio in 1970" *British Journal of Political Science* **6** 43-81

Cureton E E, 1968 "Psychometrics" in *International Encyclopedia of the Social Sciences* Volume 13, Ed. D L Sills (Free Press, New York) pp 95-112

Downing B, Hudson T, Taylor P J, Bland P, Villaneuva N, 1980 "The decline of party voting: a geographical analysis of the 1978 Massachusetts election" *Professional Geographer* **32**(4) 454-461

Dykstra R R, Reynolds D R, 1978 "In search of Wisconsin Progressivism, 1904-1952: a test of the Rogin scenario" in *The History of American Electoral Behaviour* Eds J H Silbey, A G Bogue, W H Flanigan (Princeton University Press, Princeton, NJ)

Elffers H, 1980 "On uninterpretability of factor analysis results" *Transactions, Institute of British Geographers* New Series **5** 318-329

Getis A, Boots B, 1978 *Models of Spatial Processes* (Cambridge University Press, Cambridge)

Goheen P G, 1970 *Victorian Toronto, 1850 to 1900: Pattern and Process of Growth* RP-127, Department of Geography, University of Chicago, Ill.

Gould P, 1968 "Review of B J L Berry 'Essays on commodity flows'" *Geographical Review* **58** 158-161

Gray F, 1975 "Non-explanation in urban geography" *Area* **7** 228-235

Gregory D, 1978 *Ideology, Science and Human Geography* (Hutchinson, London)

Gregory S, 1963 *Statistical Methods and the Geographer* (Longman Group, Harlow, Essex)

Gudgin G, Taylor P J, 1979 *Seats, Votes, and the Spatial Organisation of Elections* (Pion, London)

Harvey D, 1973 *Social Justice and the City* (Edward Arnold, London)

Herbert D, 1972 *Urban Geography: A Social Perspective* (David & Charles, Newton Abbot)

Johnston R J, 1971 *Urban Residential Patterns* (Bell, London)

Johnston R J, 1973 "Residential differentiation in major New Zealand urban areas: a comparative factorial ecology" in *Social Patterns in Cities* Eds B D Clark, M B Gleave Special publication number 5, Institute of British Geographers, London, pp 143-169

Johnston R J, 1977 "Urban geography: city structures" (Progress Report) *Progress in Human Geography* **1** 118-129

Johnston R J, 1978 *Multivariate Statistical Analysis in Geography* (Longman Group, Harlow, Essex)

Kamin L J, 1977 *The Science and Politics of I.Q.* (Penguin Books, Harmondsworth, Middx)

King L J, 1969 *Statistical Analysis in Geography* (Prentice Hall, Englewood Cliffs, NJ)

Kirby A M, Taylor P J, 1976 "A geographical analysis of the voting pattern in the EEC referendum, 5 June 1975" *Regional Studies* **10** 183-192

Macdonnell W R, 1902 "On criminal anthropometry and the identification of criminals" *Biometrika* **1** 177-227

McPhail I R, 1971 "The vote for Mayor of Los Angeles in 1969" *Annals, Association of American Geographers* **61** 744-758

MacRae D, Meldrum J A, 1969 "Factor analysis of aggregate voting statistics" in *Quantitative Ecological Analysis in the Social Sciences* Eds M Dogan, S Rokkan (MIT Press, Cambridge, Mass)

Miller W, 1977 *Electoral Dynamics in Britain since 1918* (St Martin's Press, New York)

Pearson K, 1901 "On lines and planes of closest fit to systems of points in space" *Philosophical Magazine* 6th Series 2 557-572

Rees P H, 1972 "Problems of classifying subareas within cities" in *City Classification Handbook* Ed. B J L Berry (John Wiley, New York) pp 265-330

Robson B T, 1973 "Foreword" in *Social Patterns in Cities* Eds B D Clark, M B Gleave, Special publication number 5, Institute of British Geographers, London, pp vii-ix

Rummel R J, 1970 *Applied Factor Analysis* (Northwestern University Press, Evanston, Ill.)

Schattsneider E E, 1959 *The Semi-sovereign People: A Realist's View of Democracy in America* (Dryden Press, Hinsdale, Ill.)

Spearman C, 1904 "General intelligence objectively determined and measured" *American Journal of Psychology* **15** 202-293

Spence N A, 1968 "A multifactor regionalization of British counties on the basis of employment data for 1961" *Regional Studies* **2** 87-104

Stevens S S, 1946 "On the theory of scales of measurement" *Science* **103** 677-680

Strotz R H, 1968 "Econometrics" in *International Encyclopedia of the Social Sciences* Ed. D L Sills, Volume 4 (Free Press, New York) pp 350-359

Taylor P J, 1976 "An interpretation of the quantification debate in British geography" *Transactions, Institute of British Geographers* New Series **1** 129-142

Taylor P J, 1978 "Political geography" (Progress Report) *Progress in Human Geography* **2** 153-162

Taylor P J, Gudgin G, 1981 "Geographical analyses of elections" in *Quantitative Geography in Britain* Eds R J Bennett, N Wrigley (Routledge and Kegan Paul, Henley-on-Thames, Oxon)

Taylor P J, Hadfield H, 1981 "Inequality in public housing and the state" in *Conflict, Politics and the Urban Scene* Eds K R Cox, R J Johnston (Longman Group, Harlow, Essex)

Timms D W G, 1970 "Comparative factorial ecology: some New Zealand examples" *Environment and Planning* **2** 455-469

Timms D W G, 1971 *The Urban Mosaic* (Cambridge University Press, Cambridge)

Warnes A M, 1973 "Residential patterns in an emerging industrial town" in *Social Patterns in Cities* Eds B D Clark, M B Gleave, Special publication number 5, Institute of British Geographers, London, pp 169-189

Williams K, 1971 "Do you seriously want to be a factor analyst?" *Area* **3** 228-229

Microcomputer graphics in multivariate analysis: an analysis of socioprofessional structure of the Rouen area

Y Guermond and the IMAGE Group †

The IMAGE group of the University of Rouen has worked for some years on the use of 'graphical microsystems' in geography. Our starting consideration has been that the traditional computer involves a limited number of peripherals. The addition of a digitizer or of a plotter has always been possible, but as a task is often limited to specialists, since it requires a complete knowledge of the hardware and software systems to be able to add the necessary subroutines. As a result, graphical representation by means of computers has been limited for a long time to the larger French laboratories. Recent improvements of microcomputers, however, now enables us to solve these problems in every department of geography.

The equipment in our department in Rouen consists mainly of a microcomputer HP 9825, a Summagraphics digitizer, a plotter HP 9872, and a Benson tracer(1). Present work on the urban structure of the Rouen region provides an illustration of the use of microcomputer graphics, which includes:

graphical printouts of factor analyses,
graphical adjustments,
mapping the residuals of a multiple linear regression,
classification and mapping of the different classes.

Mapping is carried out by the Cartovec system, developed in Rouen, which allows the printout of chloropleth maps.

The Rouen Study

The Rouen area (369793 inhabitants in 1968, and 388711 in 1975) has been chosen as a case study for our chain of microcomputer graphical programs. Data were available at 'block' level (2280 blocks on magnetic tape), which could be reconsolidated on the basis of 106 statistical districts (data were transferred onto cassette). A digital representation of the map of these 106 districts enabled us to calculate the distance to the city centre, as well as the area of each of these districts, and thus to include the population density, which we were previously unable to calculate since the census does not provide the size of each area.

† Members of the IMAGE group in the University of Rouen are as follows: B Lannuzel, A Leduc, and X Verlut (Department of Mathematics), L de Golbery, Y Guermond, P Lecarpentier, and G Pinchon (Department of Geography).

(1) HP—Hewlett Packard Company, Palo Alto, Calif.; Summagraphics Inc., 35 Brentwood Avenue, Fairfield, Conn. 06430; Benson Cie, 1 rue Jean Lemoine, 94015 Creteil, France.

Finally we had sixteen variables which were submitted to a principal components analysis. They were

DCE Distance to the city centre,
DEN Population density,
PPI Persons per room,
PMEN Persons per household,
MVA % of population less than 20 years old,
PAC % of active population in the total population,
COM % of tradesmen in the total active population,
CS % of upper management in the total active population,
CM % of middle management in the total active population,
EMP % other white collar workers in the total active population,
OUV % blue collar workers in the total active population,
ETR % nonnationals in the total population,
CFT % dwellings with amenities,
VOI % households with one car or more,
MIN % of single family dwellings,
A62 % post-1962 buildings.

Analysis of major components using the plotter printout allows us to locate the different districts in relation to the two axes of the principal

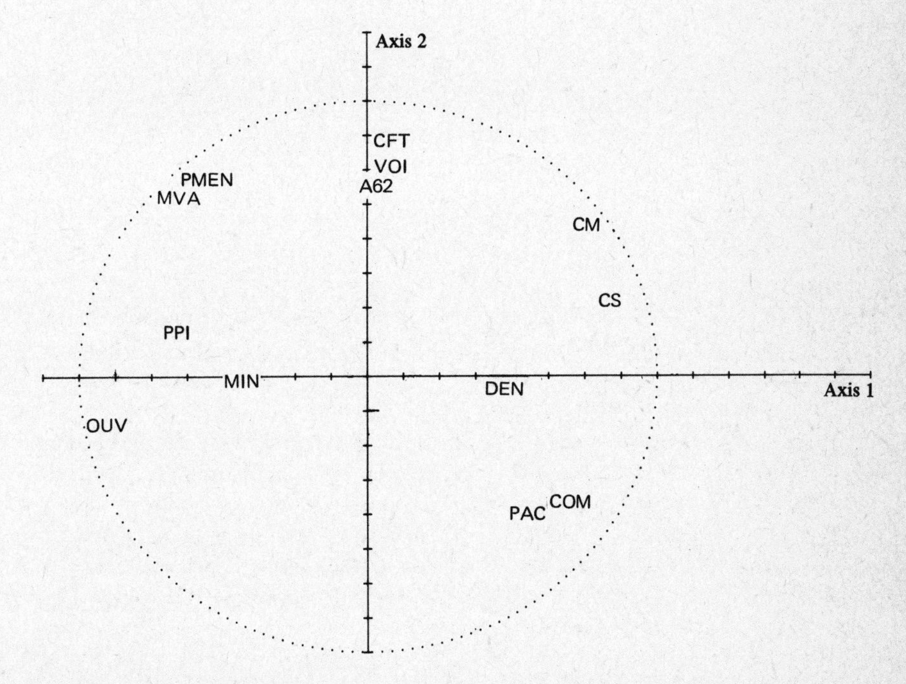

Figure 18.1. Normalised principal components analysis results (coefficient of explanation = 59·7%).

components, as shown in figure 18.1. The first three principal components account for 74% of the total variance as follows:

The first axis (35% of the variance) relates to socioprofessional structure, white collar blocks (CS, CM, COM) against working-class blocks (OUV, MVA, PPI);

The second axis (25% of the variance) relates to demographic structure, newly-established blocks of young households with a high percentage of population under twenty (MVA, PMEN, CFT, VOI, A62) against blocks of tradespeople (COM, PAC);

The third axis (14% of the variance) relates to the nature of building types, blocks of detached dwellings away from the town centre (MIN) against more crowded blocks (PPI, DEN).

Simple regression of the Rouen data

The printout by the plotter of clusters of points on a graph of two variables allows us to select one relationship from seven possible models, viz:

$$Y = aX + b\,, \tag{18.1}$$

$$Y = \frac{a}{X} + b\,, \tag{18.2}$$

$$Y = a \lg X + b\,, \tag{18.3}$$

$$Y = aX^{1/2} + b\,, \tag{18.4}$$

$$\lg Y = aX + b\,, \tag{18.5}$$

$$Y = aX^2 + bX + c\,, \tag{18.6}$$

$$\lg Y = a \lg X + b\,. \tag{18.7}$$

To describe the relationships in terms of distance from the town centre (figures 18.2 and 18.3), we generally choose model (18.3) or model (18.7). In the case of socioprofessional structure, we find a marked inverse relationship between the decrease in the percentage of high-level management with distance from the city centre ($\lg \mathrm{CS} = -0{\cdot}52 \lg \mathrm{DCE} + 0{\cdot}47$) and the increase in the percentage of blue collar workers ($\lg \mathrm{OUV} = 0{\cdot}38 \lg \mathrm{DCE} - 1{\cdot}7$).

The spatial distribution of tradespeople (figure 18.3) is complex. The relationship between their locations and distances from the town centre was found to be weak ($r = 0{\cdot}38$). Logarithmic analysis underlines the fact that many small retailers are to be found in the hypercentre (at times they represent more than 20% of the residents), but the polynomial model ($Y = aX^2 + bX + c$) highlights their small numbers in outlying residential areas (6 km to 8 km from the centre) and their reappearance in parishes of the urban periphery. The opposite results are obtained for the distribution of under-twenties (MVA, figure 18.3), as their number is high in the outlying estates. Here again, models (18.6) and (18.7) illustrate two different aspects of the same situation.

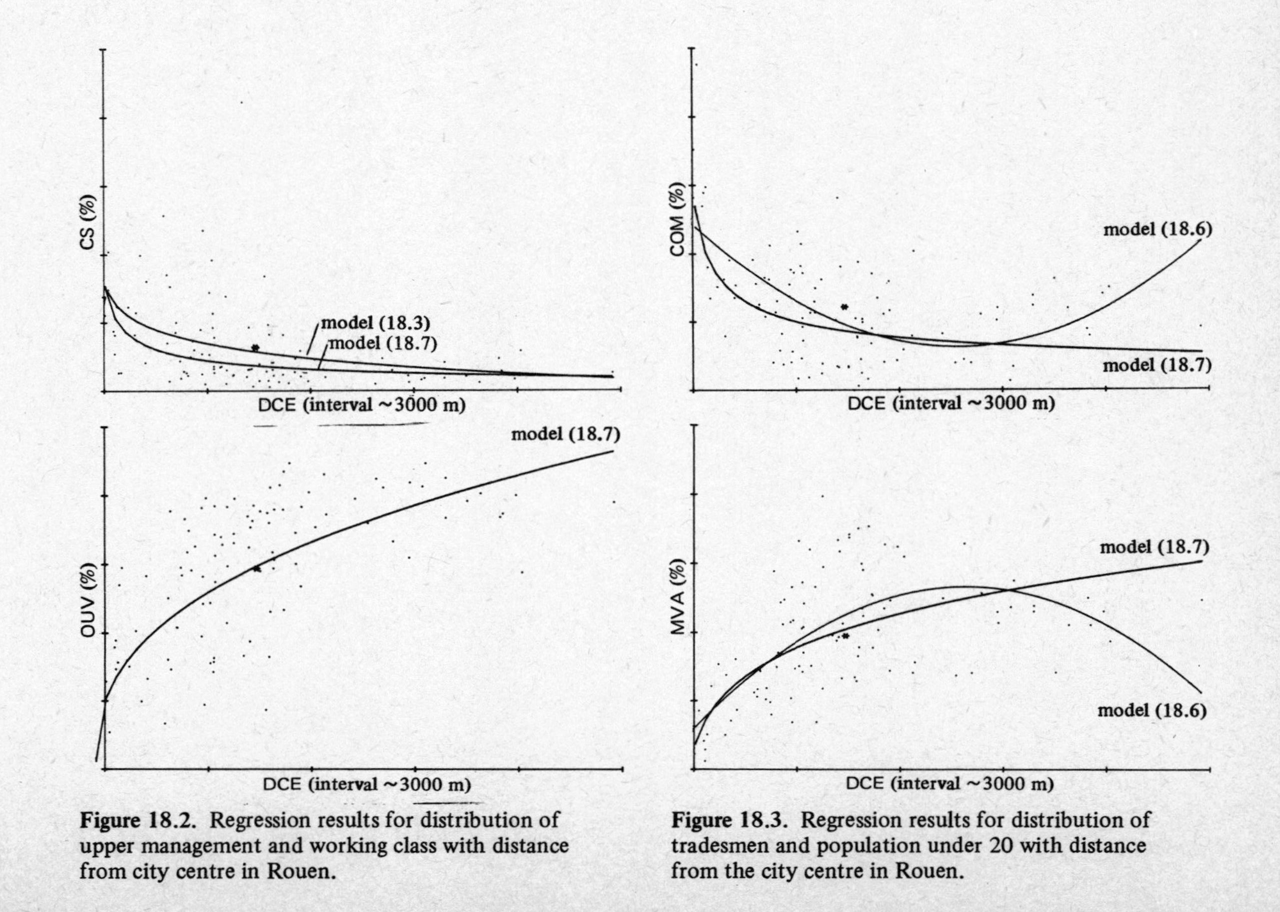

Figure 18.2. Regression results for distribution of upper management and working class with distance from city centre in Rouen.

Figure 18.3. Regression results for distribution of tradesmen and population under 20 with distance from the city centre in Rouen.

Multiple regression of the Rouen data

From results already obtained, we attempted to analyse the distribution of white collar workers by trying linear combinations with other variables. The program we initiated enabled us to modify the number of variables under consideration, as also the number of statistical observables, so as to eliminate those that did not fit within the analysis.

Our aim was to explain the distribution of the percentages of white collar workers (EMP) by means of data on the nature of the types of building identified by

distance from the town centre (DCE), $r = -0{\cdot}60$,
population density (DEN), $r = 0{\cdot}48$,
% of single family dwellings (MIN), $r = -0{\cdot}44$.

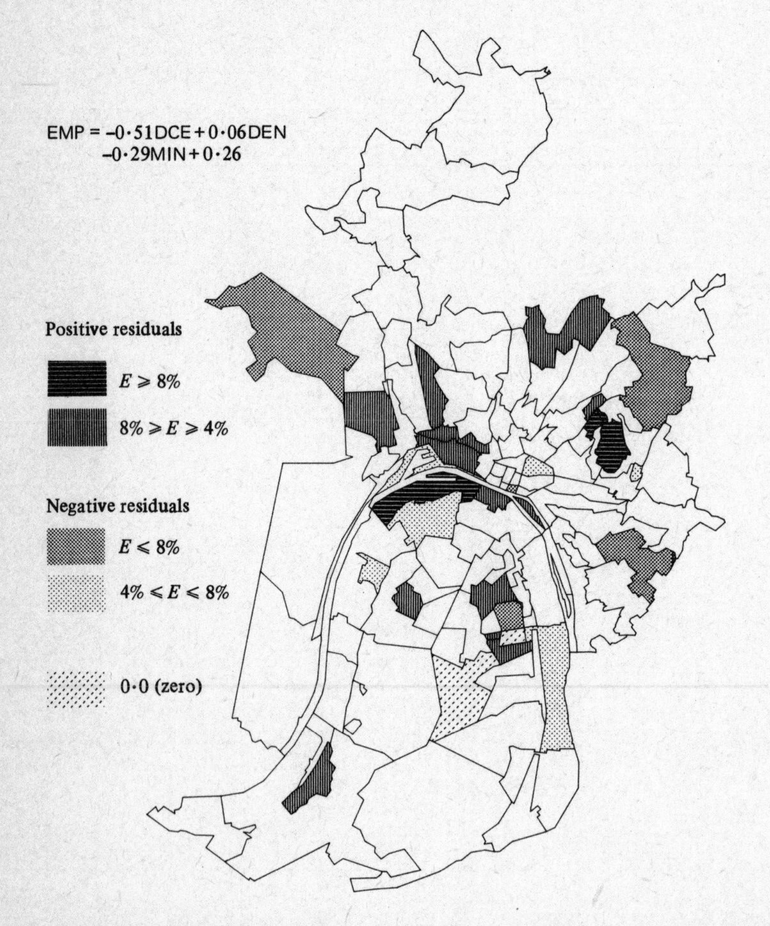

Figure 18.4. Map of the residuals for regression of the socioprofessional structure of Rouen in relation to building type.

For all districts taken together, the coefficient of multiple linear correlation between the four variables is 0·64. If we eliminate those districts with a residual value above 0·08, we significantly improve the *r* multiple, which rises to 0·72. Figure 18.4 shows the results of this calculation, obtained by using the regression

$$\text{EMP} = -0{\cdot}51\text{DCE} + 0{\cdot}06\text{DEN} - 0{\cdot}29\text{MIN} + 0{\cdot}26\,.$$

The map gives the location of residuals higher than 0·08 (eliminated after the first calculation) and then higher than 0·04. For other districts this value varies by less than 4% of the real value (the average percentage of white collar workers in the Rouen area being 18%). We can see that the districts in which the proportion of white collar workers is higher than that predicted by the model fall into three categories:
(1) suburbs close to the town centre,
(2) districts to the south, of detached housing built between the wars,
(3) recent residential estates to the northeast.

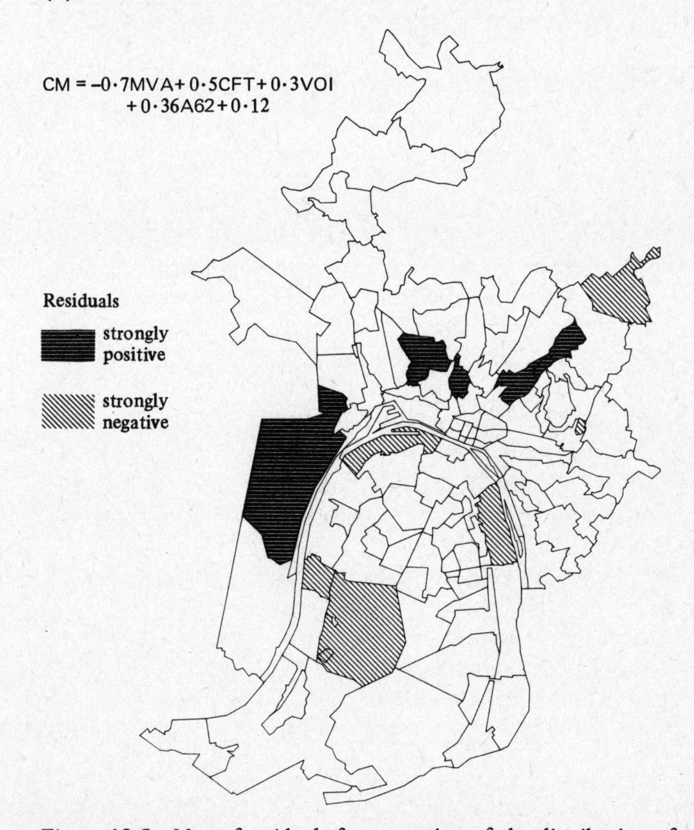

Figure 18.5. Map of residuals for regression of the distribution of middle-management groups in relation to domestic amenities, housing, and car ownership. Residuals are mapped which deviate by more than ±4% from predicted levels.

We attempted to explain a further socioprofessional variable, the percentage of middle management, by making use of those variables seemingly related (under-twenties, domestic amenities, cars, new housing). For the regression

$$CM = -0{\cdot}7MVA + 0{\cdot}5CFT + 0{\cdot}3VOI + 0{\cdot}36A62 + 0{\cdot}12 ,$$

the r multiple coefficient is 0·83. The average percentage of middle management in the Rouen area is 11%. The map (figure 18.5) shows that the location of strongly positive residuals (actual proportion at least 4% higher than the theoretical proportion) had a very clear geographical significance; most middle management were to be found to the north and west of the town, on the edges of the plateau (belvedere sites). The negative residuals were found around the port and shunting yards (Seine valley, up-river from the town centre).

Classification of Rouen social areas

From a classification tree, produced on a Benson tracer, we can propose two divisions:

4 classes, with 58% loss of detail,

9 classes, with only 36% loss of detail.

Above 9, the number of classes rises too rapidly for the division to be retained in terms of the small total number of districts. The value of the proposed divisions can be illustrated by the Fischer–Snedecor test; the homogeneity hypothesis is rejected for all variables in divisions of 7 and 9 classes. The same applies for division into 4 classes to a limit of 0·05. Below this limit, in the latter division, only variable ETR (percentage of nonnationals) is rejected; this was not unexpected considering the small number of nonnationals and their distribution in colonies or ghettos not related to the other social groups.

The averages and the standard deviations (σ) of the 15 variables are summarised in table 18.1, and this enables us to give a clear definition of each of the 9 classes. Distance from the town centre is no longer included as a variable, as the plotter's cartographic representations enable us to locate the various classes (figure 18.6).

The division into conventional zones does not apply on the right bank of the river Seine, owing to relief conditions, and even on the left bank it shows us modifications that we are able to analyse by reexamining each large grouping of districts in more detail.

Working-class suburbia

This is the area near the town centre, which includes type 7 districts, where the population under twenty is small and where we find 21·5% of white collar workers. Detached houses generally predominate (85%) and only 7% of dwellings were built later than 1962. This 'old established

Table 18.1. Mean of each category in relation to the mean of the whole area for nine types of area.

Variable	DEN	PPI	PMEN	MVA	PAC	COM	CS	CM	EMP	OUV	ETR	CFT	VOI	MIN	A62
Mean of the whole area	2448	1	3·2	35	44	7	6	11	18	47	2·5	66	58	74	8
Class 1, New working-class suburbs															
Type 1. Decaying suburbs												−			
Type 7. Old-established white collar suburbs				−											
Type 4. Mixed suburbs															
Class 3, Dense working-class suburbs															
Type 6. Subproletariat		+								+	++				
Type 5. Working class		+	+	+								+			
Class 4, Town centre															
Type 9. Town centre rebuilt since 1945	+	−	−	−	+	++	++	+		−−		+		−	−
Type 8. Old town centre	++		−	−	+	+				−			−	−	
Class 2, Prosperous districts															
Type 2. Upper middle-class housing		−					++	+		−					
Type 3. Recent middle-class housing								+	+			+		−	++

Key: ++ more than $+2\sigma$ greater than the mean of the whole area.
+ more than $+1\sigma$ greater than the mean of the whole area.
− more than -1σ less than the mean of the whole area.
−− more than -2σ less than the mean of the whole area.

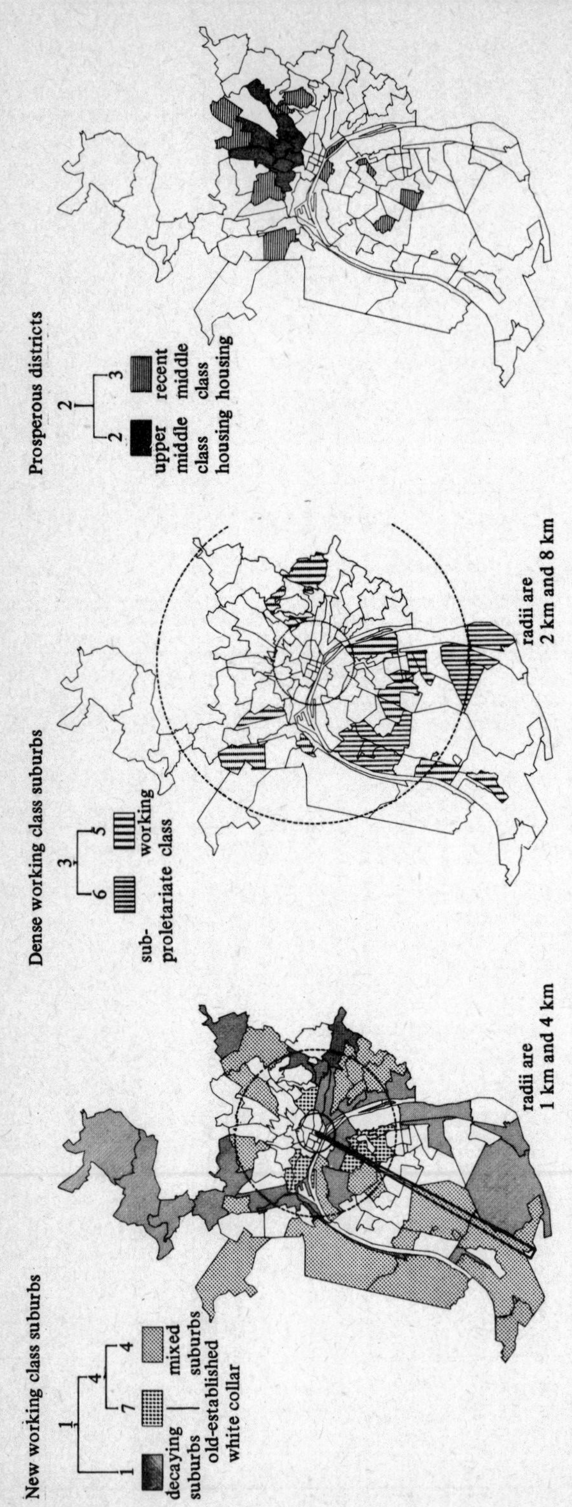

Figure 18.6. Typology of the city of Rouen in terms of social areas.

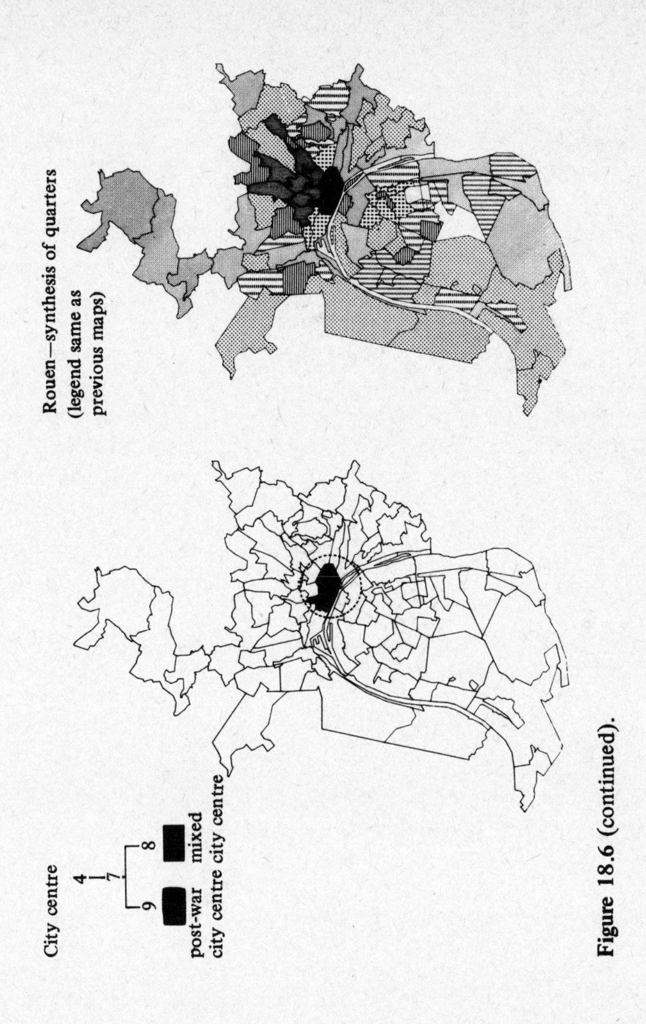

Figure 18.6 (continued).

white collar suburb' stretches as far as the warehouse district west of the town centre, near the port, and also to the south and to the northeast. These districts are located between 1 km and 4 km from the town centre.

Type 1 districts (decaying suburbs) are more widespread, stretching along the main industrial axes, the Cailly valley to the northwest, the port area up-river, and to the southeast. Here we find 60% blue collar workers ($\bar{x}+\sigma x = 64\%$), a low standard of domestic amenities, and very little recent construction.

Type 4 is a mixed suburb (all variables having a value close to the average for the whole Rouen area). We can distinguish these districts from those above by the relatively high percentage of households with a car. These districts are generally located on the urban periphery (6 km to 10 km from the town centre), but they do extend closer to the centre on the small areas of high ground that were formerly not urbanised and which do not attract upper management in the same way as do the 'belvederes', for example, to the east and northeast of the town. The three districts of this type close to the town centre on the left bank are somewhat a special case, this is an old-established working-class suburb close to the harbour and its surrounding industrial area.

The overpopulated working-class suburbs

These extend almost exclusively between 2 km and 8 km from the town centre. This area only exceeds 8 km to the southwest (working-class 'colonies' near the port) and to the southeast (industrial zone up-river). These are overpopulated suburbs in the sense that the number of persons per room is always high, although population density per square kilometre is only average and at times low.

Type 6 districts ('subproletariat') are uncommon. Here density is particularly low (with a weak standard deviation for this variable), apartments are very small (high PPI) and the percentage of blue collar workers is high. These are also the only districts where the percentage of nonnationals becomes a significant variable (above the average $+2\sigma$). We can see on the map that these districts correspond to infill areas between factories (the harbour zone to the south) or to ill-defined areas bordering a wasteland littered with dumping grounds and slums.

Type 5 districts have a higher population density and a higher standard of domestic amenities. Families are large, and hence there are large numbers of under-twenties, and the number of people per room is also high. These are generally working-class colonies established between the wars. They are located on the left bank, 4 km from the town centre (outside the old-established white collar suburbs) and elsewhere they are scattered randomly, although always 3 km or 4 km from the town centre, on any available land (that is, near the shunting yards near the Seine), as well as in the old-established industrial valleys of Robec (to the east) and Cailly (to the northwest).

The prosperous districts

These districts are generally near the town centre. Type 2 districts (large numbers of upper and middle managers, less than 7% nonnationals, a high percentage of detached houses, and many cars) are very tightly confined to the higher ground to the north.

Type 3 districts (middle management and other white collar workers) extend around the above districts on the high ground, and we also find them scattered in the urban framework; they correspond to large newly-built residential complexes. The number of persons per room is higher than we find in type 2 districts.

The town centre

The town centre was limited in 1968 to that part of the right bank located inside the boulevards. The area overlooking the Seine, which was destroyed during the Second World War, was rebuilt during the 1950s to a standard attractive to upper management. The two type 9 districts have the smallest percentage of blue collar workers in the entire Rouen area (7·7%), and the highest percentage of tradespeople.

In type 8 districts the tradespeople are equally numerous, but there is a greater social mix (we find 30% blue collar workers). Standards of living are not so uniformly high in these districts, and the standard deviation is high for this variable (only 49% of dwellings with amenities, with a standard deviation of 15%, whereas in type 9 districts we find 96% of dwellings with amenities, with a standard deviation of 0·5%). Density is high in all these areas, but reaches its peak in type 8 districts. In both types we find few under-twenties.

The compilation of the four previous maps illustrates that the traditional opposition, between the ‘right bank’ and the ‘left bank’, in the social distribution of the Rouen area, stems from the concentration of the prosperous sector on the borders of the higher plateau areas north of the centre, as well as in the centre itself. The division between the two principal groups on the classification tree has a particularly striking spatial manifestation.

In the purely descriptive work of this study, we have confined ourselves to the use of classical simple techniques. Our aim has been to develop programmes that respond to the problems of current geographical interpretation, for use on modest computer installations, and thus to demonstrate that ‘microprocessing technology’ should become a widespread and valuable aid in geographical investigation. It seems to us that one of the reasons for the slowing down of much geographical work is the intervening stage of hand-mapping, starting from computer listings that are not the ideal format for geographers. A well-conceived graphical picture, allowing the selection and the quick association of the different information, is the preferred tool for geographical work.

19

Urban development and spatial analysis: Montpellier 1954–1978

M Chesnais

The last two decades have been characterised by a large and relatively rapid expansion of our towns. This process of urbanisation has been accompanied by transformation or changing patterns of land use which are often without precedent in recent times. The study of the processes of urbanisation therefore involves aspects both of time and space. In such studies it is essential that two conditions be fulfilled. On the one hand, it is essential that the boundaries of the land chosen for the study of these processes do not change over time, either in their area or in their ease of observation through a set of reference points. In addition, observation over different periods must be consistent, so as to allow a satisfactory chronological analysis.

These problems have been tackled in a research programme investigating the growth of Montpellier and the effects of this growth on spatial transformation. It is not my intention here to present the detailed results of this work but rather to expand on them by identifying the conditions of land-use change and the resulting effects upon spatial transformation. Chronological and spatial patterns can both be investigated, as well as the various situations arising from them, so as to determine the characteristics of the process of change in state (a state being defined by a stated land use and a fixed moment of observation). Hence the intention is to analyse the degree of diversification observed during the transformation process.

In effect, the processes are least complex when the transformation is uniform; for example, all vine-growing plots (initial state) are transformed by the construction of detached houses (a single subsequent state). However, transformation is multiform when one initial state, which applies to all plots under consideration, is replaced by the largest possible variety of subsequent states—vineyards could be replaced by apartments, houses, factories, roads, etc. In the first case, diversification is minimal; in the second, it approaches a maximum. Hence, in the first case, the spatial transformation is relatively homogeneous, whilst in the second it tends to be very heterogeneous.

These two situations roughly illustrate the nature of the problem that exists; in detailed terms, we need to be more precise. We need to find a method of analysis, for example using the measurement of entropy, which allows us to appreciate the extent of diversification in the changing pattern of land use.

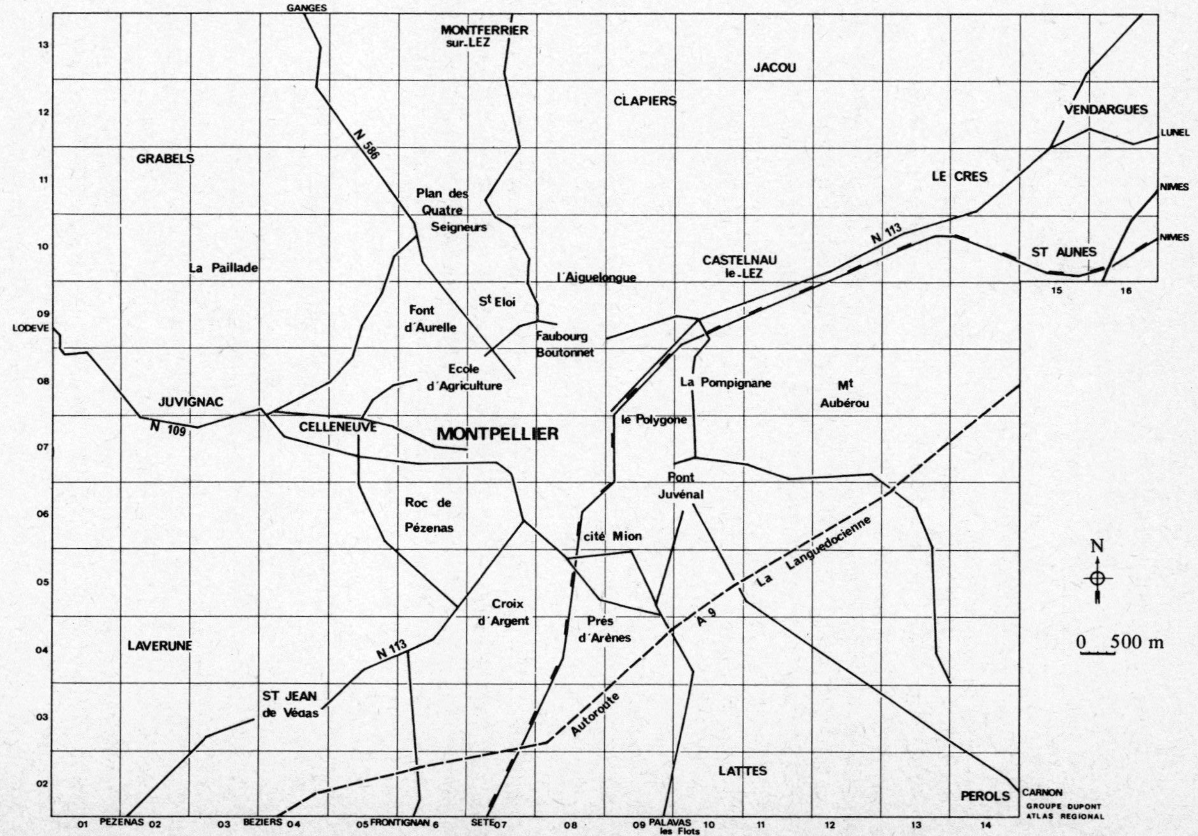

Figure 19.1.
The study area.

Conditions and processes of data compilation

The study area is centred on Montpellier and covers 176 km^2; an observation grid of 176 areas was devised. Each area comprises 64 observation districts, each of which in turn comprise a 25 m radius study zone. In this way, the data obtained correspond to a one-eighth sample survey. From this grid we have been able to obtain 11264 observations.

The data have been obtained from the study of aerial photographs taken in 1954, 1963, 1971, and 1978, the same sample grid being used so as to ensure continuity of observation at different times. In addition to Montpellier itself, the study zone covers an area characterised by uncultivated hills and vineyard slopes to the north of a NE–SW line, as shown in figure 19.1 approximately from Vendargues to Saint Joan-de-Vedas. The land to the southeast, however, comprises a plain filled with vineyards.

The detailed analysis was based upon a choice of 21 spatial variables, each defined from identifiable characteristics within the study zone. From these, the dimensions of the observation grid and the scales for the aerial photographs (between 1/14500 and 1/25000, according to the sources used) were determined.

The definition of the spatial variables was as wide ranging as possible, taking into account all identifiable forms of land use over the entire period 1954–1978, some forms not being present on all four observation dates. Hence, continuity is achieved as much through the permanence of the grid data as through the pertinence of the land uses observed. To analyse the effects of urban development on land-use transformation, and the consequent modifications, three periods were defined, 1954–1963, 1963–1971, 1971–1978, to characterise the continuity of the sequence 1954–1978.

The definition of an observation zone, at a fixed moment in time, by a particular land use, constitutes a state. Hence we can identify as many states as we can spatial variables. For each period there are two potential directions of change for each observation zone (change in state or no change), and these allow us to construct a transition matrix or table of changes in state.

Conditions of the transformation analysis

Overall view

Before undertaking any analysis it is necessary to define how the geographical dimensions of the observation grid can be related to the statistical nature of the information. In each observation zone, an area of 1 km^2, we have 64 observations (as we have 64 basic observation areas). However, the number of observations of each variable is such that any results obtained would be statistically unsatisfactory. In order to improve the analysis, two possibilities exist: either to regroup the study zones so as to create larger numbers of observations; or to regroup the spatial variables in order to reduce the number of occurrences. In the research results already published, the two possibilities have been tested by analysing groups of four study zones (hence 4 km^2) on the one hand, and by a grouping of the variables into thirteen categories on the other. If the data on the latter are relatively precise, owing to the dimension of the study zone itself, the geographical detail is considerably reduced by a factor of four.

Having taken the above results into account, priority is accorded here to geographical precision, and the 176 study zones have been retained. However, to obtain a sufficient statistical significance, we must reduce the number of spatial-variable categories to seven as shown in table 19.1. The first four represent built-up areas, and the final three unbuilt areas.

Amongst the first, two refer to residential buildings, separating apartment blocks from individual dwellings. The third represents all buildings concerned with production and services, and the fourth relates to the transport system. The unbuilt categories comprise one that can be said to represent a transitional state between built-up and unbuilt land. Earthworks, with the exception of quarries, are used for construction purposes, foreshadowing building. The fallow lands represent the abandonment of fields formerly under cultivation. However, the detailed analysis carried out in the general study has led us to reinterpret this spatial variable, in effect, waste lands can directly foreshadow a change in state to construction, but, they can also correspond to a rest period in an agricultural cycle. This cannot be determined by aerial photography alone; knowledge of the area and personal observations are required. In addition, the studies have shown that urban development has an effect on the noncultivation of land, acting as a regulator in the geographical redistribution of agricultural activity. However, generally speaking, the significance of the transitional state of the earthworks and fallow-lands category remains. The other two categories correspond to agricultural land that could generally be called 'natural', more through reference to ancient states than to the evolution of recent decades which has progressively excluded every form that can be so described. In this case, we are essentially dealing with hilly areas of sparse vegetation, some watercourses, a very narrow fringe of low lands bordering the lagoons, and wooded plains. The relative lack of precision that could result from this grouping is in fact compensated for by the precise, known location of each of the different elements.

Essentially, agriculture is characterised here by viticulture; the distinction between hillside and plain vineyards being clearly identifiable in the study area. There is a variety of ancilliary types of agriculture, which

Table 19.1. Spatial variables employed in the analysis.

1. Residential buildings	5. Earthworks and fallow lands
2. Apartment houses	6. Agricultural lands
3. Industrial, commercial, and public buildings	7. Woodlands, scrub, and water.
4. Transportation	

include cultivation under glass, open field cultivation, orchards, nurseries, and grasslands in some valleys.

We have seven classes of spatial variables and, at the four successive dates, their distribution over the study area gives us an initial idea of the evolution resulting from urban development. As shown in table 19.2, the spread of urbanisation has essentially resulted from the construction of residential buildings—particularly individual houses, which have spread more rapidly in recent years than have apartment blocks. The increase in surface area devoted to production and services has been primarily concerned with the latter: hospitals, universities, etc, and to a lesser degree with commerce, and least of all with industry. The spread of the transport infrastructure is also of significance, particularly the 'Languedocienne' motorway, which passes through the study area, and the major roads towards the coast and the airport.

As for earthworks and fallow lands, after a rapid initial expansion in the first study period, and a subsequent slight regression related to an increase in construction in the second period, their proportion has tended to stabilise. Examination shows us that urban development has primarily affected agricultural land and only recently has it affected sectors such as hillsides.

Table 19.2. Evolution of the study area in relation to the seven variables.

Spatial variable		Period			
		1954	1963	1971	1978
Built-up	1. Apartment blocks	1·8	2·7	4·4	5·3
	2. Houses	3·7	5·8	9·4	13·3
	3. Industries	1·0	1·2	3·0	4·5
	4. Transport	8·9	9·0	10·1	11·2
Intermediate	5. Earthworks and fallow lands	6·0	12·2	11·2	13·0
Unbuilt	6. Agriculture	57·3	47·9	42·4	34·9
	7. 'Natural' environment	21·3	21·1	19·4	17·7
		100%	100%	100%	100%

Table 19.3. The renewal rate for each variable used in the study, expressed as the percentage of the variable which undergoes change during the period.

Spatial variable		Period		
		1954–63	1963–71	1971–78
Built-up	1. Apartment blocks	2·9	7·9	1·8
	2. Houses	5·0	6·1	2·4
	3. Industries	4·5	4·3	1·2
	4. Transport	7·4	13·5	3·3
Intermediate	5. Earthworks and fallow lands	51·4	70·1	55·2
Unbuilt	6. Agriculture	21·2	22·9	24·3
	7. 'Natural' environment	9·0	18·8	18·3

I will end this summary with a brief analysis of the transition matrices derived from the study in order to illustrate the rate of renewal for each class of variable. Table 19.3 gives the proportion of each study zone at the beginning of each period, which subsequently undergoes a change of state during that period. If internal transformations within the urban framework are identifiable, particularly during the second period when expansion was greatest, the most notable renewal is to be found amongst the earthworks and fallow lands; more than two-thirds of the zones thus categorised at the beginning of the second period have subsequently undergone a change in state. But as shown in table 19.2, their total proportion of 12% has remained roughly stable. The renewal rate of the agricultural sector has not been spectacular, and its proportion of the total from 1954–1978 decreased from more than half the total area to around one-third. This illustrates the constant, regular effects of urban development on agricultural land and the role of the 'regulatory factor' played by the earthworks and fallow lands. In addition, the effects on the 'natural' environment are apparent from the second period.

Measurement of the diversification of the transformation process

We can utilise various approaches to analyse the processes of spatial transformation through urbanisation. Here we are concerned with an analysis of the diversification. For a given area, we have constructed a 7 x 7 matrix showing changes in state. From this we can calculate the entropy of each linear distribution (that is, how the study zones characterised by a known variable at the beginning of a period have or have not subsequently changed their state). For each row of the matrix the study zones were then divided between the six other land-use possibilities available. The closer the division between the six possibilities corresponds to an equiprobable distribution, the higher is the entropy, and the greater the diversification arising from the transformation processes. This applies to each row of the matrix and enables us to characterise the evolution of each zone, from its original state, during each period. In addition, the total of the entropy values obtained for each row allows us to quantify the extent of diversification of the whole area being studied.

In this way we obtained, for the whole study area, successive values expressed as a percentage of the maximum entropy (which remains constant over the three periods owing to the permanence of the observation grid and the number of variables). These values were:

1954–63, 67·3%; 1963–71, 78·5%; 1971–78, 83·5%.

We see that the diversification deriving from the land-use transformation processes increases in the study area over time. These are aggregate total data, but data that we can usefully supplement by considering not only the study area as a whole, but each of the 176 study zones, as discussed below.

Geographical evolution and diversification by period

For each of the 176 zones, three 7 x 7 transition matrices, characterising successive periods have been constructed. To facilitate comparisons, entropy values were converted to percentages of the maximum entropy value, which does not vary as all the zones are characterised by matrices of the same dimensions, 7 x 7, constructed on the basis of the same variables and from the same number of observations (64). As in the previous cases, only those distributions illustrating changes in state have been taken into account, thus diagonal values are excluded.

The values obtained were classified after the establishment of frequency histograms, coded numerically, and then represented automatically (figure 19.2).

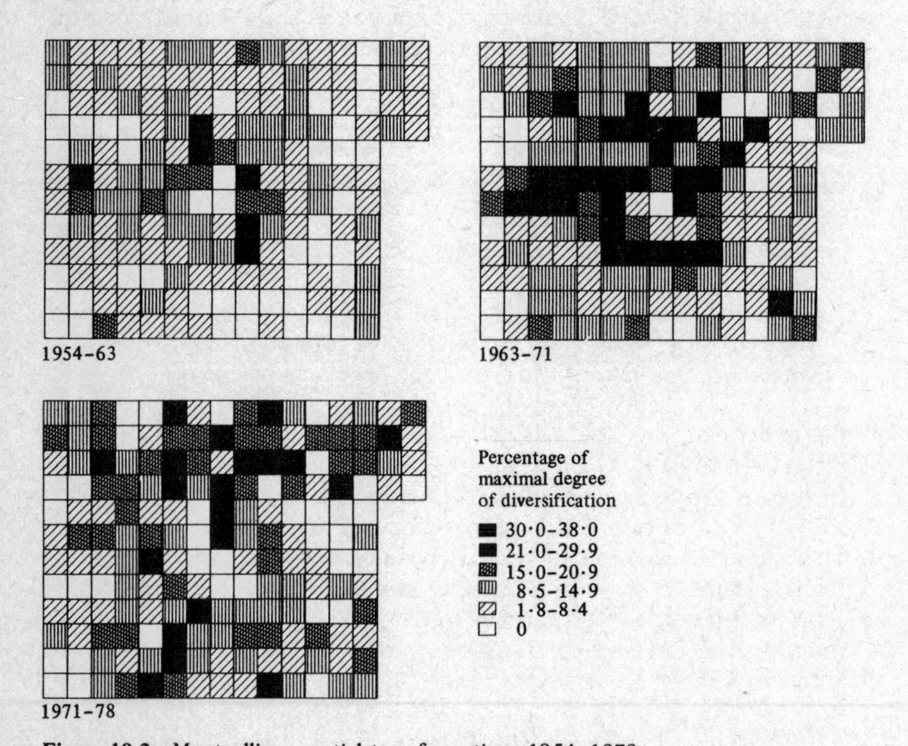

Figure 19.2. Montpellier: spatial transformations 1954-1978.

First period, 1954-1963

The biggest diversification in change of state is represented by values under 40%. This underlines the problem of keeping geographical definitions constant, since the larger the area (the entire study area) the higher the cumulative effect, and the higher the entropy.

Zones where diversification is high are few in number and are situated to the north and east of the town. In the town centre, diversification is nonexistent; this does not signify that there is no change in state, but that if change does occur it involves only two states.

Second period, 1963–1971
A larger number of zones than before are characterised by relatively high entropy values. Diversification has spread, particularly in a ring around the newly-developed centre, but also to the west, the southeast, and the northeast.

Third period, 1971–1978
Although general diversification has again increased, it is a different form of geographical distribution of land-use transformations. This is illustrated by the inclusion of zones characterised by the highest entropy values, particularly in the north of the study area, as well as in the southwest.

At this point we must compare the three periods. In effect, when considering the evolution of diversification for the entire area, we see that it has risen from 67·3% to 83·5%, and on examining its geographical distribution over the three periods, it is the second period that seems to show the greatest diversification, having the greatest number of high-entropy values. This leads to the conclusion that, in the area studied, the growth of Montpellier involves an increasing diversification in land-use transformations; embryonic in the first period, geographically concentrated around the town in the second period, and more widely spread in the final period.

Although the urbanisation process is accompanied by an increase in the diversification of land-use patterns resulting from changes in land use, a diffusion phenomenon has developed across almost the entire area being studied.

Evolution over the three-period sequence
The presence of a diffusion phenomenon should not detract from the considerable geographical diversity in the distribution of the profiles characterising each study zone. In effect, an automatic classification process, using a table 176 x 3, identifies fifteen different types shown on figure 19.3. This tends in part to cloud the findings that we would expect to encounter here, namely that urbanisation comprised a diffusion of diversification processes in successive waves. Nothing to this effect appears in figure 19.3 in adjacent zones, although we must remember that in the space of one kilometre situations vary greatly, and the classification process will have reduced some of these effects.

The urban development of Montpellier and its surrounding areas has therefore been accompanied by a considerable degree of variability in geographical order, which could lead us to question oversimplistic urban diffusion models founded solely on the centre–periphery relationship.

The results obtained illustrate much more than would at first appear; they also allow us to appreciate the different stages of development of spatial transformations.

However, we can attempt to arrange this information geographically by regrouping the fifteen categories given in figure 19.3 into five groups which then express spatial behaviour tendencies. Figure 19.4 illustrates the resulting distribution. Certain zones are characterised by an increasing diversification during the study periods and reach relatively high values during the last period. Their distribution is characteristic both in the northern and in the southern areas. In contrast, the central area, widening to the west, comprises zones where generally the degree of diversification has remained just about stable, has diminished, or is very low or non-existent.

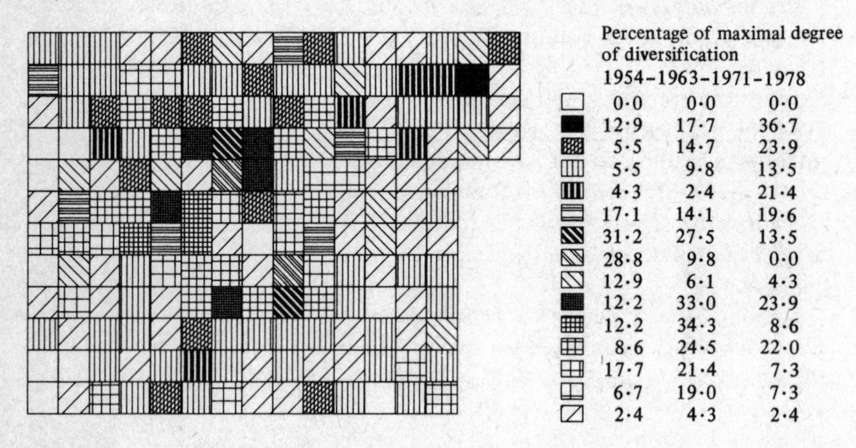

Figure 19.3. Types of spatial transformations.

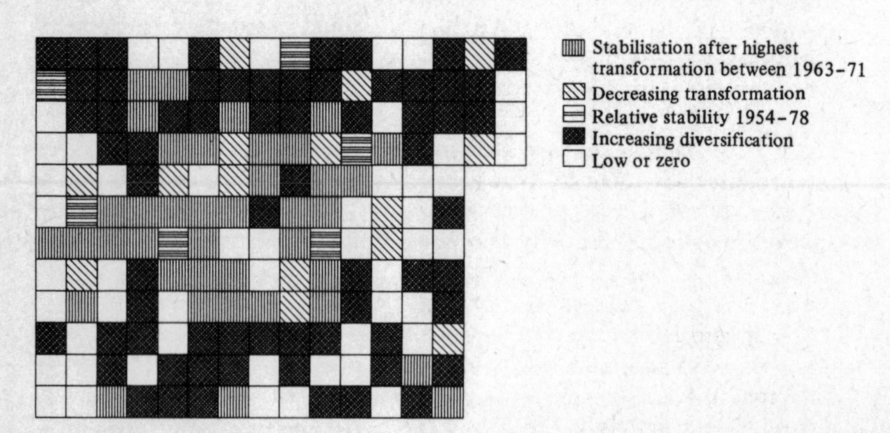

Figure 19.4. The main types of spatial transformation (derived from groupings of the categories in figure 19.3).

The representation discussed above illustrates the effects of urban development in a structural dimension and, through the above distribution, enables us to formulate a hypothesis: that the north, and in the south, factors intervene to create the widest variety of new land uses in relation to preexisting conditions. This is a very simple hypothesis. Other analyses will have to determine to what extent, apart from strictly periurban growth, the attraction of the hills in the north and the coastal plain in the south is the generator of a greater diversity of spatial changes and whether this is a superficial or a deep-seated modification of land use during the processes of an urban concentration.

Conclusion: the operational character of a spatial application of the measurement of entropy

Already used in the course of our research, the measurement of entropy has enabled us to reach levels of analysis not originally envisaged. Through its ability to describe the essence of a distribution, through its guaranteed accuracy in use when information is correctly presented, this technique allows us to identify the largest or smallest regularity within a distribution, either of a group of variables or of their interrelations. In the case of this study, the use of entropy measurement has enabled us to isolate the surface aspects in the organisation of spatial transformation processes, and above all to clarify how these processes have operated.

Acknowledgements. This chapter is based on part of a study undertaken by F Auriac, M C Bernard, M Chesnais, R Ferras, M C Maurel, and M Vigouroux (members of the Dupont Group) on the subject "Montpellier 1954–1978: process of spatial transformation and urban development".

An automatic method of data simplification for area mapping (choropleth)

J P Grimmeau

Introduction

The author of a choropleth map with thirty or more elementary areas does not generally return to the original values when he carries out an analysis. Instead he tends to work on the simplified data which are on the map and assumes that (1), if two neighbouring areas are in the same class, they are similar to each other, and (2), if they are in two different classes, they are dissimilar. The classical methods of choropleth mapping do not warrant these two assumptions, and this chapter is aimed at developing a method of data simplification which is accurate as far as possible. Taking a subjective method proposed by Annaert (1968) as a starting point, I have tried to develop an automatic method, using cluster analysis, which will always bring similar neighbouring areas into the same class, so that the boundaries shown on the map will be significant (Grimmeau, 1977a; 1977b). Under these circumstances the class limits cannot always be strictly respected. I have therefore reworked the method by introducing measures of similarity and dissimilarity based on probability. The principles and first results of the new version are presented in this chapter and are discussed more fully in Grimmeau (1980).

Method

The grouping method employed here for the mapping proceeds by using four steps of analysis. The first step employs a single-link cluster analysis with a spatial constraint to obtain a division of 'regions' bounded by spatial discontinuities. Here I propose to use a probabilistic distance measure such that the areas will be grouped together subject to the likelihood of the statistical null hypothesis that the difference between two values occurred by chance. The single-link method has the disadvantage that it does not guarantee the internal homogeneity of the groups—quite different values can be aggregated because they are linked together by a chain of intermediate values. Hence it is necessary to introduce a supplementary criterion of 'region' homogeneity which can also be chosen probabilistically, and a 'region' will be taken as homogeneous depending on the likelihood of the hypothesis that all the differences between the local values and the average value occurred by chance. The single-link analysis works as well upwards as downwards, and the supplementary criterion can easily be introduced into a downwards procedure. In this case at each step of the division process the internal homogeneity of the groups is tested, and only the heterogeneous ones are subdivided further.

The second step in the analysis aims to bring together similar neighbouring groups which are separated by a local discontinuity. It is necessary to proceed upwards with a centroid method, spatial constraints, and a probabilistic distance. Nonneighbouring similar groups are brought together by the method, but without spatial constraint as a third step.

At this stage of the analysis the less coherent parts of the data, which are attributed to chance, are eliminated though the data have not yet been really simplified. The result obtained here is governed by the chosen significance level and by the adopted neighbourhood network (see Grimmeau, 1979).

If the number of groups is still too large for mapping, a fourth step is to use the centroid upwards analysis without neighbourhood constraint. The most suitable distance criterion then seems to be the decrease in variance from one level of grouping to the next. The method can be adapted to the problem under consideration by choosing an appropriate variance. This variance may be used weighted or unweighted, for example, by the number of aggregated areas, by the denominator of the mapped ratio (if the data are in such a form), or by the proportion of the surface covered by each group on the map. Moreover, the variance can be computed only by using the original numbers or by using transformed values (say by taking logarithms). The four steps of the analysis are shown in table 20.1.

Table 20.1. The method of grouping employed.

Step	Process	Spatial constraint	Link	Distance	Supplementary criterion
1	downwards	yes	single	probabilistic	homogeneity
2	upwards	yes	centroid	probabilistic	-
3	upwards	no	centroid	probabilistic	-
4	upwards	no	centroid	loss of variance	-

An application of the method

The method has so far been applied to the mapping of population density, the proportion of commuters using a train, and the employment ratio. All data are for Belgium for the smallest territorial divisions (the communes), and only the train-commuter pattern is discussed here. This data set is widely influenced by chance; in some parts of Belgium the numbers are very low: 50% of the Belgian communes have no more than 25 commuters using a train.

In the first step of the analysis, the 2663 communes were merged into 943 larger regions; the homogeneity of the regions was tested with α set at 1%; the last division separates two communes whose standardised difference was no larger than 0·6. Of course, the mapped boundaries correspond with larger standardised differences.

In the second step of the analysis, the comparison of neighbouring regions, the number was reduced to 700.

The third step of the analysis, the comparison of the regions without taking their neighbourhood into consideration, allowed the retention of only 54 groups. At this point it was considered that no information had been lost, only the noise has been eliminated.

The fourth step of the analysis (simplification) was the identification of 7 classes. This number was chosen by taking into account the possibilities of efficient cartographic representation (see Robinson et al, 1978, page 172). The loss of information produced by this stage is 2%, and the whole process causes a 2·5% loss of information. Computing time on a CDC6500 was 6 minutes, drawing included.

At the level of the 700 regions defined by the first two steps, the groups do not overlap. Limits can thus be defined at a middle distance between the values of the last region of one group and the first region of the next group. At the level of the communes the groups do overlap. The cumulated frequency curves of the values within each group are S shaped, which corresponds to the expectation of the probabilistic nature of the method.

Discussion of the results

For a better evaluation of the solution, a second map was constructed with the same data, strictly within the limits defined above. This map is a classical one in that the limiting values of the different types of shading are respected, but we must keep in mind that they derive from the cluster analysis and are thus less 'arbitrary' than in a traditional solution. I shall call these two maps the typological map and the classes map. The two maps are shown in figures 20.1.

First we must examine whether the overlap between the two maps is too large to be admissible. If we compare the two solutions (table 20.2) we see that 82·2% of the communes are classified in the same way on both maps, 17·2% with a shift of one row, and 0·6% (16 communes) with a shift of two rows. If we consider the differences between the values of the 475 communes classified differently in the two solutions and the nearest boundary of the group they belong to, these differences extend from −20% to +12%, and the median difference (without taking the sign into account) is about 2%. Since the cumulated frequency curve of the standardised differences is not so far from the unit normal distribution, the deviations can be largely attributed to chance.

As the first aim of the method is to obtain a clearer definition on the map, let us see if there is any noticeable improvement. Table 20.3 compares the neighbourhood links in the two solutions, distinguishing them according to whether they connect communes classified in the same way or not. Among the 7805 neighbourhood relations, 84% are from the same type in both maps, and 16% are from the two opposite types; 1051

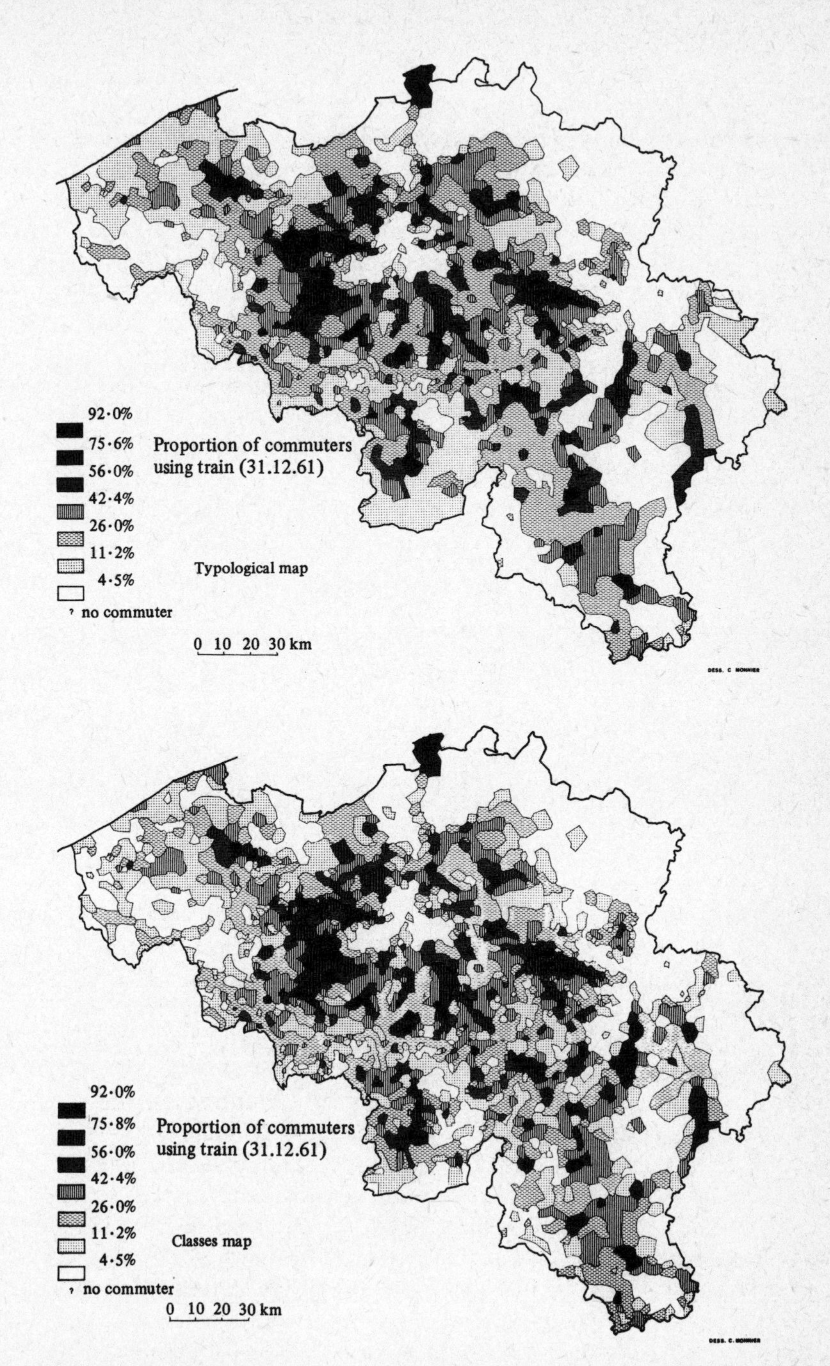

Figure 20.1. The proportions of commuters using the train, 31.12.1961, (a) shown on a typological map, (b) shown on a classes map.

relations connect communes from different classes but from the same typological group (I shall call them 'tied' relations) against 198 connecting communes from the same class but from different typological groups ('untied' relations). The number of tied relations is far larger than the untied relations, so that there are more relations between communes from the same group (48·8%) than between communes from the same class (37·8%).

Table 20.2. Comparison of the classification of the communes on the two maps.

		Typological map							
		1	2	3	4	5	6	7	Total
Classes map	1	**644**	102	6					752
	2	71	**333**	71					475
	3	8	42	**512**	42	2			606
	4			42	**308**	34			384
	5				22	**172**	6		200
	6					13	**162**	7	182
	7						7	**56**	63
	Total	723	477	631	372	221	175	63	2662

Table 20.3. Comparisons of the neighbourhood links on the two maps according to whether they connect communes classified in the same way (equal) or not classified in the same way (not equal), numbers and percentages.

		Typological map					
		equal		not equal		total	
Classes map	equal	2755	35·3%	198	2·5%	2953	37·8%
	not equal	1051	13·5%	3801	48·7%	4852	62·2%
	total	3806	48·8%	3999	51·2%	7805	100%

Figure 20.2 shows the cumulative frequency distributions of the chi-squared statistic associated with the neighbourhood relations according to whether they are connecting communes from the same group or not, and from the same class or not. We can see that similar neighbouring communes are less often classified in the different groups than in different classes. But all the differences between neighbouring communes from the same group cannot be attributed to chance; we must not forget that at the end of the third step 54 significantly different groups remain, and that to obtain a readable map we must proceed to a simplification, after which significantly different units go into the same group (among which some can be neighbours).

As there are more neighbourhood relations between communes of the same group than between communes of the same class, the typological map must contain less areas than the classes map. There are 484 and 739 areas respectively; the number of areas is thus lowered by the classification typology from 35%. This remarkable cartographical simplification causes a loss of only 0·4% information (2·5% for the typological map, 2·1% for the other one).

The typological map is simpler and better from a statistical point of view. However, is it better from a geographical point of view? This is more difficult to answer. What we can say is that the typological map comes closer to the railway network, the primary explanatory factor of the geographical contrasts. A large proportion of the points of the classes map that do not correspond with the railway network vanish. The simplification is particularly strong in the regions where the numbers of

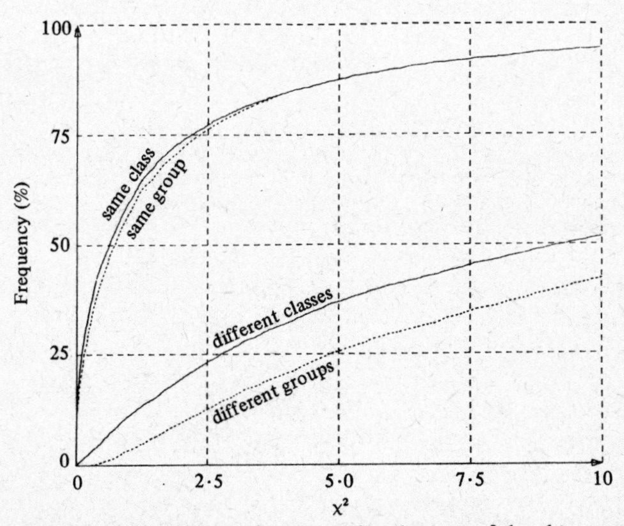

Figure 20.2. Cumulative frequency distributions of the chi-squared statistic associated with the neighbourhood relations, according to whether they are connecting communes from the same group or not, and from the same class or not.

commuters by train are very small. This spares us the need to search for a geographical explanation for what may be due to chance.

References

Annaert J, 1968 *Carte des densités par régions homogènes en Belgique, au 31-12-61* Institut de Géographie, Bruxelles (unpublished)

Grimmeau J P, 1977a *Contribution méthodologique à la mise en oeuvre de techniques quantitatives dans un but cartographique—application à l'étude par communes de la mobilité de main-d'oeuvre en Belgique au 31-12-61* unpublished Ph D, University of Bruxelles

Grimmeau J P, 1977b "Cartographie par plages et discontinuités spatiales" *Espace Géographique* **6** 49–58

Grimmeau J P, 1979 "Les réseaux de voisinage: critères de définition et utilisations géographiques" Besançon, *8e colloque sur les méthodes mathématiques appliquées à la géographie* to be published in: *Cahiers de géographie de Besançon, Notes et séminaires de Recherche*

Grimmeau J P, 1980 "Pour une cartographie probabilistie" Besançon, *9e colloque sur les méthodes appliquées à la géographie* to be published in: *Cahiers de géographie de Besançon, Notes et séminaires de Recherche*

Robinson A, Sale R V, Morrison J, 1978 *Elements of Cartography* (John Wiley, New York)

21

A definition of Venice in a metropolitan context

G Zanetto

Introduction

Since the 19th century efforts have been made to adapt Venice to the requirements of industrial development, but it was only after the First World War that they took a radically new direction. At that time an industrial port was built on that part of the dry land which looks onto the lagoon. Consequently in 1926 the commune of Venice was extended to a larger territory, including the new port and also a small town on the dry land (Mestre) which was to become, some years later, a twin town larger than Venice itself (see Zanetto and Lando, 1980).

This was the beginning of the integration of Venice into an urban system (400000 inhabitants today) whose elements have been ceaselessly specialising in the performance of their functions. Venice has therefore lost the characteristic of a town complete in itself and has become an area specialised in particular functions, with a massive displacement of its people (see Lando, 1978; Zanetto and Lando, 1978).

We can affirm without doubt that commerce and tourism are strongly concentrated in Venice, but it is more difficult to state which specialisation involves Venice as a residential area, since its population has moved rapidly inside the conurbation, induced both by economic convenience (Zanetto et al, 1980) and by cultural choice (see Lando and Zanetto, 1978; 1979a; Zanetto, 1977; 1980). As a result we now have to ask ourselves what characteristics distinguish the population living in the isle of Venice from the population of the whole commune, and what is the importance of such diversity in the total socioeconomic variance of the area? A principal component analysis based on the 1971 census data is used here to answer this question.

The analysis

After some adjustments, the principal components analysis was carried out using the thirty-four variables defined in table 21.1 for the 344 census areas of the commune of Venice in 1971. After Varimax rotation of the axes, three significant components are extracted whose loadings are listed in table 21.2.

According to many studies on the subject, the first component is a good quanitifier of the socioeconomic status of the commune (see Lando and Zanetto, 1979b), as defined by employment, occupation, and educational variables, as well as by the size of dwellings. The scores of this component enable us to deduce a status hierarchy of urban quarters.

These hierarchies are organised around the three highest scores which are found in the centres of Mestre, Venice, and Lido, each of which is surrounded by poor quarters. Venice is shown by the first component as a town containing a complete range of social strata, regularly arranged around the high-status centre. Since the same situation can be found on the dry land and on the islands of the lagoon, nothing characterises Venice from the status point of view.

In addition to this first dimension, the scores of the second principal component, which is nearly as important as the first (22·5% compared with 24·2% of variance), are distributed in a way which is perfectly in tune with expectations. Except for a few and justifiable exceptions, the areas related to Venice present positive scores, whereas all the other areas

Table 21.1. Socioeconomic variables from the 1971 census of the municipality of Venice.

01	mean number of persons per family	
02	% of the population aged 65 and over	
03	% of the population aged 14 and under	
04	replacement (population aged 15 to 39)/(population aged 40 to 65)	
05	% of working women in the female population	
06	professionals and undertakers	% in the active population
07	craftsmen and independent workers	
08	employees and managers	
09	blue collar workers	
	total workers in	
10	manufacture	% in the active population
11	commerce	
12	transport	
13	banking and insurance	
14	service industry	
15	public administration	
16	agriculture and fishing	
17	building	
18	% of 'elementary' education in population over 10	
19	% of 'middle' education in population over 13	
20	% of 'secondary' education in population over 18	
21	% of graduates in population over 24	
22	% of illiterates in population over 45	
23	density of population	
24	% of dwellings occupied by owners	
25	mean number of rooms per person	
26	% of dwellings with a bathroom	
27	% of dwellings connected to the gas network	
28	% of dwellings with central heating	
29	housing defects (% of dwellings lacking comforts)	
30	% of one- or two-room dwellings	
31	% of three- or four-room dwellings	
32	% of five- or more room dwellings	
33	% of houses built before 1945	
34	% of houses built after 1946	

Variables have been standardised before the principal components analysis.

have negative scores. Exceptions are represented by one unique area among the 139 of Venice—an island drained and built up in the 1950s; and by twenty areas among the 205 of the rest of the commune, which correspond to ancient lodgings in public ownership, and were thus rescued from urban renewal. Only six areas among them present scores higher than 0·20, and none reaches 0·50. We can therefore state with reasonable certainty that it is this second dimension which distinguishes Venice from the rest of its commune, and points to its distinctive specialisation.

Table 21.2. Loadings of three principal components (Varimax solution) for the municipality of Venice. (Note: only loadings greater than 0·40 are shown.)

Variable	Principal component		
	1	2	3
01	-	−0·810	-
02	+0·451	+0·629	-
03	−0·502	−0·714	-
04	−0·453	−0·657	-
05	+0·618	-	-
06	+0·795	-	-
07	-	-	-
08	+0·523	-	−0·476
09	−0·706	-	-
10	−0·412	−0·584	-
11	-	+0·662	-
12	−0·492	-	-
13	+0·539	+0·464	-
14	+0·618	+0·506	-
15	-	-	-
16	-	−0·402	-
17	-	−0·444	-
18	−0·781	-	-
19	+0·400	-	−0·669
20	+0·740	-	-
21	+0·842	-	-
22	-	-	+0·569
23	-	+0·794	-
24	-	−0·595	-
25	+0·712	+0·531	-
26	-	-	−0·878
27	-	+0·791	-
28	-	-	−0·744
29	-	−0·507	+0·634
30	-	-	+0·801
31	−0·823	-	-
32	+0·853	-	-
33	-	+0·671	+0·481
34	-	−0·685	−0·557
Eigenvalue	8·24	7·63	5·38
% of variance	24·2	22·5	15·8

Population living in the isle of Venice is characterised, above all, from a demographical point of view: we find an aged population and that the mean family size is smaller than elsewhere (variables 01, 02, 03, 04). Regarding occupations, employment in commerce and services has a strong concentration to the detriment of industry as a whole (variables 11, 14, 10). We find an obvious concentration of old dwellings and a lack of new ones; houses are particularly small, crowded, and are occupied by their owners more than is found elsewhere; moreover, they are not comfortable from a modern point of view (variables 24, 25, 29, 33, 34). Finally, Venice is characterised by a general proliferation of the gas network, as well as by a high density of population (variables 27, 23).

In short, the second component points out a compact and ancient town, whose population is specialised in services, but beaten by successive selective migration which has been caused by the obvious difficulties of urban renewal. However, one cannot believe that this fact is totally characteristic of Venice. The third principal component defines the quality of dwellings more closely. The scores on this component show that some areas of Venice retain a great variety of comforts and that the impact of renewal has been all but negligible (Costa et al, 1978).

Maybe the key to a correct interpretation is the drop in the Venetian population (from 184000 in 1950, to 94000 at present) in the face of the growth of tourism and a way of living to which Venice is not completely suited.

References

Costa P, Lando F, Zanetto G, 1978 "Venezia: rinnovo urbano 1967-1976" *Città Classe* **15/16** 46-49

Lando F, 1978 "La struttura socio-economica veneziana: un tentativo di analisi" *La Rivista Veneta* **28/29** 125-140

Lando F, Zanetto G, 1978 "Geografia e percezione dello spazio" in *La Percezione dell'ambiente: l'Esperimento di Venezia* Eds P Balboni, A De Marchi, F Lando, G Zanetto (Ciedart, Venice) pp 11-50

Lando F, Zanetto G, 1979a "Venise: le milieu lagunaire dans la perception de ses habitants" *L'Espace Géographique* 3 153-155

Lando F, Zanetto G, 1979b "La compléxité urbaine de la terre ferme vénitienne" *Revue d'Analyse Spatiale Quantitative et Appliquée* 7 1-14

Zanetto G, (Ed.), 1977 *The Lagoon as it is seen by its Users* symposium on "Environment, Participation, and Quality of Life", Venice; Council of Europe, Fondazione Cini, pp 31 + XX

Zanetto G, 1980 "Percezione ambientale: una ricerca, un progetto e qualche perplessità" in *Ricerca geografica e Percezione ambientale* Eds M Cesa-Bianchi, R Geipel (Unicopli, Milan) pp 275-287

Zanetto G, Costa P, Lando F, 1980 "Rinnovo urbano e trasformazione sociali nel centro storico di Venezia" *Sistemi Urbani* (Gruida, Naples) forthcoming

Zanetto G, Lando F, 1978 "Le migrazioni da e per Venezia insulare" in *Italiani in Movimento* Ed. G Valussi (Geap, Pordenone) pp 327-330

Zanetto G, Lando F, 1980 "Mestre: analisi tipologica di una struttura urbana" *Bollettino della Società Geografica Italiana* 7/9 213-255

Delimiting geographical units and selecting the indicators: correspondence analysis as an aid

M Vigouroux

In principal components analysis, without rotation of the axes, the effect of classification is well known; the first axis or general dimension, for the most part, classifies objects according to their size. However, in the urban framework, these divisions are very irregular: large housing estates, vast expanses of suburbia, diverse small blocks in the town centre, and a mosaic of 19th century dwellings. We must also add to this the conflict between the declining town centre and the growth of the periphery.

The use of percentage variables has often been tried but this does not solve the problem. Percentage methods and a more detailed division of the town centre gives priority to the old and to tradespeople, to the detriment of the younger generations and the blue collar workers of the suburban estates, segregated into larger districts.

In addition, if the most numerous categories are formed by blue and white collar workers, we would usually hesitate to fragment the upper-income categories (although these two are very diversified: the small and large businessman, the professional person, the executive). With the process of increase of indicators, they eventually become correlated and this tends to affect the results of the analysis.

This chapter examines the utility of correspondence analysis. As developed by J P Benzecri's team (Benzecri et al, 1973), it enables us to avoid many of these difficulties. The original model functions on a grid of two variables divided into modalities [in this example, the grouping of districts (the town) and the grouping of socioprofessional categories (the working population of the town)]. From the contingency table of frequencies, considered as probabilities, the analysis processes raw data, and the distributional equivalence renders the results insensitive to division into classes. We therefore work on separate classes that will be regrouped, as required, and on a detailed network of locations that can be simplified, if necessary. In addition, from one observation date to another, work on the raw data clearly allows comparisons between analyses; for example, that of a town centre which is declining differently in terms of districts and classes. Thus the technique permits a purely *a posteriori* approach and facilitates the subsequent establishment of variables and of experimentally devised districts.

Example

The example developed here is for Perpignan and uses 1975 census data. The original set of data contains thirty-seven census tracts described by nine occupation groups (that is the labour force of the set of thirty-seven tracts). The nine groups are: industrial and large traders, IGC; independent enterprises, APC; professional groups and managers, PLC; middle white collar workers, CMO; lower white collar workers, EMP; upper blue collar workers, COQ; lower blue collar workers, OSM; service workers, PSE; others, DIV.

Comparison is made with the same data set with thirty-seven tracts and ten occupation groups, for which PLC is divided into LIB (professional) and CSU (manager).

From these two data sets, the results of two correspondence analyses give us the following results. First the cumulative proportion of variance is given by:

Component	1	2	3	4	5
37 x 9 data set	65·2%	78·5%	86·2%	92·5%	95·2%
37 x 10 data set	62·9%	76·0%	83·4%	90·0%	93·2%

Note that the difference is normal when the number of indicators increases.

The loadings of the indicators onto the components give, for the eight common indicators in the two analyses, very similar results on components 1, 2, 3, but quite different on components 4 and 5, see table 22.1.

From these results it should be noted that,
first, the loadings of PLC are nearly the same as the arithmetic mean of loadings of CSU and LIB;
second, the weight of PLC in the set of indicators is very close to the sum of weights of CSU and LIB; and
third, on component 2, LIB and CSU are slightly divergent and they are classifying more precisely the census tracts into LIB dominant or CSU dominant. In Perpignan, PLC is homogeneous enough and it is not necessary to differentiate between LIB and CSU.

Table 22.1. Loadings and aids for interpretation (Lebart and Fenelon, 1973, page 249).

Data set	Group	Component 1			Component 2			Component 3		
		L	C	W	L	C	W	L	C	W
37 x 10	LIB	+851	697	103	+53	3	2	+338	110	139
	CSU	+772	709	161	−121	18	19	+293	102	198
	LIB + CSU	(+811)		(264)	(−34)		(21)	(+316)		(337)
37 x 9	PLC	+800	823	265	−53	4	6	+316	129	353

Notes: L is the loading; C is the contribution; W is the weight.

In addition, it should be noted that the census tract scores are very similar on each of the three components, *even on the second.*

Some conclusions

On the three components, which account for 83% to 86% of the variance, the data set is stable with nine or ten indicators, especially for the census tracts. The variation in the number of indicators has consequences *only* on the loadings of the modified indicators. Dividing PLC into LIB and CSU gives a better explanation of the classification of the census tracts. Another good use of the property ('distributional equivalence') would be to divide or to sum some census tracts, the number of indicators being kept constant.

The brief results presented in this chapter demonstrate that before analysis of a data set itself, the use of correspondence analysis allows the selection of pertinent indicators from among census data, and to have some answer to the agglomeration problem of geographical units. Strictly, indicators must be one qualitative variable (for instance, occupation, economic activity, land-use type, and so on). In geography, the set of geographical units is always a qualitative variable (city, region, set of towns, etc), and hence this method of analysis has considerable potential.

References

Benzecri J P, and others, 1973 *L'Analyse des Données* (Dunod, Paris)
Lebart L, Fenelon J P, 1973 *Statistique et Informatique Appliquées* (Dunod, Paris)

23

Some aspects of the multivariate analysis of spatial structures

G Löffler

If we characterise the result of a regionalisation or cluster analysis of choristically defined criteria as a generalised description of the spatial structure, there are a number of spatial structures of different significance corresponding to the various theoretical concepts which are operationalised by groups of variables. Distance matrices, used for a lot of classification strategies that take into account the differences between spatial units, can be regarded as quantitative descriptions of these spatial structures.

In a similar manner to Berry's field theory (1968) in which the relationship between such spatial structure and an interaction structure, operationalised by a distance matrix and an interaction matrix, is tested, we can measure the relationship between two different spatial structures by computing the "cophenetic correlation coefficient" r_c (Sokal and Rohlf, 1962) and making use of distance matrices (Löffler, 1981a). When measuring the degree of spatial congruence of two spatial structures in this way, we may ascertain whether a correlation exists between them or not. As the cophenetic correlation coefficient can be handled in the same way as the product-moment correlation coefficient, we are allowed to use it for the same statistical techniques. The use of differences or distances between objects or spatial units is not new. Apart from its use by Berry, who employed it in geography for the first time, this strategy is normally used in the social sciences to compute correlation, regression, and path coefficients for nonmetric data (Holm, 1977; Leitner and Wohlschlägl, 1980). In geography, the application of differences between variables measured in spatial units means that we do not compare only the variability of two phenomena in all geographical units independently, but we also compare the spatial structures of the phenomena as they are described within the distance matrices.

Another advantage of this approach to the analysis of spatial structures is the fact that we are able to compute distance-oriented and direction-oriented correlation coefficients. Since matrices of this kind are comparable with those of kilometre distances or those of neighbourhood situations (that is, the number of spatial units located between two given units) we have the opportunity of testing hypotheses on the spatial autocorrelation or spatial cross-correlation of phenomena and on the range and direction of their influences.

If we choose a neighbourhood matrix and a distance matrix, the range of spatial autocorrelation can be tested in the following way:
A step-by-step reduction of both matrices, beginning with the largest size or order of neighbourhood, permits one to compute correlation coefficients

between the differences of the phenomenon and the orders of neighbourhood. The results of this computation are n correlation coefficients for n orders of neighbourhood. These enable us to reject or confirm hypotheses on the existence of spatial autocorrelation for special distances. If we want to test a hypothesis on the existence of spatial cross-correlation for a special distance, we have only to compute the correlation between two phenomena (1, 2) and the situation of neighbourhood (3). By the use of the partial correlation coefficient $r_{1,2.3}$ we can see whether a spatial cross-correlation exists or not. An increase of the partial correlation coefficient confirms the hypothesis that a spatial cross-correlation exists between the two phenomena. By testing all neighbourhood orders it will easily be seen at which distance a maximum can be found.

Hypotheses on the directions and/or distances of influences can be tested by the choice of spatial differences and kilometre distances. For example consider testing the range or direction of the influence of urban agglomerations on their 'umland'. Only the operation of differencing the phenomena between these agglomerations and the other spatial units (or their kilometre distances or situations of neighbourhood relative to the agglomerations) need be derived from the complete matrices. On the one hand, the correlation between the chosen kilometre distances and the differences enables us to interpret the influence of the urban agglomeration relative to other areas with regard to their distance and/or direction. On the other hand, we can show in which way correlations between two phenomena are varying depending upon the distance to the urban agglomerations.

Full details of empirical examples of the use of all these variants of the (multivariate) analysis of spatial structures can be found in Löffler (1981b).

References

Berry B J L, 1968 "A synthesis of formal and functional regions using a general field theory of spatial behaviour" in *Spatial Analysis* Eds B J L Berry, D F Marble (Prentice Hall, Englewood Cliffs, NJ) pp 419-430

Holm K, (Ed.), 1977 *Die Befragung* **5** Uni Taschenbücher Nr 435 (Francke, Munich)

Leitner H, Wohlschlägl H, 1980 "Metrische und ordinale Pfadanalyse: Ein Verfahren zur Testung komplexer Kausalmodelle in der Geographie" *Geographische Zeitschrift* **68**(2) 81-106

Löffler G, 1981a "Cluster-Strukturvergleiche—Wahl der Klassenzahl und 'natürliche' Klassifikation" *Theorie und Quantitative Methodik in der Geographie. Tagungsband Zürich 26-28 Mars 1980* Ed. D Steiner (Geographisches Institut der ETH, Zürich) (forthcoming)

Löffler G, 1981b "A multivariate analysis of the relationship between different spatial structures—some aspects of developing regional space-(time)-models of Swedish agriculture" in *German Quantitative Geography* Papers presented at the Second European Conference on Theoretical and Quantitative Geography in Cambridge, 11-14 September 1980, Eds G Bahrenberg, U Streit, Münstersche Geographische Arbeiten (Schöningh, Paderborn) (forthcoming)

Sokal R R, Rohlf F J, 1962 "The comparison of dendrograms by objective methods" *Taxon* **11** 33-40